H. Haken Synergetics—An Introduction

Springer Series in Synergetics

Synergetics, an interdisciplinary field of research, is concerned with the cooperation of individual parts of a system that produces macroscopic spatial, temporal or functional structures. It deals with deterministic as well as stochastic processes.

Volume 1	**Synergetics,** An Introduction 2nd Edition By H. Haken
Related Title	**Synergetics,** A Workshop Editor: H. Haken

Hermann Haken

Synergetics

An Introduction

Nonequilibrium Phase Transitions and Self-Organization in Physics, Chemistry and Biology

Second Enlarged Edition

With 152 Figures

Springer-Verlag Berlin Heidelberg New York 1978

Professor Dr. Hermann Haken

Institut für Theoretische Physik der Universität Stuttgart
Pfaffenwaldring 57/IV, D-7000 Stuttgart 80, Fed. Rep. of Germany

ISBN 3-540-08866-0 2. Auflage Springer-Verlag Berlin Heidelberg New York
ISBN 0-387-08866-0 2nd edition Springer-Verlag New York Heidelberg Berlin

ISBN 3-540-07885-1 1. Auflage Springer-Verlag Berlin Heidelberg New York
ISBN 0-387-07885-1 1st edition Springer-Verlag New York Heidelberg Berlin

Library of Congress Cataloging in Publication Data. Haken, H. Synergetics : an introduction. Bibliography: p. Includes index. 1. System theory. I. Title. Q295.H35 1978 003 78-17686

Offset printing and bookbinding: Konrad Triltsch, Graphischer Betrieb, Würzburg

2153/3130 – 543210

To the Memory of

Maria and Anton Vollath

Preface to the Second Edition

The publication of this second edition was motivated by several facts. First of all, the first edition had been sold out in less than one year. It had found excellent critics and enthusiastic responses from professors and students welcoming this new interdisciplinary approach. This appreciation is reflected by the fact that the book is presently translated into Russian and Japanese also.

I have used this opportunity to include some of the most interesting recent developments. Therefore I have added a whole new chapter on the fascinating and rapidly growing field of chaos dealing with irregular motion caused by deterministic forces. This kind of phenomenon is presently found in quite diverse fields ranging from physics to biology. Furthermore I have included a section on the analytical treatment of a morphogenetic model using the order parameter concept developed in this book. Among the further additions, there is now a complete description of the onset of ultrashort laser pulses. It goes without saying that the few minor misprints or errors of the first edition have been corrected.

I wish to thank all who have helped me to incorporate these additions.

Stuttgart, July 1978 Hermann Haken

Preface to the First Edition

The spontaneous formation of well organized structures out of germs or even out of chaos is one of the most fascinating phenomena and most challenging problems scientists are confronted with. Such phenomena are an experience of our daily life when we observe the growth of plants and animals. Thinking of much larger time scales, scientists are led into the problems of evolution, and, ultimately, of the origin of living matter. When we try to explain or understand in some sense these extremely complex biological phenomena it is a natural question, whether processes of self-organization may be found in much simpler systems of the unanimated world.

In recent years it has become more and more evident that there exist numerous examples in physical and chemical systems where well organized spatial, temporal, or spatio-temporal structures arise out of chaotic states. Furthermore, as in living organisms, the functioning of these systems can be maintained only by a flux of energy (and matter) through them. In contrast to man-made machines, which are devised to exhibit special structures and functionings, these structures develop spontaneously—they are self-organizing. It came as a surprise to many scientists that numerous such systems show striking similarities in their behavior when passing from the disordered to the ordered state. This strongly indicates that the functioning of such systems obeys the same basic principles. In our book we wish to explain such basic principles and underlying conceptions and to present the mathematical tools to cope with them.

This book is meant as a text for students of physics, chemistry and biology who want to learn about these principles and methods. I have tried to present mathematics in an elementary fashion wherever possible. Therefore the knowledge of an undergraduate course in calculus should be sufficient. A good deal of important mathematical results is nowadays buried under a complicated nomenclature. I have avoided it as far as possible though, of course, a certain number of technical expressions must be used. I explain them wherever they are introduced. Incidentally, a good many of the methods can also be used for other problems, not only for self-organizing systems. To achieve a self-contained text I included some chapters which require some more patience or a more profound mathematical background of the reader. Those chapters are marked by an asterisk. Some of them contain very recent results so that they may also be profitable for research workers.

The basic knowledge required for the physical, chemical and biological systems is, on the average, not very special. The corresponding chapters are arranged in such a way that a student of one of these disciplines need only to read "his" chapter. Nevertheless it is highly recommended to browse through the other chapters just to get a feeling of how analogous all these systems are among each other. I have called this discipline "synergetics". What we investigate is the joint action of many subsystems (mostly of the same or of few different kinds) so as to produce structure and functioning on a macroscopic scale. On the other hand, many different disciplines cooperate here to find general principles governing self-organizing systems.

I wish to thank Dr. Lotsch of Springer-Verlag who suggested writing an extended version of my article "Cooperative phenomena in systems far from thermal equilibrium and in nonphysical systems", in Rev. Mod. Phys. (1975). In the course of writing this "extension", eventually a completely new manuscript evolved. I wanted to make this field especially understandable to students of physics, chemistry and biology. In a way, this book and my previous article have become complementary.

It is a pleasure to thank my colleagues and friends, especially Prof. W. Weidlich, for many fruitful discussions over the years. The assistance of my secretary, Mrs. U. Funke, and of my coworker Dr. A. Wunderlin was an enormous help for me in writing this book and I wish to express my deep gratitude to them. Dr. Wunderlin checked the formulas very carefully, recalculating many of them, prepared many of the figures, and made valuable suggestions how to improve the manuscript. In spite of her extended administrative work, Mrs. U. Funke has drawn most of the figures and wrote several versions of the manuscript, including the formulas, in a perfect way. Her willingness and tireless efforts encouraged me again and again to complete this book.

Stuttgart, November 1976 Hermann Haken

Contents

* Sections with an asterisk in the heading may be omitted during a first reading.

1. Goal

Why You Might Read This Book

1.1 Order and Disorder: Some Typical Phenomena

Let us begin with some typical observations of our daily life. When we bring a cold body in contact with a hot body, heat is exchanged so that eventually both bodies acquire the same temperature (Fig. 1.1). The system has become completely homogeneous, at least macroscopically. The reverse process, however, is never observed in nature. Thus there is a unique direction into which this process goes.

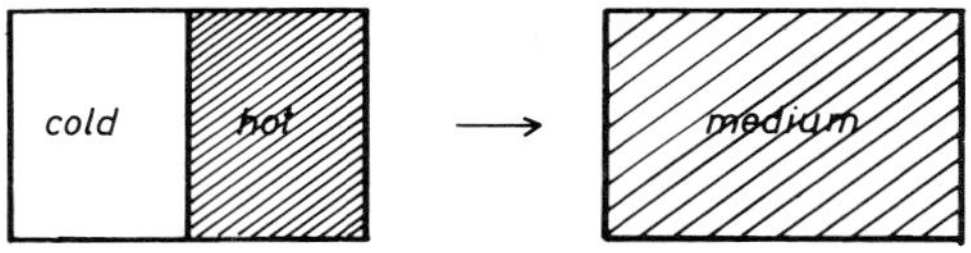

Fig. 1.1. Irreversible heat exchange

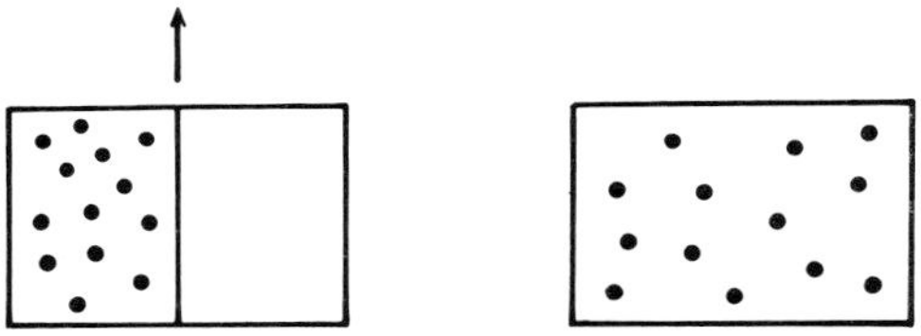

Fig. 1.2. Irreversible expansion of a gas

Fig. 1.3. Drop of ink spreading in water

When we have a vessel filled with gas atoms and remove the piston the gas will fill the whole space (Fig. 1.2). The opposite process does not occur. The gas by itself does not concentrate again in just half the volume of the vessel. When we put a drop of ink into water, the ink spreads out more and more until finally a homogeneous distribution is reached (Fig. 1.3). The reverse process was never observed. When an airplane writes words in the sky with smoke, the letters become more and more diffuse and disappear (Fig. 1.4). In all these cases the systems develop to a unique final state, called a state of thermal equilibrium. The original structures disappear being replaced by homogeneous systems. When analysing these phenomena on the microscopic level considering the motion of atoms or molecules one finds that disorder is increased.

Fig. 1.4. Diffusion of clouds

Let us conclude these examples with one for the degradation of energy. Consider a moving car whose engine has stopped. At first the car goes on moving. From a physicist's point of view it has a single degree of freedom (motion in one direction) with a certain kinetic energy. This kinetic energy is eaten up by friction, converting that energy into heat (warming up the wheels etc.). Since heat means thermal motion of many particles, the energy of a single degree of freedom (motion of the car) has been distributed over many degrees of freedom. On the other hand, quite obviously, by merely heating up the wheels we cannot make a vehicle go.

In the realm of thermodynamics, these phenomena have found their proper description. There exists a quantity called entropy which is a measure for the degree of disorder. The (phenomenologically derived) laws of thermodynamics state that in a closed system (i.e., a system with no contacts to the outer world) the entropy ever increases to its maximal value.

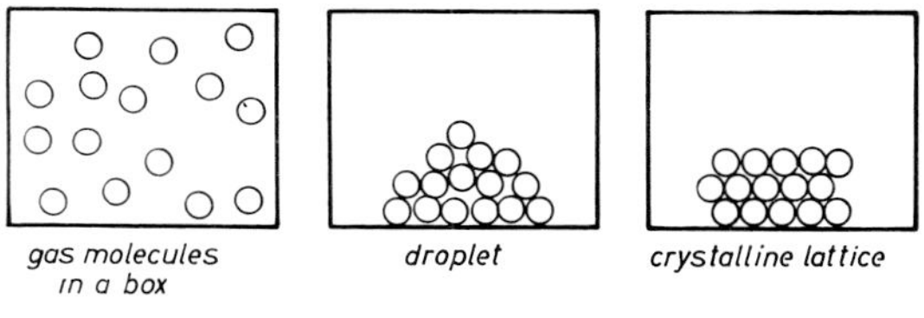

Fig. 1.5. Water in its different phases

On the other hand when we manipulate a system from the outside we can change its degree of order. Consider for example water vapor (Fig. 1.5). At elevated temperature its molecules move freely without mutual correlation. When temperature is lowered, a liquid drop is formed, the molecules now keep a mean distance between each other. Their motion is thus highly correlated. Finally, at still lower temperature, at the freezing point, water is transformed into ice crystals. The molecules are now well arranged in a fixed order. The transitions between the different aggregate states, also called phases, are quite abrupt. Though the same kind of molecules are involved all the time, the macroscopic features of the three phases differ drastically. Quite obviously, their mechanical, optical, electrical, thermal properties differ wildly.

Another type of ordering occurs in ferromagnets (e.g., the magnetic needle of a compass). When a ferromagnet is heated, it suddenly loses its magnetization. When temperature is lowered, the magnet suddenly regains its magnetization (cf. Fig. 1.6). What happens on a microscopic, atomic level, is this: We may visualize the magnet as being composed of many, elementary (atomic) magnets (called spins). At elevated temperature, the elementary magnets point in random directions (Fig. 1.7). Their

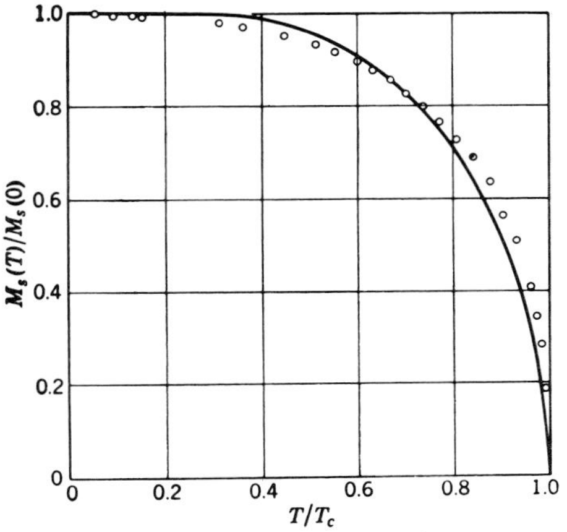

Fig. 1.6. Magnetization of ferromagnet as function of temperature (After C. Kittel: *Introduction to Solid State Physics*. Wiley Inc., New York 1956)

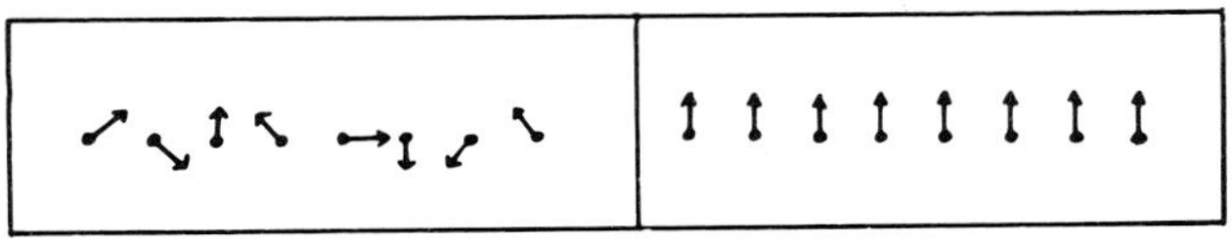

Fig. 1.7. Elementary magnets pointing into random directions ($T > T_c$) (lhs); aligned elementary magnets ($T < T_c$) (rhs)

magnetic moments, when added up, cancel each other and no macroscopic magnetization results. Below a critical temperature, T_c, the elementary magnets are lined up, giving rise to a macroscopic magnetization. Thus the order on the microscopic level is a cause of a new feature of the material on the macroscopic level. The change of one phase to the other one is called phase transition. A similarly dramatic phase transition is observed in superconductors. In certain metals and alloys the electrical resistance completely and abruptly vanishes below a certain temperature (Fig. 1.8). This phenomenon is caused by a certain ordering of the metal electrons. There are numerous further examples of such phase transitions which often show striking similarities.

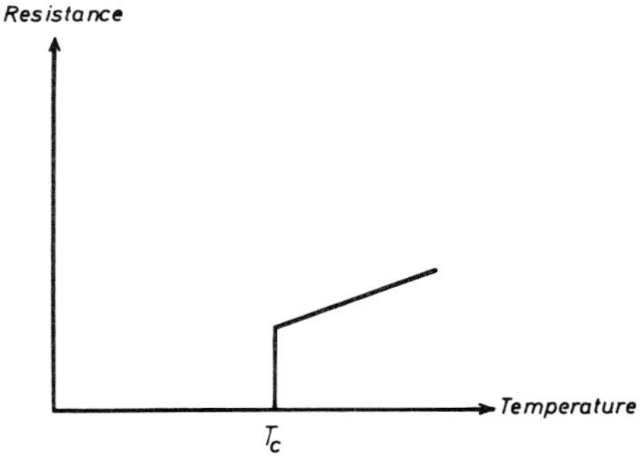

Fig. 1.8. Resistance of a superconductor as a function of temperature (schematic)

While this is a very interesting field of research, it does not give us a clue for the explanation of any biological processes. Here order and proper functioning are not achieved by lowering the temperature, but by maintaining a flux of energy and matter through the system. What happens, among many other things, is this: The energy fed into the system in the form of chemical energy, whose processing involves many microscopic steps, eventually results in ordered phenomena on a macroscopic scale: formation of macroscopic patterns (morphogenesis), locomotion (i.e., few degrees of freedom!) etc.

In view of the physical phenomena and thermodynamic laws we have mentioned above, the possibility of explaining biological phenomena, especially the creation of order on a macroscopic level out of chaos, seems to look rather hopeless. This has led prominent scientists to believe that such an explanation is impossible. However, let us not be discouraged by the opinion of some authorities. Let us rather reexamine the problem from a different point of view. The example of the car teaches us that it is possible to concentrate energy from many degrees of freedom into a single degree of freedom. Indeed, in the car's engine the chemical energy of gasoline is first essentially transformed into heat. In the cylinder the piston is then pushed into a single prescribed direction whereby the transformation of energy of many degrees of freedom into a single degree of freedom is accomplished. Two facts are important to recall here:

1) The whole process becomes possible through a man-made machine. In it we have established well-defined constraints.
2) We start from a situation far from thermal equilibrium. Indeed pushing the piston corresponds to an approach to thermal equilibrium under the given constraints.

The immediate objection against this machine as model for biological systems lies in the fact that biological systems are self-organized, not man-made. This leads us to the question if we can find systems in nature which operate far from thermal equilibrium (see 2 above) and which act under *natural* constraints. Some systems of this kind have been discovered quite recently, others have been known for some while. We describe a few typical examples:

A system lying on the border line between a natural system and a man-made device is the *laser*. We treat here the laser as a device, but laser action (in the microwave region) has been found to take place in interstellar space. We consider the

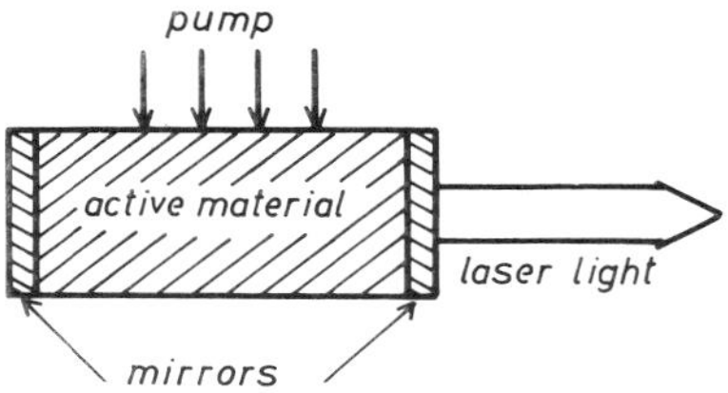

Fig. 1.9. Typical setup of a laser

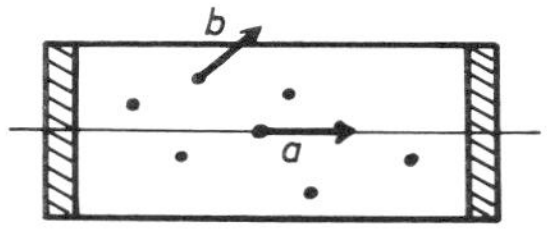

Fig. 1.10. Photons emitted in axial direction (a) have a much longer lifetime t_0 in the "cavity" than all other photons (b)

solid state laser as an example. It consists of a rod of material in which specific atoms are embedded (Fig. 1.9). Usually mirrors are fixed at the end faces of the rod. Each atom may be excited from the outside, e.g., by shining light on it. The atom then acts as a microscopic antenna and emits a light-wavetrack. This emission process lasts typically 10^{-8} second and the wavetrack has a length of 3 meters. The mirrors serve for a selection of these tracks: Those running in the axial direction are reflected several times between the mirrors and stay longer in the laser, while all other tracks leave it very quickly (Fig. 1.10). When we start pumping energy into the laser, the following happens: At small pump power the laser operates as a lamp. The atomic antennas emit independently of each other, (i.e., randomly) light-wavetracks. At a certain pump power, called laser threshold, a completely new phenomenon occurs. An unknown demon seems to let the atomic antennas oscillate in phase. They emit now a single giant wavetrack whose length is, say 300,000 km! (Fig. 1.11). The emitted light intensity, (i.e., the output power) increases drastically with further increasing input (pump) power (Fig. 1.12). Evidently the macroscopic properties of the laser have changed dramatically in a way reminescent of the phase transition of for example, the ferromagnet.

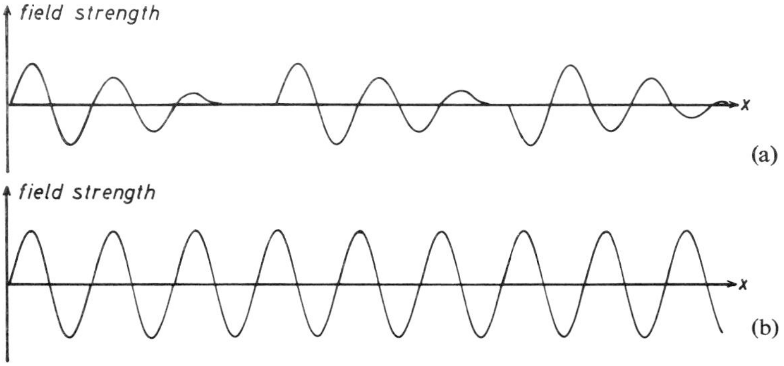

Fig. 1.11. Wave tracks emitted (a) from a lamp, (b) from a laser

As we shall see later in this book, this analogy goes far deeper. Obviously the laser is a system far from thermal equilibrium. As the pump energy is entering the system, it is converted into laser light with its unique properties. Then this light leaves the laser. The obvious question is this: What is the demon that tells the sub-

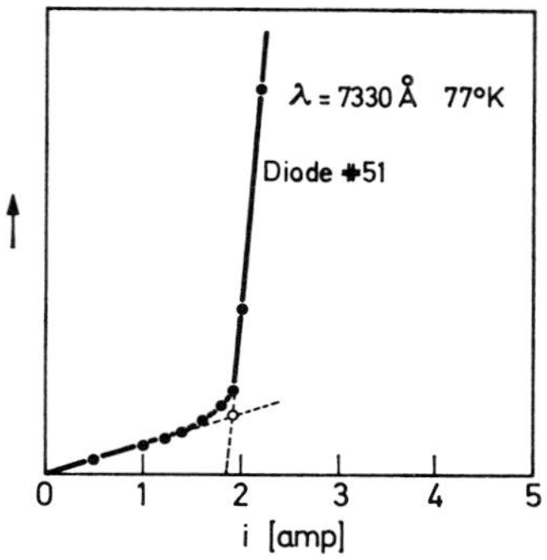

Fig. 1.12. Output power versus input power of a laser below and above its threshold (After M. H. Pilkuhn, unpubl. result)

systems (i.e., the atoms) to behave in such a well organized manner? Or, in more scientific language, what mechanisms and principles are able to explain the self-organization of the atoms (or atomic antennas)? When the laser is pumped still higher, again suddenly a completely new phenomenon occurs. The rod regularly emits light flashes of extremely short duration, say 10^{-12} second. Let us consider as a second example *fluid dynamics*, or more specifically, the flow of a fluid round a cylinder. At low speed the flow portrait is exhibited in Fig. 1.13(a). At higher speed,

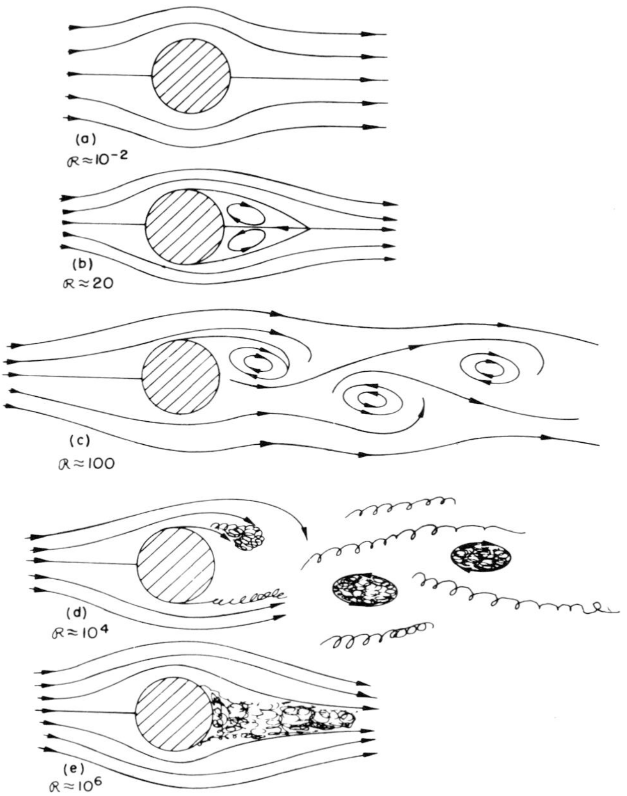

Fig. 1.13. Flow of a fluid round a cylinder for different velocities (After R. P. Feynman, R. B. Leighton, M. Sands: *The Feynman Lectures of Phys.*, Vol. II. Addison-Wesley 1965)

suddenly a new, static pattern appears: a pair of vortices (b). With still higher speed, a dynamic pattern appears, the vortices are now oscillating (c). Finally at still higher speed, an irregular pattern called turbulent flow arises (e). While we will not treat this case in our book, the following will be given an investigation.

The *convection instability* (Bénard instability). We consider a fluid layer heated from below and kept at a fixed temperature from above (Fig. 1.14). At a small temperature difference (more precisely, gradient) heat is transported by heat conduction and the fluid remains quiescent. When the temperature gradient reaches a critical value, the fluid starts a macroscopic motion. Since heated parts expand, these parts move up by buoyancy, cool, and fall back again to the bottom. Amazingly, this motion is well regulated. Either rolls (Fig. 1.15) or hexagons (Fig. 1.16) are observed. Thus, out of a completely homogeneous state, a dynamic well-ordered spatial pattern emerges. When the temperature gradient is further increased, new phenomena occur. The rolls start a wavy motion along their axes. Further patterns are exhibited in Fig. 1.17. Note, that the spokes oscillate temporarily. These phenomena play a fundamental role for example in meteorology, determining the motion of air and the formation of clouds (see e.g., Fig. 1.18).

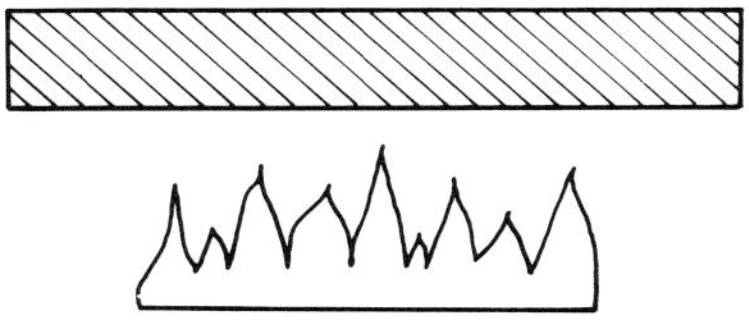

Fig. 1.14. Fluid layer heated from below for small Rayleigh numbers. Heat is transported by conduction

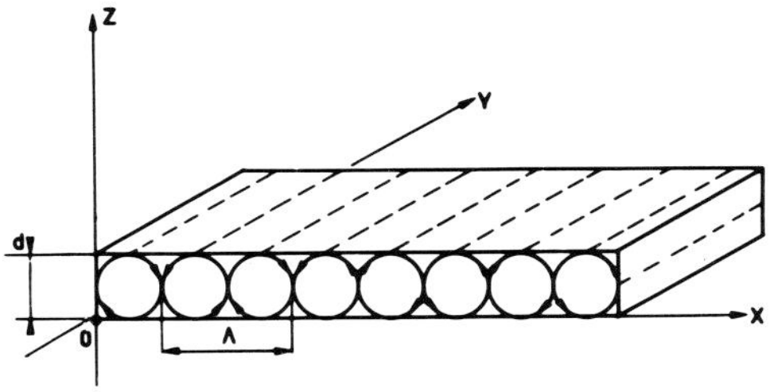

Fig. 1.15. Fluid motion in form of roles for Rayleigh numbers somewhat bigger than the critical

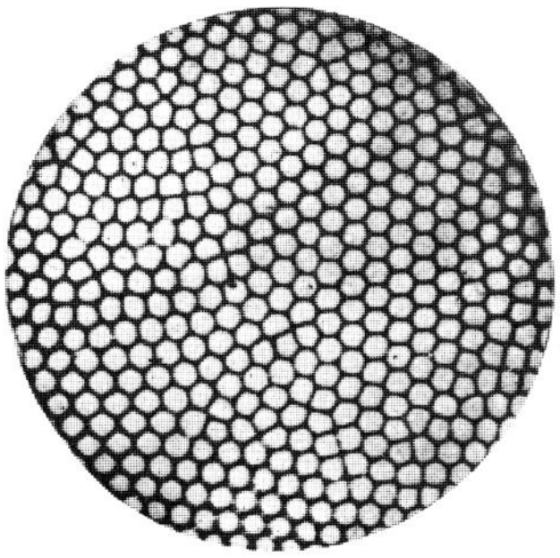

Fig. 1.16. The cell structure of the Bénard instability as seen from above (After S. Chandrasekhar: *Hydrodynamic and Hydromagnetic Stability.* Clarendon Press, Oxford 1961)

Fig. 1.17. Pattern of fluid motion at elevated Rayleigh number (After F. H. Busse, J. A. Whitehead: J. Fluid Mech. *47*, 305 (1971))

Fig. 1.18. A typical pattern of cloud streets (After R. Scorer: *Clouds of the World.* Lothian Publ. Co., Melbourne 1972)

A closely related phenomenon is the *Taylor instability*. Here a fluid is put between two rotating coaxial cylinders. Above a critical rotation speed, Taylor vortices occur. In further experiments also one of the cylinders is heated. Since in a number of cases, stars can be described as rotating liquid masses with thermal gradients, the impact of this and related effects on astrophysics is evident. There are numerous further examples of such ordering phenomena in physical systems far from thermal equilibrium. However we shall now move on to *chemistry*.

In a number of chemical reactions, spatial, temporal or spatio-temporal patterns occur. An example is provided by the Belousov-Zhabotinsky reaction. Here $Ce_2(SO_4)_3$, $KBrO_3$, $CH_2(COOH)_2$, H_2SO_4 as well as a few drops of Ferroine (redox indicator) are mixed and stirred. The resulting homogeneous mixture is then put into a test tube, where immediately temporal oscillations occur. The solution changes color periodically from red, indicating an excess of Ce^{3+}, to blue, indicating an excess of Ce^{4+} (Fig. 1.19). Since the reaction takes place in a closed system, the system eventually approaches a homogeneous equilibrium. Further examples of developing chemical structures are represented in Fig. 1.20. In later chapters of this book we will treat chemical reactions under *steady-state conditions*, where, nevertheless, spatio-temporal oscillations occur. It will turn out that the onset of the occurrence of such structures is governed by principles analogous to those governing disorder-order transitions in lasers, hydrodynamics, and other systems.

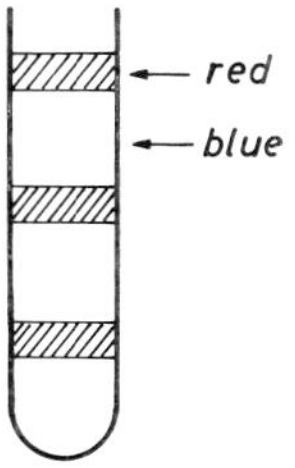

Fig. 1.19. The Belousov-Zhabotinsky reaction showing a spatial pattern (schematic)

Our last class of examples is taken from *biology*. On quite different levels, a pronounced spontaneous formation of structures is observed. On the global level, we observe an enormous variety of species. What are the factors determining their distribution and abundance? To show what kind of correlations one observes, consider Fig. 1.21 which essentially presents the temporal oscillation in numbers of snowshoe hares and lynx. What mechanism causes the oscillation? In evolution, selection plays a fundamental role. We shall find that selection of species obeys the same laws as for example, laser modes.

Let us turn to our last example. In developmental physiology it has been known long since that a set of equal (equipotent) cells may self-organize into structures with well-distinguished regions. An aggregation of cellular slime mold (*Dictyostelium disciodeum*) may serve as model for cell interactions in embryo genesis. Dictyostelium forms a multi-cellular organism by aggregation of single cells. During its growth phase the organism exists in the state of single amoeboid cells. Several hours after the end of growth, these cells aggregate forming a polar body along

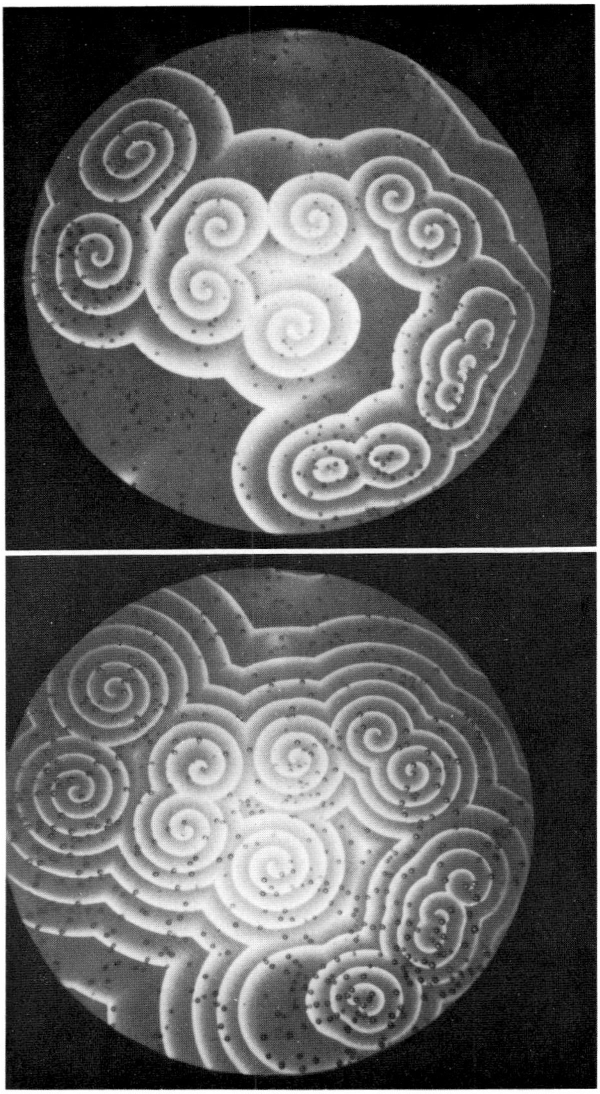

(a)

(b)

Fig. 1.20 a and b. Spirals of chemical activity in a shallow dish. Wherever two waves collide head on, both vanish. Photographs taken by A. T. Winfree with Polaroid Sx·70.

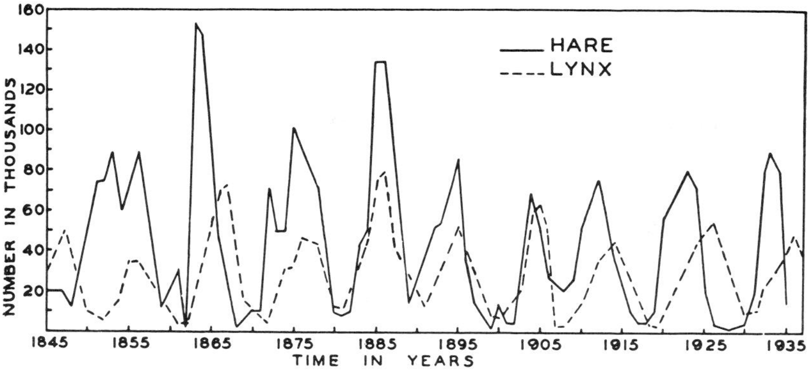

Fig. 1.21. Changes in the abundance of the lynx and the snowshoe hare, as indicated by the number of pelts received by the Hudson Bay Company (After D. A. McLulich: *Fluctuations in the Numbers of Varying Hare*. Univ. of Toronto Press, Toronto 1937)

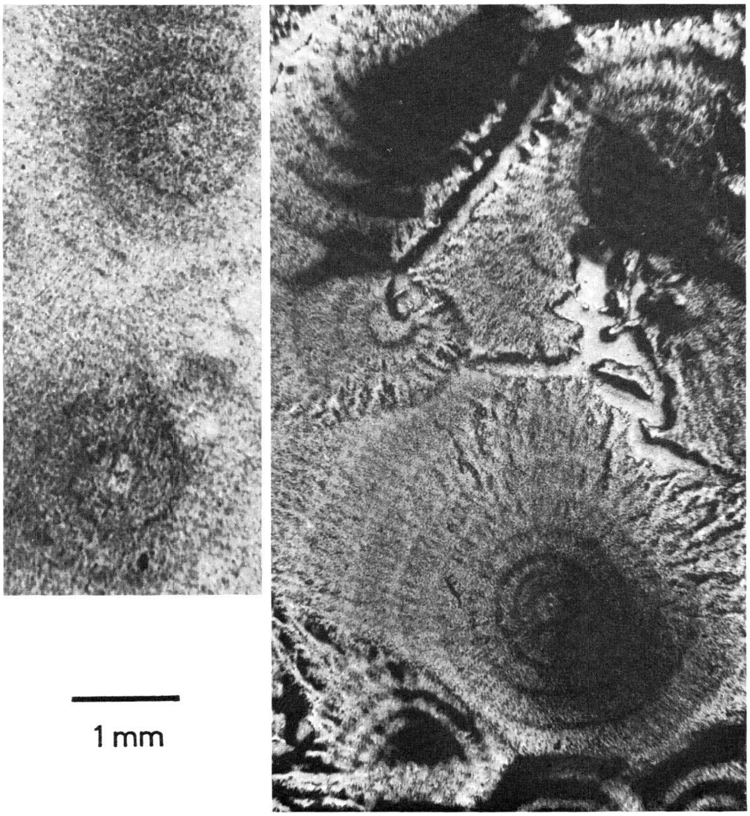

Fig. 1.22. Wave pattern of chemotactic activity in dense cell layers of slime mold (after Gerisch et al.)

which they differentiate into other spores or stalk cells the final cell types constituting the fruiting body. The single cells are capable of spontaneously emitting a certain kind of molecules called cAMP (cyclic Adenosin 3′5′Monophosphate) in the form of pulses into their surroundings. Furthermore cells are capable of amplifying cAMP pulses. Thus they perform spontaneous and stimulated emission of chemicals (in analogy to spontaneous and stimulated emission of light by laser atoms). This leads to a collective emission of chemical pulses which migrate in the form of concentration waves from a center causing a concentration gradient of cAMP. The single cells can measure the direction of the gradient and migrate towards the center with help of pseudopods. The macroscopically resulting wave patterns (spiral or concentric circles) are depicted in Fig. 1.22 and show a striking resemblance to the chemical concentration waves depicted in Fig. 1.20.

1.2 Some Typical Problems and Difficulties

In the preceding section we presented several typical examples of phenomena some of which we want to study. The first class of examples referred to closed systems. From these and numerous other examples, thermodynamics concludes that in closed systems the entropy never decreases. The proof of this theorem is left to statistical mechanics. To be quite frank, in spite of many efforts this problem is not completely solved. We will touch on this problem quite briefly, but essentially take a somewhat different point of view. We do not ask *how* can one *prove quite generally* that entropy ever increases but rather, *how* and *how fast* does the entropy increase in a given system? Furthermore it will transpire that while the entropy concept and related concepts are extremely useful tools in thermostatics and in so-called irreversible thermodynamics, it is far too rough an instrument to cope with self-organizing structures. In general in such structures entropy is changed only by a tiny amount. Furthermore, it is known from statistical mechanics that fluctuations of the entropy may occur. Thus other approaches are required. We therefore try to analyze what features are common to the non-equilibrium systems we have described above, e.g., lasers, fluid dynamics, chemical reactions, etc. In all these cases, the total system is composed of very many subsystems, e.g., atoms, molecules, cells, etc. Under certain conditions these subsystems perform a well organized collective motion or function.

To elucidate some of the central problems let us consider a string whose ends are kept fixed. It is composed of very many atoms, say 10^{22}, which are held together by forces. To treat this problem let us make a model. A chain of point masses coupled by springs (Fig. 1.23). To have a "realistic" model, let us still take an appreciable number of such point masses. We then have to determine the motion of very many interacting "particles" (the point masses) or "subsystems". Let us take the following attitude: To solve this complicated many-body problem, we use a computer into which we feed the equations of motion of the point masses and a "realistic" initial condition, e.g., that of Fig. 1.24. Then the computer will print for us large tables with numbers, giving the positions of the point masses as a function of time. Now the first essential point is this: These tables are rather useless until our brain

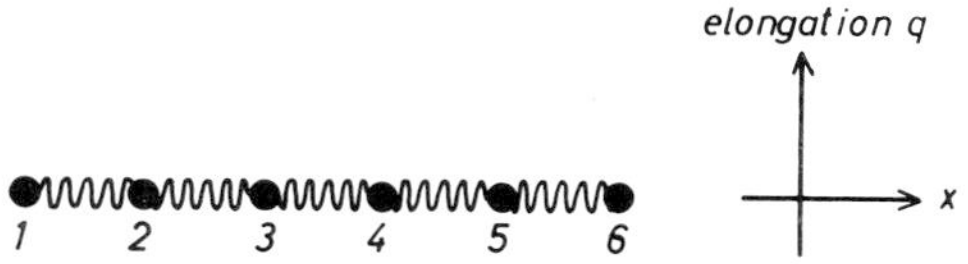

Fig. 1.23. Point masses coupled by springs

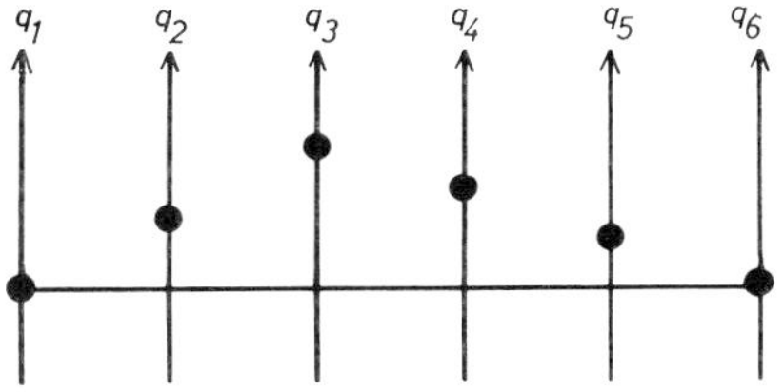

Fig. 1.24. An initial configuration of point masses

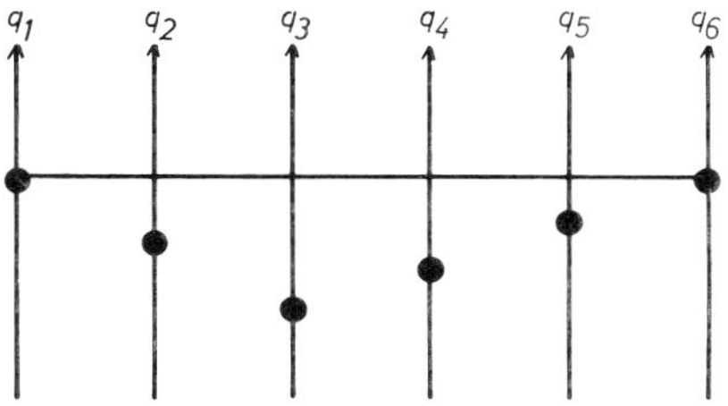

Fig. 1.25. Coordinates of point masses at a later time

selects certain "typical features". Thus we shall discover that there are correlations between the positions of adjacent atoms (Fig. 1.25). Furthermore when looking very carefully we shall observe that the motion is periodic in time. However, in this way we shall never discover that the natural description of the motion of the string is by means of a spatial sine wave (cf. Fig. 1.26), unless we already know this answer and feed it as initial condition into the computer. Now, the (spatial) sine wave is characterized by quantities i.e., the wavelength and the amplitude which are completely unknown on the microscopic (atomic) level. The essential conclusion from our example is this: To describe collective behavior we need entirely new concepts compared to the microscopic description. The notion of wavelength and amplitude is entirely different from that of atomic positions. Of course, when we know the sine wave we can deduce the positions of the single atoms.

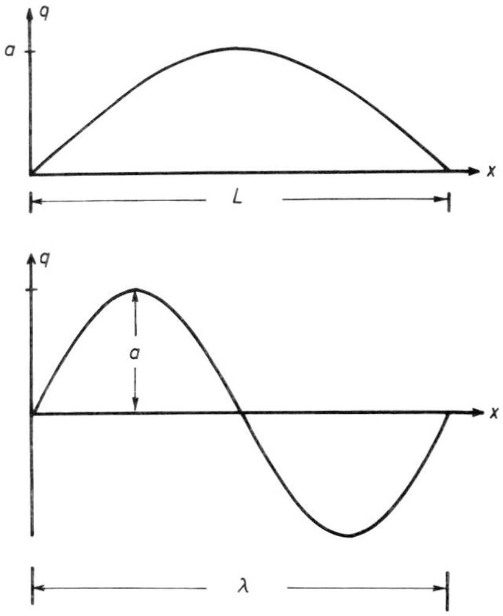

Fig. 1.26. Examples of sine-waves formed by strings. $q=\xi \sin(2\pi x/\lambda)$, ξ: amplitude, λ: wavelength. $\lambda=2L/n$, n: integer

In more complicated systems quite other "modes" may appropriately describe spatio-temporal patterns or functionings. Therefore, our mechanical example must be taken as an allegory which, however, exhibits the first main problem of multi-component systems: What is the adequate description in "macroscopic" terms, or, in what "modes" does the system operate? Why didn't the computer calculation of our example lead to these modes? The reason lies in the linearity of the corresponding equations of motion which permits that any superposition of solutions is again a solution of these equations. It will turn out that equations governing self-organization are intrinsically nonlinear. From those equations we shall find in the following that, "modes" may either *compete*, so that only one "survives", or *coexist* by stabilizing each other. Apparently the mode concept has an enormous advantage over the microscopic description. Instead of the need to know all "atomic" coordinates of very many degrees of freedom we need to know only a single or very few parameters, e.g., the mode amplitude. As we will see later, the mode amplitudes determine the kind and degree of order. We will thus call them *order parameters* and establish a connection with the idea of order parameters in phase transition theory. The mode concept implies a scaling property. The spatio-temporal patterns may be similar, just distinguished by the size (scale) of the amplitude (By the way, this "similarity" principle plays an important role in pattern recognition by the brain, but no mechanism is so far known to explain it. Thus for example a triangle is recognized as such irrespective of its size and position).

So far we have demonstrated by an *allegory* how we may be able to describe macroscopic ordered states possibly by very few parameters (or "degrees of freedom"). In our book we will devise several methods of how to find equations for these order parameters. This brings us to the last point of our present section. Even if we have such "parameters", how does self-organization, for example a spontaneous pattern formation, occur. To stay in our allegory of the chain, the question would be as follows. Consider a chain at rest with amplitude $\xi = 0$. Suddenly it starts moving in a certain mode. This is, of course, impossible, contradicting fundamental physical laws, for example that of energy conservation. Thus we have to feed energy into the system to make it move and to compensate for friction to keep it moving. The amazing thing in self-organizing systems, such as discussed in Section 1.1, is now this. Though energy is fed into the system in a *completely random* fashion, the system forms a well-defined macroscopic mode. To be quite clear, here our mechanical string model fails. A randomly excited, damped string oscillates randomly. The systems we shall investigate organize themselves *coherently*. In the next section we shall discuss our approach to deal with these puzzling features.

Let us conclude this section with a remark about the interplay between microscopic variables and order parameters, using our mechanical chain as example: The behavior of the coordinate of the μ'th point mass ($\mu = 1, 2, 3, \ldots$) is described and prescribed by the sine-wave and the size of the order-parameter: Thus the order-parameter tells the atoms how to behave. On the other hand the sine wave becomes only possible by the corresponding collective motion of the atoms. This example provides us with an allegory for many fields. Let us take here an extreme case, the brain. Take as subsystems the neurons and their connections. An enormous number of microscopic variables may describe their chemical and electrical activities. The order-parameters are ultimately the thoughts. Both systems necessitate each other. This brings us to a final remark. We have seen above that on a macroscopic level we need concepts completely different from those on a microscopic level. Consequently it will never suffice to describe only the electro-chemical processes of the brain to describe its functioning properly. Furthermore, the ensemble of thoughts forms again a "microscopic" system, the macroscopic order parameters of which we do not know. To describe them properly we need new concepts going beyond our thoughts, for us an unsolvable problem. For lack of space we cannot dwell here on these problems which are closely tied up with deep-lying problems in logic and which, in somewhat different shape, are well known to mathematicians, e.g., the Entscheidungsproblem. The systems treated in our book will be of a simpler nature, however, and no such problems will occur here.

1.3 How We Shall Proceed

Since in many cases self-organization occurs out of chaotic states, we first have to develop methods which adequately describe such states. Obviously, chaotic states bear in themselves an uncertainty. If we knew all quantities we could at least list them, find even some rules for their arrangement, and could thus cope with chaos. Rather we have to deal with uncertainties or, more precisely speaking, with probabilities. Thus our first main chapter will be devoted to probability theory. The next question is how to deal with systems about which very little is known. This leads us in a very natural way to basic concepts of information theory. Applying it to physics we recover basic relations of thermodynamics, so to speak, as a by-product. Here we are led to the idea of entropy, what it does, and where the problems still are. We then pass over to dynamic processes. We begin with simple examples of processes caused by random events, and we develop in an elementary but thorough way the mathematical apparatus for their proper treatment. After we have dealt with "chance", we pass over to "necessity", treating completely deterministic "motion". To it belong the equations of mechanics; but many other processes are also treated by deterministic equations. A central problem consists in the determination of equilibrium configurations (or "modes") and in the investigation of their stability. When external parameters are changed (e.g., pump power of the laser, temperature gradient in fluids, chemical concentrations) the old configuration (mode) may become unstable. This instability is a prerequisite for the occurrence of new modes. Quite surprisingly, it turns out that often a situation occurs which requires a random event to allow for a solution. Consider as a static example a stick under a load (cf. Fig. 1.27). For small loads, the straight position is still stable. However, beyond a critical load the straight position becomes unstable and two new equivalent equilibrium positions appear (cf. Fig. 1.27). Which one the stick acquires cannot be decided in a purely deterministic theory (unless asymmetries are admitted). In reality the development of a system is determined both by deterministic and random causes ("forces"), or, to use Monod's words, by "chance and necessity". Again we explain by means of the simplest cases the basic concepts

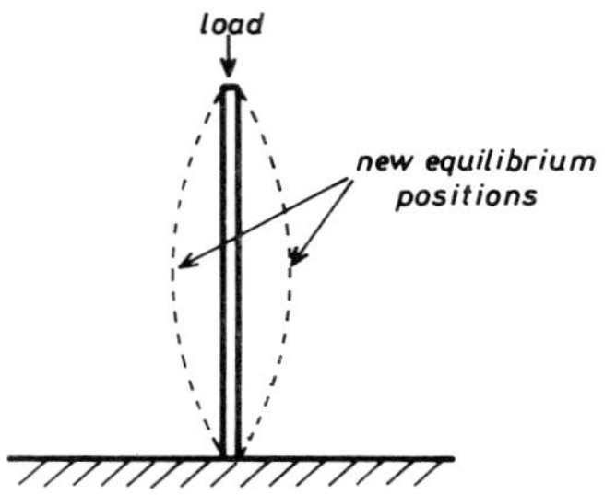

Fig. 1.27. Deformation of stick under load

and mathematical approaches. After these preliminaries we come in Chapter 7 to the central question of self-organization. We will discover how to find order parameters, how they "slave" subsystems, and how to obtain equations for order parameters. This chapter includes methods of treating continuously extended media. We will also discuss the fundamental role of fluctuations in self-organizing systems. Chapters 8 to 10 are devoted to a detailed treatment of selected examples from physics, chemistry, and biology. The logical connection between the different chapters is exhibited in Table 1.1. In the course of this book it will transpire that seemingly quite different systems behave in a completely analogous manner. This behavior is governed by a few fundamental principles. On the other hand, admittedly, we are *searching* for such analogies which show up in the essential gross features of our systems. When each system is analyzed in more and more detail down to the subsystems, quite naturally more and more differences between these systems may show up.

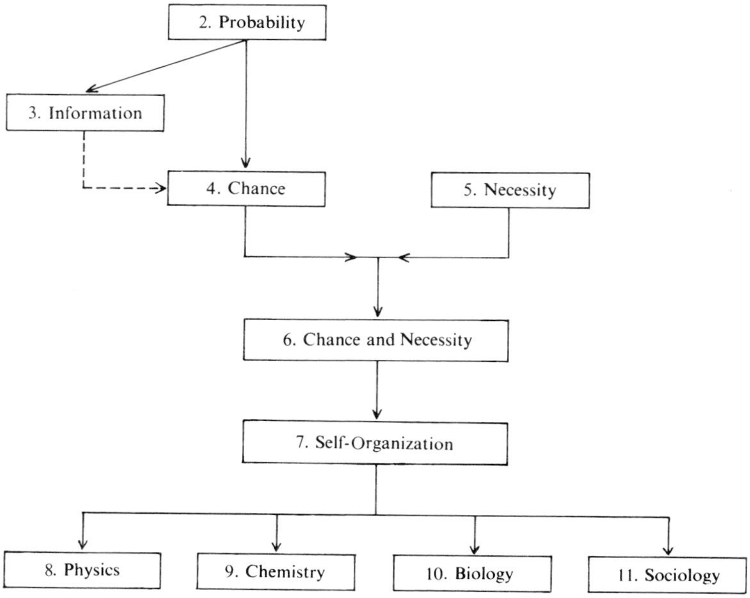

Table 1.1

2. Probability

What We Can Learn From Gambling

2.1 Object of Our Investigations: The Sample Space

The objects we shall investigate in our book may be quite different. In most cases, however, we shall treat systems consisting of very many subsystems of the same kind or of very few kinds. In this chapter we deal with the subsystems and define a few simple relations. A single subsystem may be among the following:

atoms	plants
molecules	animals
photons (light quanta)	students
cells	

Let us consider specifically a group of students. A single member of this group will be denoted by a number $\omega = 1, 2, \ldots$ The individual members of the group under consideration will be called the sample points. The total group or, mathematically stated, the total set of individuals will be called sample space (Ω) or sample set. The set Ω of sample points $1, 2, \ldots, M$ will be denoted by $\Omega = \{1, 2, \ldots, M\}$. The word "sample" is meant to indicate that a certain subset of individuals will be sampled (selected) for statistical purposes. One of the simplest examples is tossing a coin. Denoting its tail by zero and its head by one, the sample set of the coin is given by $\Omega = \{0, 1\}$. Tossing a coin now means to sample 0 or 1 at random. Another example is provided by the possible outcomes when a die is rolled. Denoting the different faces of a die by the numbers $1, 2, \ldots, 6$ the sample set is given by $\Omega = \{1, 2, 3, 4, 5, 6\}$. Though we will not be concerned with games (which are, nevertheless a very interesting subject), we shall use such simple examples to exhibit our basic ideas. Indeed, instead of rolling a die we may do certain kinds of experiments or measurements whose outcome is of a probabilistic nature. A sample point is also called a simple event, because its sampling is the outcome of an "experiment" (tossing a coin, etc.).

It will be convenient to introduce the following notations about sets. A collection of ω's will be called a subset of Ω and denoted by $A, B, \ldots$ The empty set is written as $\emptyset$, the number of points in a set S is denoted by $|S|$. If all points of the set A are contained in the set B, we write

$$A \subset B \quad \text{or} \quad B \supset A. \tag{2.1}$$

If both sets contain the same points, we write

$$A = B. \tag{2.2}$$

The union

$$A \cup B = \{\omega \mid \omega \in A \quad \text{or} \quad \omega \in B\} \qquad (2.3)^{1}$$

is a new set which contains all points contained either in A or B. (cf. Fig. 2.1). The intersection

$$A \cap B = \{\omega \mid \omega \in A \quad \text{and} \quad \omega \in B\} \tag{2.4}$$

is a set which contains those points which are both contained in A and B (cf. Fig. 2.2). The sets A and B are disjoint if (cf. Fig. 2.3).

$$A \cap B = \emptyset. \tag{2.5}$$

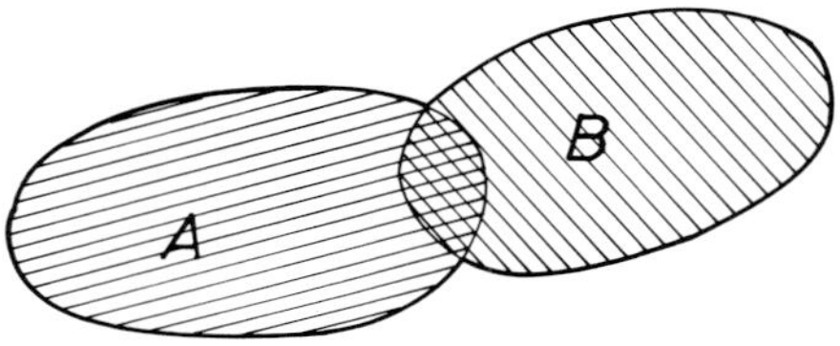

Fig. 2.1. The union $A \cup B$ of the sets A and B comprises all elements of A and B. (To visualize the relation 2.3 we represent the sets A and B by points in a plane and not along the real axis)

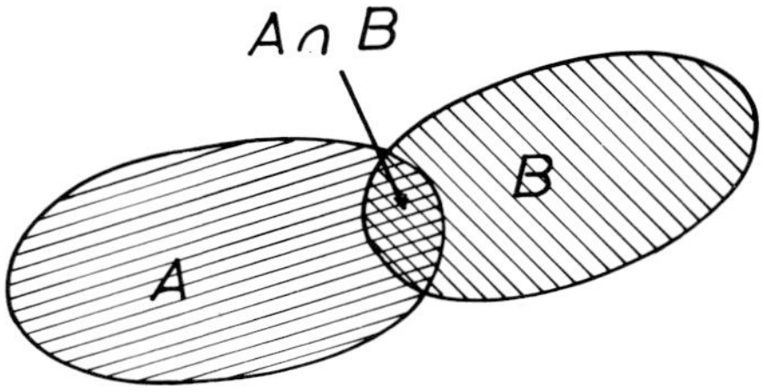

Fig. 2.2. The intersection $A \cap B$ of the sets A and B

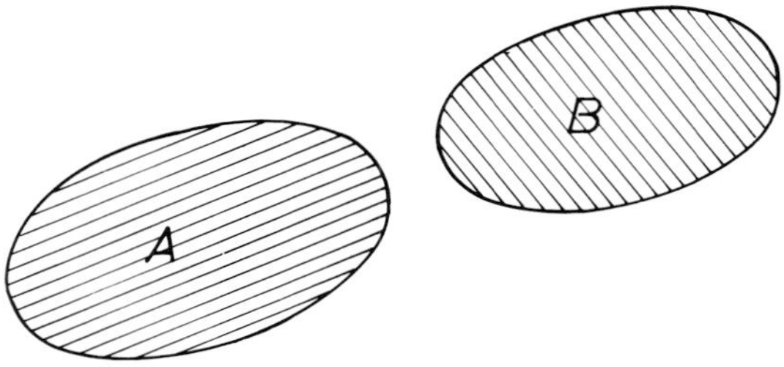

Fig. 2.3. Disjoint sets have no elements in common

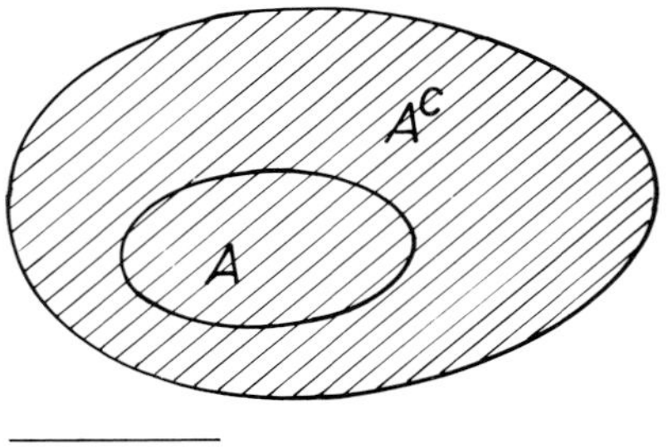

Fig. 2.4. is decomposed into A and its complement A^c

[1] Read the rhs of (2.3) as follows: all ω's, for which ω is either element of A *or* of B.

All sample points of Ω which are not contained in A form a set called the complement A^c of A (cf. Fig. 2.4). While the above-mentioned examples imply a countable number of sample points, $\omega = 1, 2, \ldots, n$ (where n may be infinite), there are other cases in which the subsystems are continuous. Think for example of an area of a thin shell. Then this area can be further and further subdivided and there are continuously many possibilities to select an area. If not otherwise noted, however, we will assume that the sample space Ω is discrete.

Exercises on 2.1

Prove the following relations 1)–4):

1) if $A \subset B$, $B \subset C$, then $A \subset C$;
   if $A \subset B$ and $B \subset A$, then $A = B$

2) $(A^c)^c = A$, $\Omega^c = \emptyset$, $\emptyset^c = \Omega$

3) a) $(A \cup B) \cup C = A \cup (B \cup C)$ (associativity)
   b) $A \cup B = B \cup A$ (commutativity)
   c) $(A \cap B) \cap C = A \cap (B \cap C)$
   d) $A \cap B = B \cap A$ (commutativity)

   e) $A \cap (B \cup C) = (A \cap B) \cup (A \cap C)$ (distributivity)
   f) $A \cup A = A \cap A = A$, $A \cup \emptyset = A$, $A \cap \emptyset = \emptyset$
   g) $A \cup \Omega = \Omega$, $A \cap \Omega = A$, $A \cup A^c = \Omega$, $A \cap A^c = \emptyset$
   h) $A \cup (B \cap C) = (A \cup B) \cap (A \cup C)$ (distributivity)
   i) $(A \cap B)^c = A^c \cup B^c$; $(A \cup B)^c = A^c \cap B^c$

   Hint: Take an arbitrary element ω of the sets defined on the lhs, show that it is contained in the set on the rhs of the corresponding equations. Then do the same with an arbitrary element ω' of the rhs.

4) de Morgan's law:
   a) $(A_1 \cup A_2 \cup \cdots \cup A_n)^c = A_1^c \cap A_2^c \cap \cdots \cap A_n^c$
   b) $(A_1 \cap A_2 \cap \cdots \cap A_n)^c = A_1^c \cup A_2^2 \cup \cdots \cup A_n^c$

   Hint: Prove by complete induction.

2.2 Random Variables

Since our ultimate goal is to establish a quantitative theory, we must describe the properties of the sample points quantitatively. Consider for example gas atoms which we label by the index ω. At each instant the atom ω has a certain velocity which is a measurable quantity. Further examples are provided by humans who might be classified according to their heights; or people who vote yes or no, and the numbers 1 and 0 are ascribed to the votes yes and no, respectively. In each case we can ascribe a numerically valued function X to the sample point ω so that we have for each ω a number $X(\omega)$ or, in mathematical terms, $\omega \to X(\omega)$ (cf. Fig. 2.5).

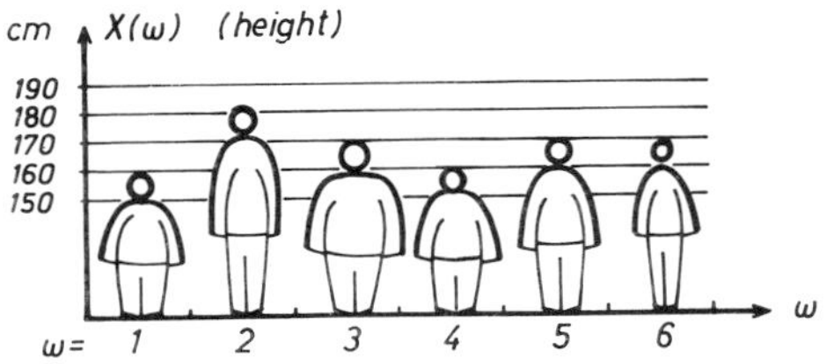

Fig. 2.5. Random variable $X(\omega)$ (heights of persons)

This function X is called a random variable, simply because the sample point ω is picked at random: Making a velocity measurement of a single gas molecule or the rolling of a die. Once the sample point is picked, $X(\omega)$ is determined. It is, of course, possible to ascribe several numerically valued functions $X_1, X_2, \ldots$ to the sample points ω, e.g. molecules might be distinguished by their weights, their velocities, their rotational energies etc. We mention a few simple facts about random variables. If X and Y are random variables, then also the linear combination $aX + bY$, the product XY, and the ratio X/Y $(Y \neq 0)$, are also random variables. More generally we may state the following: If φ is a function of two ordinary variables and X and Y are random variables, then $\omega \to \varphi(X(\omega), Y(\omega))$ is also a random variable. A case of particular interest is given by the sum of random variables

$$S_n(\omega) = X_1(\omega) + \cdots + X_n(\omega). \tag{2.6}$$

Because later on we want to treat the joint action of many subsystems (distinguished by an index $i = 1, 2, \ldots, n$), we have often to deal with such a sum.

Examples are provided by the weight of n persons in a lift, the total firing rate of n neurons, or the light wave built up from n wavetracks emitted from atoms. We shall see later that such functions as (2.6) reveal whether the subsystems (persons, neurons, atoms) act independently of each other, or in a well-organized way.

2.3 Probability

Probability theory has at least to some extent its root in the considerations of the outcomes of games. Indeed if one wants to present the basic ideas, it is still advantageous to resort to these examples. One of the simplest games is tossing a coin where one can find head or tail. There are two outcomes. However, if one bets on head there is only one favorable outcome. Intuitively it is rather obvious to define as probability for the positive outcome the ratio between the number of positive outcomes 1 divided by the number of all possible outcomes 2, so that we obtain $P = 1/2$. When a die is thrown there are six possible outcomes so that the sample space $\Omega = \{1, 2, 3, 4, 5, 6\}$. The probability to find a particular number $k = 1, 2, 3, 4, 5, 6$ is $P(k) = 1/6$. Another way of reaching this result is as follows. When, for symmetry reasons, we ascribe the same probability to all six outcomes and demand that the sum of the probabilities of these outcomes must be one, we again obtain

$$P(k) = \tfrac{1}{6},\ k = 1, 2, 3, 4, 5, 6. \tag{2.7}$$

Such symmetry arguments also play an important role in other examples, but in many cases they require far-reaching analysis. First of all, in our example it is assumed that the die is perfect. Each outcome depends, furthermore, on the way the die is thrown. Thus we must assume that this is again done in a symmetric way which after some thinking is much less obvious. In the following we shall assume that the symmetry conditions are approximately realized. These statements can be reformulated in the spirit of probability theory. The six outcomes are treated as equally likely and our assumption of this "equal likelihood" is based on symmetry.

In the following we shall call $P(k)$ a probability but when k is considered as varying, the function P will be called a probability measure. Such probability measures may not only be defined for single sample points ω but also for subsets A, B, etc. For example in the case of the die, a subset could consist of all even numbers 2, 4, 6. The probability of finding an even number when throwing a die is evidently $P(\{2,4,6\}) = 3/6 = 1/2$ or in short $P(A) = 1/2$ where $A = \{2,4,6\}$.

We now are able to define a probability measure on the sample space Ω. It is a function on Ω which satisfies three axioms:

1) For every set $A \subset \Omega$ the value of the function is a nonnegative number $P(A) \geq 0$.
2) For any two disjoint sets A and B, the value of the function of their union $A \cup B$ is equal to the sum of its values for A and B,

$$P(A \cup B) = P(A) + P(B), \text{ provided } A \cap B = \emptyset. \tag{2.8}$$

3) The value of the function for Ω (as a subset) is equal to 1

$$P(\Omega) = 1. \tag{2.9}$$

2.4 Distribution

In Section 2.2 we introduced the concept of a random variable which relates a certain quantity X (e.g., the height of a person) to a sample point ω (e.g., the person). We now consider a task which is in some sense the inverse of that relation. We prescribe a certain range of the random variable, e.g., the height between 160 and 170 cm and we ask, what is the probability of finding a person in the population whose height lies in this range (cf. Fig. 2.6). In abstract mathematical terms our problem is as follows: Let X be a random variable and a and b two constant numbers; then we are looking for the subset of sample points ω for which $a \leq X(\omega) \leq b$ or in short

$$\{a \leq X \leq b\} = \{\omega \mid a \leq X(\omega) \leq b\}. \tag{2.10}$$

We again assume that the total set Ω is countable. We have already seen in Section

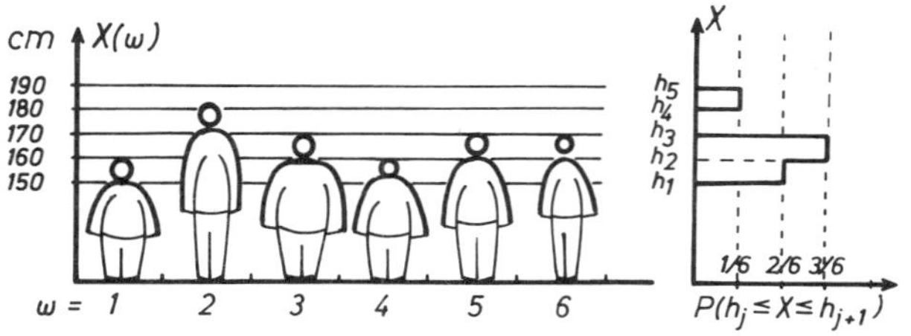

Fig. 2.6. The probability P to find a person between heights h_j and h_{j+1} (h_1= 150 cm, h_2=160 cm, etc.)

2.3 that in this case we may assign a probability to every subset of Ω. Thus we may assign to the subset defined by (2.10) a probability which we denote by

$$P(a \leq X \leq b). \tag{2.11}$$

To illustrate this definition we consider the example of a die. We ask for the probability that, if the die is rolled once, the number of spots lies in between 2 and 5. The subset of events which have to be counted according to (2.10) is given by the numbers of 2, 3, 4, 5. (Note that in our example $X(\omega) = \omega$). Because each number appears with the probability 1/6, the probability to find the subset 2, 3, 4, 5 is $P = 4/6 = 2/3$. This example immediately lends itself to generalization. We could equally well ask what is the probability of throwing an odd number of spots, so that $X = 1, 3, 5$. In that case, X does not stem from an interval but from a certain well-defined subset, A, so that we define now quite generally

$$P(X \in A) = P(\{\omega \mid X(\omega) \in A\}) \tag{2.12}$$

where A is a set of real numbers. As a special case of the definition (2.11) or (2.12), we may consider that the value of the random variable X is given: $X = x$. In this case (2.11) reduces to

$$P(X = x) = P(X \in \{x\}). \tag{2.13}$$

Now we come to a general rule of how to evaluate $P(a \leq X \leq b)$. This is suggested by the way we have counted the probability of throwing a die with a resulting number of spots lying inbetween 2 and 5. We use the fact that $X(\omega)$ is countable if Ω is countable. We denote the distinct values of $X(\omega)$ by $v_1, v_2, \ldots, v_n, \ldots$ and the set $\{v_1, v_2, \ldots\}$ by V_x. We further define that $P(X = x) = 0$ if $x \notin V_x$. We may furthermore admit that some of the v_n's have zero probabilities. We abbreviate $p_n = P(X = v_n)$ (cf. Fig. 2.7a, b). Using the axioms on page 21 we may deduce that the probability $P(a \leq X \leq b)$ is given by

$$P(a \leq X \leq b) = \textstyle\sum_{a \leq v_n \leq b} p_n, \tag{2.14}$$

and more generally the probability $P(X \in A)$ by

$$P(X \in A) = \textstyle\sum_{v_n \in A} p_n. \tag{2.15}$$

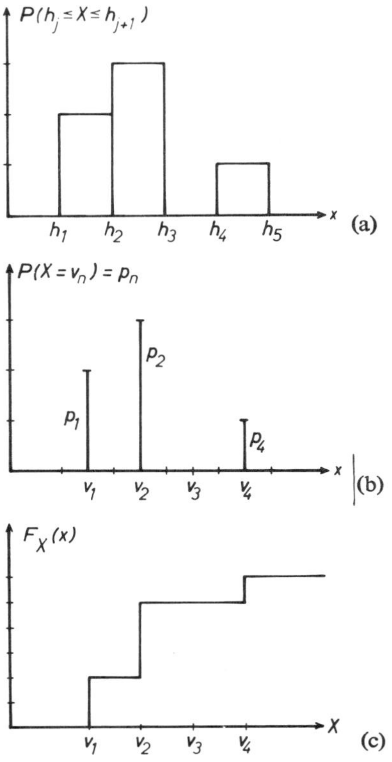

Fig. 2.7. (a) Same as rhs of Fig. 2.6, but abscissa and ordinate exchanged. (b) Probability measure p_n. We made the heights discrete by taking the average value $v_n = 1/2(h_n + h_{n+1})$ of a person's height in each interval $h_n \dots h_{n+1}$. (c) Distribution function $F_X(x)$ corresponding to *b*)

If the set A consists of all real numbers X of the interval of $-\infty$ until x we define the so called *distribution function* of X by (cf. Fig. 2.7c)

$$F_X(x) = P(X \le x) = \sum_{v_n \le x} p_n. \tag{2.16}$$

The p_n's are sometimes called elementary probabilities. They have the properties

$$p_n \ge 0 \quad \text{for all } n, \tag{2.17}$$

and

$$\sum_n p_n = 1 \tag{2.18}$$

again in accordance with the axioms. The reader should be warned that "distribution" is used with two different meanings:

I	II
"Distribution function" defined by (2.16)	"Probability distribution" (2.13) or "probability density"

2.5 Random Variables with Densities

In many practical applications the random variable X is not discrete but continuous. Consider, for instance, a needle spinning around an axis. When the needle comes to rest (due to friction), its final position may be considered as a random variable. If we describe this position by an angle ψ it is obvious that ψ is a continuous random variable (cf. Fig. 2.8). We have therefore to extend the general formulation of the foregoing section to this more general case. In particular, we wish to extend (2.14) in a suitable manner, and we may expect that this generalization consists in replacing the sum in (2.14) by an integral. To put this on mathematically safe ground, we first consider a mapping $\xi \to f(\xi)$ where the function $f(\xi)$ is defined on the real axis for $-\infty$ to $+\infty$. We require that $f(\xi)$ has the following properties

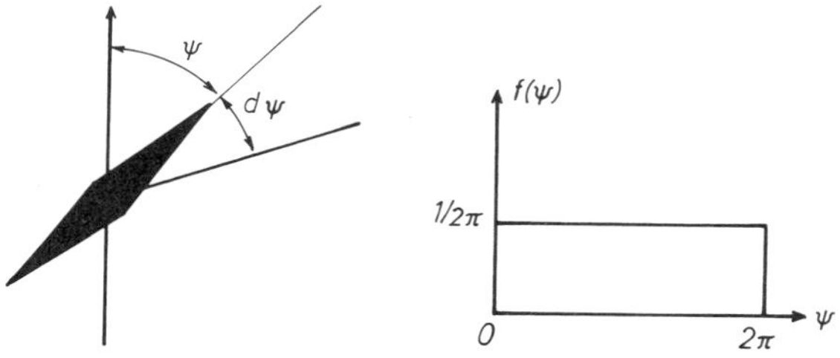

Fig. 2.8. Angle ψ and its density function $f(\psi)$

1) $\forall \xi : f(\xi) \geq 0,$ (2.19)[2]

2) $\int_{-\infty}^{\infty} f(\xi)\, d\xi = 1.$ (2.20)

(The integral is meant as Riemann's integral). We will call f a density function and assume that it is piecewise continuous so that $\int_a^b f(\xi)\, d\xi$ exists. We again consider a random variable X defined on Ω with $\omega \to X(\omega)$. We now describe the probability by (cf. Fig. 2.9)

$$P(a \leq X \leq b) = \int_a^b f(\xi)\, d\xi. \quad (2.21)^3$$

The meaning of this definition becomes immediately evident if we think again of the needle. The random variable ψ may acquire continuous values in the interval between 0 and 2π. If we exclude, for instance, gravitational effects so that we may assume that all directions have the same probability (assumption of equal like-

[2] Read $\forall \xi$: "for all ξ's".

[3] More precisely, $f(\xi)$ should carry the suffix X to indicate that $f_X(\xi)$ relates to the random variable X (cf. (2.16)).

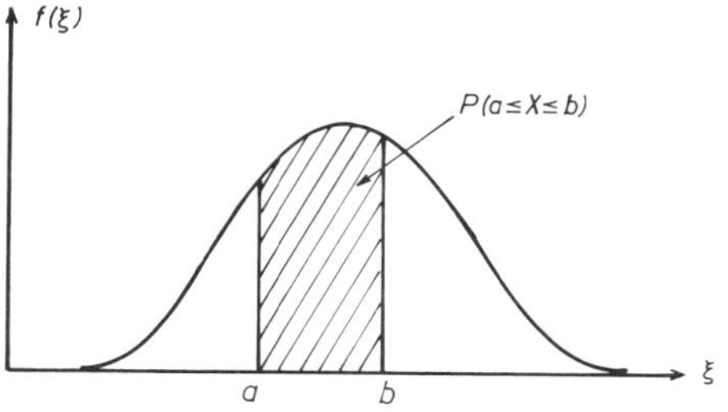

Fig. 2.9. $P(a \leq X \leq b)$ as area

lihood!), the probability of finding the needle in a certain direction in the interval ψ_0 and $\psi_0 + d\psi$ is $(1/2\pi)d\psi$. Then the probability of finding ψ in the interval ψ_1 and ψ_2 will be given by

$$P(\psi_1 \leq \psi \leq \psi_2) = \int_{\psi_1}^{\psi_2} \frac{1}{2\pi} d\psi = \frac{1}{2\pi}(\psi_2 - \psi_1).$$

Note that $1/(2\pi)$ stems from the normalization condition (2.20) and that in our case $f(\xi) = 1/(2\pi)$ is a constant. We may generalize (2.21) if A consists of a union of intervals. We then define accordingly

$$P(X \in A) = \int_A f(\xi)\, d\xi. \tag{2.22}$$

A random variable with density is sometimes also called a continuous random variable. Generalizing the definition (2.16) to continuous random variables we obtain the following: If $A = (-\infty, x)$, then the distribution function $F_X(x)$ is given by

$$F_X(x) \equiv P(X \leq x) = \int_{-\infty}^{x} f(\xi)\, d\xi. \tag{2.23}$$

In particular if f is continuous we find

$$F'_X(x) = f_X(x). \tag{2.24}$$

Exercise on 2.5

1) Dirac's δ-function is defined by:

$$\delta(x - x_0) = 0 \quad \text{for} \quad x \neq x_0$$

and

$$\int_{x_0-\varepsilon}^{x_0+\varepsilon} \delta(x - x_0)\, dx = 1 \quad \text{for any} \quad \varepsilon > 0.$$

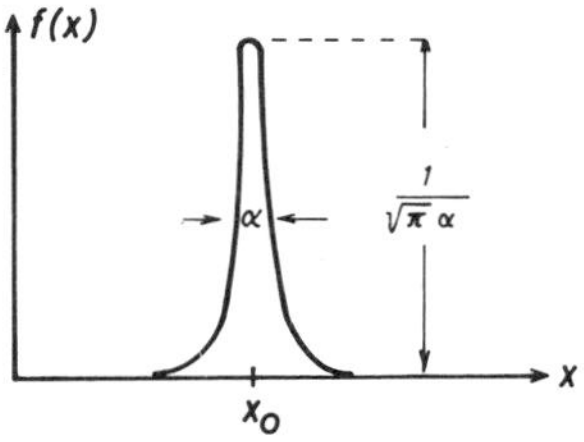

Fig. 2.10. The δ-function may be considered as a Gaussian function $\alpha^{-1}\pi^{-1/2}\exp[-(x-x_0)^2/\alpha^2]$ in the limit $\alpha \to 0$

Show: The distribution function

$$f(x) = \textstyle\sum_{j=1}^{n} p_j \delta(x - x_j)$$

allows us to write (2.16) in the form (2.23).

2) a) Plot $f(x) = \alpha \exp(-\alpha x)$ and the corresponding $F_X(x)$, $x \geq 0$ versus x.
b) Plot $f(x) = \beta(\exp(-\alpha x) + \gamma\delta(x - x_1))$

and the corresponding $F_X(x)$; $x \geq 0$, versus x. Determine β from the normalization condition (2.20).

Hint: The δ-function is "infinitely high." Thus indicate it by an arrow.

2.6 Joint Probability

So far we have considered only a single random variable, i.e., the height of persons. We may ascribe to them, however, simultaneously other random variables, e.g., their weights, color, etc. (Fig. 2.11). This leads us to introduce the "joint probability" which is the probability of finding a person with given weight, a given height, etc. To cast this concept into mathematical terms we treat the example of two random variables. We introduce the set S consisting of all pairs of values (u, v) that the

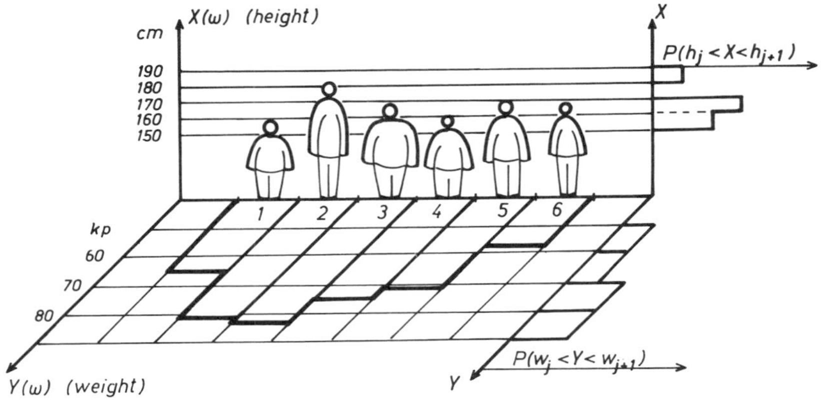

Fig. 2.11. The random variables X (height) and Y (weight). On the rhs the probability measures for the height (irrespective of weight) and for the weight (irrespective of height) are plotted

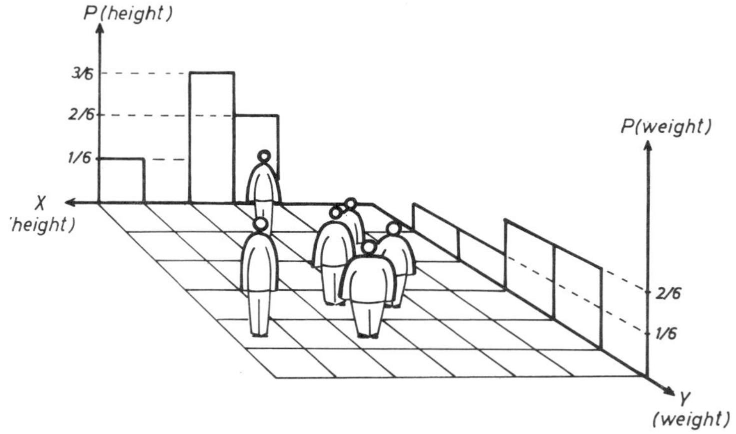

Fig. 2.12. The persons are grouped according to their weights and heights

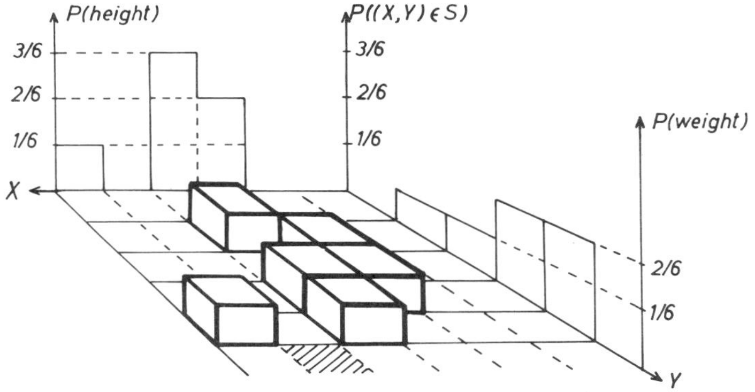

Fig.2.13. The joint probability $P((X, Y) \in S)$ plotted over X and Y. The subsets S are the single squares (see for example shaded square)

random variables (X, Y) acquire. For any subset $S' \in S$ we then define the joint probability, generalizing (2.12), as (cf. Fig. 2.13)

$$P((X, Y) \in S') = P(\{\omega \mid (X(\omega), Y(\omega)) \in S'\}). \tag{2.25}$$

Labelling the distinct values of $X(\omega) = u_m$ by m and those of $Y(\omega) = v_n$ by n, we put

$$P(X = u_m, Y = v_n) = p_{mn}. \tag{2.26}$$

Using the axioms of Section 2.3, we may easily show that the probability $P(X = u_m)$ i.e. the probability that $X(\omega) = u_m$ irrespective of which values Y acquires, is given by

$$P(X = u_m) = \sum_n P(X = u_m, Y = v_n). \tag{2.27}$$

The joint probability is sometimes also called multi-variate probability.

Exercise on 2.6

1) Generalize the definitions and relations of Sections 2.4 and 2.5 to the present case.
2) Generalize the above definitions to several random variables $X_1(\omega)$, $X_2(\omega)$, . . . , $X_N(\omega)$.

2.7 Mathematical Expectation E(X), and Moments

The quantity $E(X)$ we shall define is also called expected value, the mean, or, the first moment. We consider a random variable X which is defined on a countable space Ω. Consider as an example the die. We may ask, what is the mean value of spots which we obtain when throwing the die many times. In this case the mean value is defined as the sum over 1, 2, . . . , 6, divided by the total number of possible throws. Remembering that 1/(number of possible throws) is equal to the probability of throwing a given number of spots, we are led to the following expression for the mean:

$$E(X) = \sum_{\omega \in \Omega} X(\omega) P(\{\omega\}) \tag{2.28}$$

where $X(\omega)$ is the random variable, P is the probability of the sample point ω and the sum runs over all points of the sample set Ω. If we label the sample points by integers 1, 2, . . . , n we may use the definition of Section 2.4 and write

$$E(X) = \sum_n p_n v_n. \tag{2.29}$$

Because each function of a random variable is again a random variable we can readily generalize (2.28) to the mean of a function $\varphi(X)$ and find in analogy to (2.28)

$$E(\varphi(X)) = \sum_{\omega \in \Omega} \varphi(X(\omega)) P(\{\omega\}), \tag{2.30}$$

and in analogy to (2.29),

$$E(\varphi(X)) = \sum_n p_n \varphi(v_n). \tag{2.31}$$

The definition of the mean (or mathematical expectation) may be immediately generalized to a continuous variable so that we only write down the resulting expression

$$E(\varphi(X)) = \int_{-\infty}^{\infty} \varphi(\xi) f(\xi)\, d\xi. \tag{2.32}$$

When we put $\varphi(X) = X^r$ in (2.30), or (2.31), or (2.32) the mathematical expectation

$E(X^r)$ is called the rth moment of X. We further define the variance by

$$E\{(X - E(X))^2\} = \sigma^2. \tag{2.33}$$

Its square root, σ, is called standard deviation.

2.8 Conditional Probabilities

So far we have been considering probabilities without further conditions. In many practical cases, however, we deal with a situation, which, if translated to the primitive game of rolling a die, could be described as follows: We roll a die but now ask for the probability that the number of spots obtained is three *under the condition* that the number is odd. To find a proper tool for the evaluation of the corresponding probability, we remind the reader of the simple rules we have established earlier. If Ω is finite and all sample points have the same weight, then the probability of finding a member out of a set A is given by

$$P(A) = \frac{|A|}{|\Omega|} \tag{2.34}$$

where $|A|$, $|\Omega|$ denote the number of elements of A or Ω. This rule can be generalized if Ω is countable and each point ω has the weight $P(\omega)$. Then

$$P(A) = \frac{\sum_{\omega \in A} P(\omega)}{\sum_{\omega \in \Omega} P(\omega)}. \tag{2.35}$$

In the cases (2.34) and (2.35) we have admitted the total sample space. As an example for (2.34) and (2.35) we just have to consider again the die as described at the beginning of the section.

We now ask for the following probability: We restrict the sample points to a certain subset $S(\omega)$ and ask for the proportional weight of the part of A in S relative to S. Or, in the case of the die, we admit as sample points only those which are odd. In analogy to formula (2.35) we find for this probability

$$P(A \mid S) = \frac{\sum_{\omega \in A \cap S} P(\omega)}{\sum_{\omega \in S} P(\omega)}. \tag{2.36}$$

Extending the denominator and numerator in (2.36) by $1/\sum_{\omega \in \Omega} P(\omega)$ we may rewrite (2.36) in the form

$$P(A \mid S) = \frac{P(A \cap S)}{P(S)}. \tag{2.37}$$

This quantity is called the *conditional probability* of A relative to S. In the literature other terminologies are also used, such as "knowing S", "given S", "under the hypothesis of S".

Exercises on 2.8

1) A die is thrown. Determine the probability to obtain 2 (3) spots under the hypothesis that an odd number is thrown.
   Hint: Verify $A = \{2\}$, or $= \{3\}$, $S = \{1, 3, 5\}$. Determine $A \cap S$ and use (2.37).
2) Given are the stochastic variables X and Y with probability measure

$$P(m, n) \equiv P(X = m, Y = n).$$

Show that

$$P(m \mid n) = \frac{P(m, n)}{\sum_m P(m, n)}.$$

2.9 Independent and Dependent Random Variables

For simplicity we consider countable, valued random variables, though the definition may be readily generalized to noncountable, valued random variables. We have already mentioned that several random variables can be defined simultaneously, e.g., the weight and height of humans. In this case we really expect a certain relation between weight and height so that the random variables are not independent of each other. On the other hand when we roll two dice simultaneously and consider the number of spots of the first die as random variable, X_1, and that of the second die as random variable, X_2, then we expect that these random variables are independent of each other. As one verifies very simply by this example, the joint probability may be written as a product of probabilities of each individual die. Thus we are led to define the following quite generally: Random variables $X_1, X_2, \ldots, X_n$ are independent if and only if for any real numbers $x_1, \ldots, x_n$ we have

$$P(X_1 = x_1, \ldots, X_n = x_n) = P(X_1 = x_1)P(X_2 = x_2) \cdots P(X_n = x_n). \tag{2.38}$$

In a more general formulation which may be derived from (2.38) we state that the variables vary independently if and only if for arbitrary countable sets $S_1, \ldots, S_n$, the following holds:

$$P(X_1 \in S_1, \ldots, X_n \in S_n) = P(X_1 \in S_1) \cdots P(X_n \in S_n). \tag{2.39}$$

We can mention a consequence of (2.38) with many practical applications. Let $\varphi_1, \ldots, \varphi_n$ be arbitrary, real valued functions defined on the total real axis and $X_1, \ldots, X_n$ independent random variables. Then the random variables $\varphi_1(X_1), \ldots, \varphi_n(X_n)$ are also independent. If random variables are not independent of each other, it is desirable to have a measure for the degree of their independence or, in a more positive statement, about their correlation. Because the expectation value of the product of independent random variables factorizes (which follows from (2.38)), a measure for the correlation will be the deviation of $E(XY)$ from $E(X)E(Y)$. To avoid having large values of the random variables X, Y, with a small correlation mimic a large correlation, one normalizes the difference

$$E(XY) - E(X)E(Y) \tag{2.40}$$

by dividing it by the standard deviations $\sigma(X)$ and $\sigma(Y)$. Thus we define the so-called correlation

$$\rho(X, Y) = \frac{E(XY) - E(X)E(Y)}{\sigma(X)\sigma(Y)} \tag{2.41}$$

Using the definition of the variance (2.33), we may show that

$$\sigma^2(X + Y) = \sigma^2(X) + \sigma^2(Y) \tag{2.42}$$

for independent random variables X and Y (with finite variances).

Exercise on 2.9

The random variables X, Y may both assume the values 0 and 1. Check if these variables are statistically independent for the following joint probabilities:

a)	b)	c)
$P(X = 0, Y = 0) = \frac{1}{4}$	$= \frac{1}{2}$	$= 1$
$P(X = 1, Y = 0) = \frac{1}{4}$	$= 0$	$= 0$
$P(X = 0, Y = 1) = \frac{1}{4}$	$= 0$	$= 0$
$P(X = 1, Y = 1) = \frac{1}{4}$	$= \frac{1}{2}$	$= 0$

Visualize the results by coin tossing.

2.10* Generating Functions and Characteristic Functions

We start with a special case. Consider a random variable X taking only non-negative integer values. Examples for X are the number of gas molecules in a cell of a given volume out of a bigger volume, or, the number of viruses in a volume of blood out of a much larger volume. Let the probability distribution of X be given by

$$P(X = j) = a_j, \quad j = 0, 1, 2, \ldots \tag{2.43}$$

We now want to represent the distribution (2.43) by means of a single function. To this end we introduce a dummy variable z and define a generating function by

$$g(z) = \sum_{j=0}^{\infty} a_j z^j. \tag{2.44}$$

We immediately obtain the coefficients a_j with aid of the Taylor expansion of the function $g(z)$ by taking the j's derivative of $g(z)$ and dividing it by $j!$

$$a_j = \frac{1}{j!}\frac{d^j g}{dz^j}\bigg|_{z=0} \tag{2.45}$$

The great advantage of (2.44) and (2.45) rests in the fact that in a number of important practical cases $g(z)$ is an explicitly defined function, and that by means of $g(z)$ one may easily calculate expectation values and moments. We leave it to the reader how one may derive the expression of the first moment of the random variables X by means of (2.44). While (2.45) allows us to determine the probability distribution (2.43) by means of (2.44), it is also possible to derive (2.44) in terms of a function of the random variable X, namely,

$$g(z) = E(z^X). \tag{2.46}$$

This can be shown as follows: For each z, $\omega \to z^{X(\omega)}$ is a random variable and thus according to formulas (2.30), (2.43) we obtain

$$E(z^X) = \sum_{j=0}^{\infty} P(X = j) z^j = g(z). \tag{2.47}$$

We mention an important consequence of (2.44). If the random variables $X_1, \ldots, X_n$ are independent and have $g_1, \ldots, g_n$ as their generating functions, then the generating function of the sum $X_1 + X_2 + \cdots + X_n$ is given by the product $g_1 g_2 \cdots g_n$. The definition of the generating function of the form (2.46) lends itself to a generalization to more general random variables. Thus one defines for non-negative variables a generating function by replacing z by $e^{-\lambda}$

$$\text{Laplace transform } (z \to \mathrm{e}^{-\lambda});\ E(z^X) \to E(\mathrm{e}^{-\lambda X}), \tag{2.48}$$

and for arbitrary random variables by replacing z by $e^{i\Theta}$

$$\text{Fourier transform } (z \to \mathrm{e}^{i\Theta}),\ E(z^X) \to E(\mathrm{e}^{i\Theta X}). \tag{2.49}$$

The definitions (2.48) and (2.49) can be used not only for discrete values of X but also for continuously distributed values X. We leave to the reader the detailed formulation as an exercise.

Exercise on 2.10

Convince yourself that the derivatives of the characteristic function

$$\Phi_X(\Theta) = E\{\mathrm{e}^{i\Theta X}\}$$

yield the moments:

$$\left.\frac{d^n\Phi_X(\Theta)}{d\Theta^n}\right|_{\Theta=0} = i^n E(X^n).$$

2.11 A Special Probability Distribution: Binomial Distribution

In many practical cases one repeats an experiment many times with two possible outcomes. The simplest example is tossing a coin n times. But quite generally we may consider two possible outcomes as success or failure. The probability for success of a single event will be denoted by p, that of failure by q, with $p + q = 1$. We will now derive an expression for the probability that out of n trials we will have k successes. Thus we are deriving the probability distribution for the random variable X of successes where X may acquire the values $k = 0, 1, \ldots, n$. We take the example of the coin and denote success by 1 and failure by 0. If we toss the coin n times, we get a certain sequence of numbers 0, 1, 0, 0, 1, 1, . . . , 1, 0. Since the subsequent events are independent of each other, the possibility of finding this specific sequence is simply the product of the corresponding probabilities p or q. In the case just described we find $P = qpqqpp \cdots pq = p^k q^{n-k}$. By this total trial we have found the k successes, but only in a *specific sequence*.

In many practical applications we are not interested in a specific sequence but in all sequences giving rise to the same number of successes. We therefore ask the question how many different sequences of zero's and one's we can find with the same amount, k, of numbers "1". To this end we consider n boxes each of which can be filled with one number, unity or zero. (Usually such a box model is used by means of a distribution of black and white balls. Here we are taking the numbers zero and unity, instead). Because this kind of argument will be repeated very often, we give it in detail. Let us take a number, 0 or 1. Then we have n possibilities to put that number into one of the n boxes. For the next number we have $n - 1$ possibilities (because one box is already occupied), for the third $n - 2$ possibilities and so on. The total number of possibilities of filling up the boxes is given by the product of these individual possibilities which is

$$n(n-1)(n-2)\cdots 2\cdot 1 = n!$$

This filling up of the boxes with numbers does not lead in all cases to a different configuration because when we exchange two units or several units among each other we obtain the same configuration. Since the "1"'s can be distributed in $k!$ different manners over the boxes and in a similar way the "0"'s in $(n - k)!$ manners, we must divide the total number $n!$ by $k!(n - k)!$. The expression $n!/k!(n - k)!$ is denoted by $\binom{n}{k}$ and is called a binomial coefficient. Thus we find the total probability distribution in the following manner: There are $\binom{n}{k}$ different sequences each having the probability $p^k q^{n-k}$. Because for different subsets of events the probabilities are

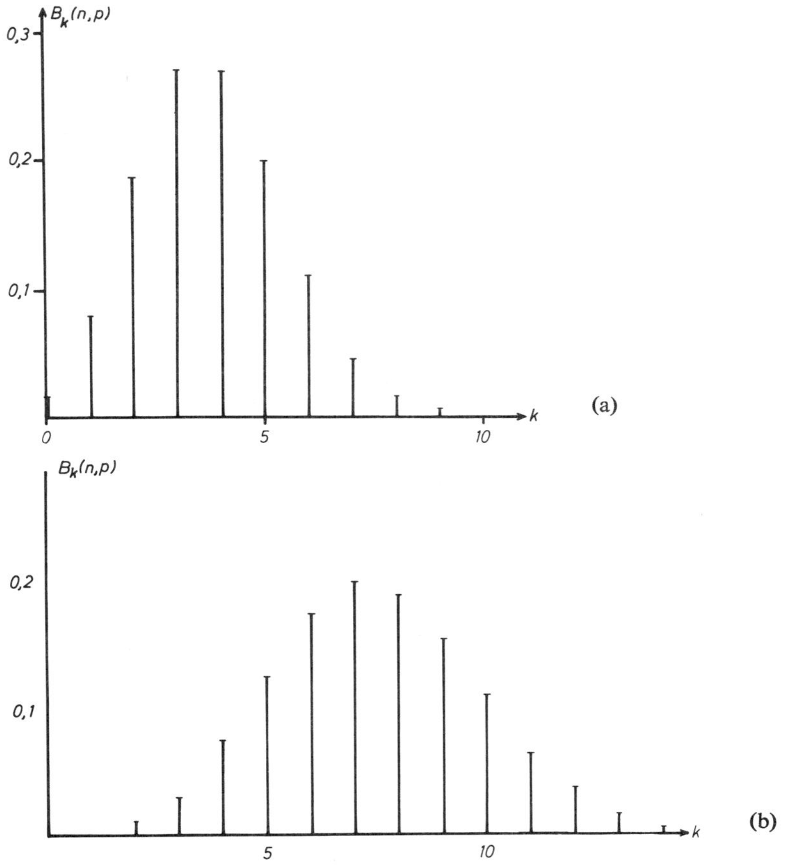

Fig. 2.14a and b. Binomial distribution $B_k(n, p)$ as function of k for $p = 1/4$ and $n = 15$ (a), $n = 30$ (b)

additive, the probability of finding k successes irrespective of the sequence, is given by

$$B_k(n, p) = \binom{n}{k} p^k q^{n-k} \tag{2.50}$$

(*binomial distribution*). Examples are given in Fig. 2.14. The mean value of the binomial distribution is

$$E(X) = np, \tag{2.51}$$

the variance

$$\sigma^2 = n \cdot p \cdot q. \tag{2.52}$$

We leave the proof as an exercise to the reader. For large n it is difficult to evaluate B. On the other hand in practical cases the number of trials n may be very large. Then (2.50) reduces to certain limiting cases which we shall discuss in subsequent sections.

The great virtue of the concept of probability consists in its flexibility which allows for application to completely different disciplines. As an example for the binomial distribution we consider the distribution of viruses in a blood cell under a microscope (cf. Fig. 2.15). We put a square grid over the cell and

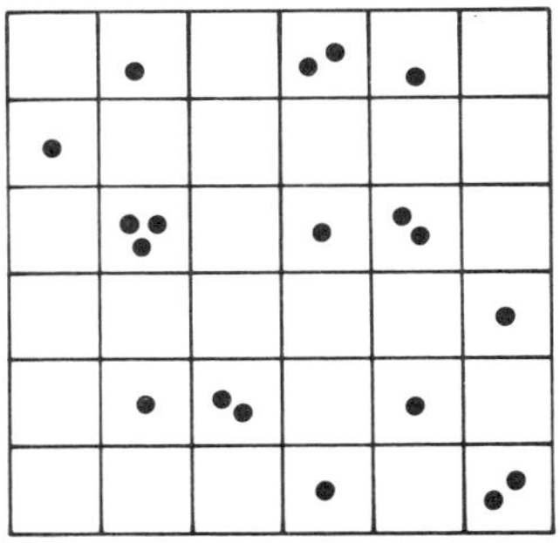

Fig. 2.15. Distribution of particles under a grid

ask for the probability distribution to find a given number of particles in a square (box). If we assume N squares and assume $n = \mu N$ particles present, then μ is the average number of particles found in a box. The problem may be mapped on the preceding one on coin tossing in the following manner: Consider a specific box and first consider only one virus. Then we have a positive event (success) if the virus is in the box and a negative event (failure) when it is outside the box. The probability for the positive event is apparently $p = 1/N$. We now have to distribute $\mu N = n$ particles into two boxes, namely into the small one under consideration and into the rest volume. Thus we make n trials. If in a "Gedankenexperiment", we put one virus after the other into the total volume, this corresponds exactly to tossing a coin. The probability for each special sequence with k successes, i.e., k viruses in the box under consideration, is given as before by $p^k q^{n-k}$. Since there are again $\binom{n}{k}$ different sequences, the total probability is again given by (2.50) or using the specific form $n = \mu N$ by

$$B_k(n, p) = \binom{n}{k} p^k q^{n-k} = \binom{\mu N}{k} \left(\frac{1}{N}\right)^k \left(1 - \frac{1}{N}\right)^{\mu N - k} \tag{2.53}$$

In practical applications one may consider N and thus also n as very large numbers, while μ is fixed and p tends to zero. Under the condition $n \to \infty$, μ fixed, $p \to 0$ one may replace (2.50) by the so-called Poisson distribution.

2.12 The Poisson Distribution

We start from (2.53) which we write using elementary steps in the form

$$B_k(n, p) = \frac{n(n-1)(n-2)\cdots(n-k+1)}{k!}\cdot\left(\frac{\mu}{n}\right)^k\left(1-\frac{\mu}{n}\right)^n\left(1-\frac{\mu}{n}\right)^{-k}$$

$$= \underbrace{\frac{\mu^k}{k!}}_{1}\underbrace{\left(1-\frac{\mu}{n}\right)^n}_{2}\underbrace{\left[\left(1-\frac{1}{n}\right)\left(1-\frac{2}{n}\right)\cdots\left(1-\frac{k-1}{n}\right)\left(1-\frac{\mu}{n}\right)^{-k}\right]}_{3} \tag{2.54}$$

For μ, k fixed but $n \to \infty$ we find the following expressions for the factors 1, 2, 3 in (2.54). The first factor remains unchanged while the second factor yields the exponential function

$$\lim_{n\to\infty}\left(1-\frac{\mu}{n}\right)^n = e^{-\mu}. \tag{2.55}$$

The third factor reduces in a straightforward way to

$$\lim_{n\to\infty}[\cdots] = 1. \tag{2.56}$$

We thus obtain the Poisson distribution

$$\pi_{k,\mu} \equiv \lim_{n\to\infty} B_k(n, p) = \frac{\mu^k}{k!}e^{-\mu}. \tag{2.57}$$

Examples are given in Fig. 2.16. Using the notation of Section 2.10 we may also write

$$a_k = \frac{\mu^k}{k!}e^{-\mu}. \tag{2.58}$$

We thus find for the generating function

$$g(z) = \sum_{k=0}^{\infty} a_k z^k = e^{(z-1)\mu}. \tag{2.59}$$

The straightforward evaluation of the mathematical expectation and of the variance (which we leave to the reader as an exercise) yields

$$E(X) = g'(1) = \mu, \tag{2.60}$$

$$\sigma^2(X) = \mu. \tag{2.61}$$

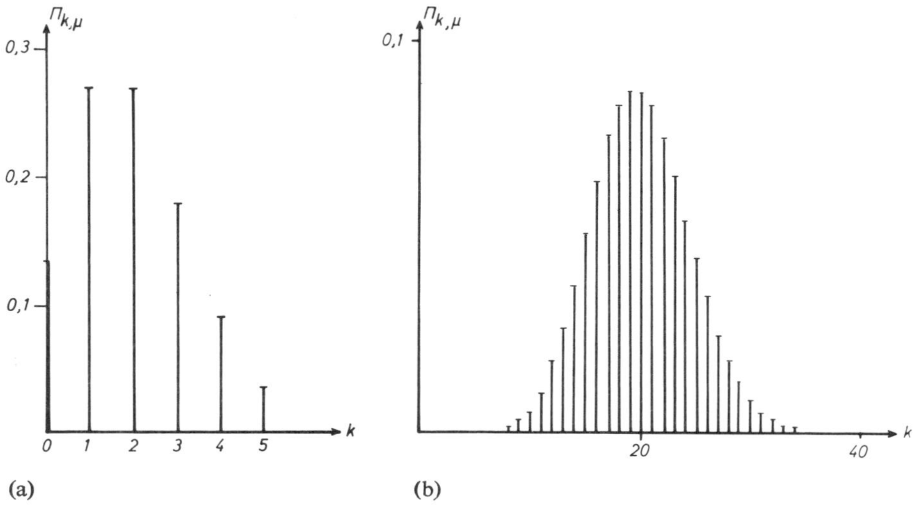

Fig. 2.16a and b. Poisson distribution $\pi_{k,\mu}$ as function of k for $\mu = 2$ (a) and $\mu = 20$ (b)

Exercise on 2.12

Prove: $E(X(X-1)\cdots(X-l+1)) = \mu^l$

Hint: Differentiate $g(z)$, (2.59), l times.

2.13 The Normal Distribution (Gaussian Distribution)

The normal distribution can be obtained as a limiting case of the binomial distribution, again for $n \to \infty$ but for p and q not small, e.g., $p = q = 1/2$. Because the presentation of the limiting procedure requires some more advanced mathematics which takes too much space here, we do not give the details but simply indicate the spirit. See also Section 4.1. We first introduce a new variable, u, by the requirement that the mean value of $k = np$ corresponds to $u = 0$; we further introduce a new scale with $k \to k/\sigma$ where σ is the variance. Because according to formula (2.52) the variance tends to infinity with $n \to \infty$, this scaling means that we are eventually replacing the discrete variables k by a continuous variable. Thus it is not surprising that we obtain instead of the original distribution function B, a density function, φ where we use simultaneously the transformation $\varphi = \sigma B$. Thus more precisely we put

$$\varphi_n(u) = \sigma B_k(n, p). \tag{2.62}$$

We then obtain the normal distribution by letting $n \to \infty$ so that we find

$$\lim_{n\to\infty} \varphi_n(u) = \varphi(u) = \frac{1}{\sqrt{2\pi}} e^{-\frac{1}{2}u^2} \tag{2.63}$$

$\varphi(u)$ is a density function (cf. Fig. 2.17). Its integral is the normal (or Gaussian) distribution

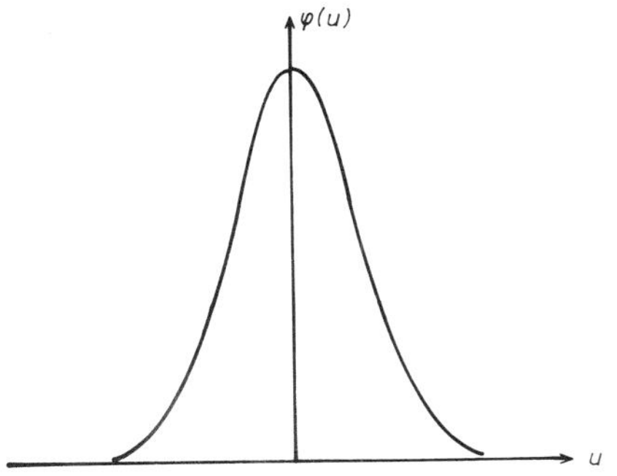

Fig. 2.17. Gaussian distribution $\varphi(u)$ as function of u. Note the bell shape

$$F(x) = \frac{1}{\sqrt{2\pi}} \int_{-\infty}^{x} e^{-\frac{1}{2}u^2}\, du. \tag{2.64}$$

Normal (or Gaussian) Probability Density of Several Variables

When several random variables are present in practical applications often the following joint Gaussian probability density appears

$$f_{\boldsymbol{X}}(\boldsymbol{x}) = (2\pi)^{-n/2} |Q|^{-1/2} \exp\left[-\tfrac{1}{2}(\boldsymbol{x} - \boldsymbol{m})^T Q^{-1} (\boldsymbol{x} - \boldsymbol{m})\right]. \tag{2.65}$$

$\boldsymbol{X}$ is a vector composed of the random variables $X_1, \ldots, X_n$, $\boldsymbol{x}$ is the corresponding vector the random variable $\boldsymbol{X}$ can acquire. $Q(\mu_{ij})$ is n by n matrix. We assume that its inverse Q^{-1} exists. $|Q| = \det Q$ denotes the determinant of Q. $\boldsymbol{m}$ is a given constant vector, T denotes the transposed vector. The following relation holds: The mean of the random variable X_j is equal to the j component of $\boldsymbol{m}$:

$$m_j = E\{X_j\}, j = 1, \ldots, n. \tag{2.66}$$

Furthermore the components of the variance μ_{ij} as defined by

$$\mu_{ij} = E\{(X_i - m_i)(X_j - m_j)\} \tag{2.67}$$

coincide with the matrix elements of $Q = (Q_{ij})$; i.e., $\mu_{ij} = Q_{ij}$.

Exercises on 2.13

1) Verify that the first moments of the Gaussian density

$$f(x) = \sqrt{\frac{\alpha}{\pi}} \exp(-\alpha x^2)$$

are given by $m_1 = 0$, $m_2 = 1/(2\alpha)$
2) Verify that the characteristic function of the Gaussian density is given by

$$E\{\exp(i\Theta X)\} = \exp(-\Theta^2/(4\alpha)).$$

3) Verify (2.66) and (2.67).
Hint: Introduce new variables $y_j = x_j - m_j$ and diagonalize Q by new linear combinations $\xi_k = \sum a_{kj} y_j$. Q is a symmetric matrix.
4) Verify

$$E\{X_1^{l_1} X_2^{l_2} \cdots X_n^{l_n}\} = i^{-(l_1+l_2+\cdots+l_n)} \left\{ \frac{\partial^{l_1+l_2+\cdots+l_n}}{\partial\Theta_1^{l_1}\partial\Theta_2^{l_2}\cdots\partial\Theta_n^{l_n}} \Phi_X(\Theta) \right\}_{\Theta=0}.$$

2.14 Stirling's Formula

For several applications in later chapters we have to evaluate $n!$ for large values of n. This is greatly facilitated by use of Stirling's formula

$$n! = \left(\frac{n}{e}\right)^n \sqrt{2\pi n}\, e^{\omega(n)} \quad \text{where} \quad \frac{1}{12(n+\frac{1}{2})} < \omega(n) < \frac{1}{12n}. \tag{2.68}$$

Since in many practical cases, $n \gg 1$, the factor $\exp \omega(n)$ may be safely dropped.

2.15* Central Limit Theorem

Let X_j, $j \geq 1$ be a sequence of independent and identically distributed random variables. We assume that the mean m and the variance σ^2 of each X_j is finite. The sum $S_n = X_1 + \cdots + X_n$, $n \geq 1$, is again a random variable (compare Section 2.2). Because the random variables X_j are independent we find for the mean

$$E(S_n) = n \cdot m \tag{2.69}$$

and for the variance

$$\sigma^2(S_n) = n\sigma^2. \tag{2.70}$$

Subtracting $n \cdot m$ from S_n and dividing the difference by $\sigma\sqrt{n}$ we obtain a new random variable

$$Y_n = \frac{S_n - n \cdot m}{\sigma\sqrt{n}} \tag{2.71}$$

which has zero mean and unit variance. The central limit theorem then makes a statement about the probability distribution of Y_n in the limit $n \to \infty$. It states

$$\lim_{n\to\infty} P(a < Y_n \le b) = \frac{1}{\sqrt{2\pi}} \int_a^b \mathrm{e}^{-\xi^2/2}\, d\xi. \tag{2.72}$$

Particularly in physics, the following form is used: We put $a = x$, and $b = x + dx$, where dx denotes a small interval. Then the integral on the rhs of (2.72) can be approximately evaluated and the result reads

$$\lim_{n\to\infty} P(x < Y_n < x + dx) = \frac{1}{\sqrt{2\pi}} \mathrm{e}^{-x^2/2}\, dx \tag{2.73}$$

or, verbally, in the limit $n \to \infty$, the probability that Y_n lies in the interval $x, \ldots,$ $x + dx$ is given by a Gaussian density multiplied by the interval dx. Note that the interval dx must not be dropped. Otherwise errors might occur if coordinate transformations are performed. For other practical applications one often uses (2.72) in a rough approximation in the form

$$P(x_1\sigma\sqrt{n} < S_n - mn < x_2\sigma\sqrt{n}) \approx \Phi(x_2) - \Phi(x_1) \tag{2.74}$$

where

$$\Phi(x) = \frac{1}{\sqrt{2\pi}} \int_{-\infty}^{x} \mathrm{e}^{-\xi^2/2}\, d\xi. \tag{2.75}$$

The central limit theorem plays an enormous role in the realm of synergetics in two ways: it applies to all cases where one has no correlation between different events but in which the outcome of the different events piles up via a sum. On the other hand, the breakdown of the central limit theorem in the form (2.72) will indicate that the random variables X_j are no more independent but *correlated.* Later on, we will encounter numerous examples of such a *cooperative* behavior.

3. Information

How to Be Unbiased

3.1 Some Basic Ideas

In this chapter we want to show how, by some sort of new interpretation of probability theory, we get an insight into a seemingly quite different discipline, namely information theory. Consider again the sequence of tossing a coin with outcomes 0 and 1. Now interpret 0 and 1 as a dash and dot of a Morse alphabet. We all know that by means of a Morse alphabet we can transmit messages so that we may ascribe a certain meaning to a certain sequence of symbols. Or, in other words, a certain sequence of symbols carries information. In information theory we try to find a *measure for the amount of information.*

Let us consider a simple example and consider R_0 different possible events ("realizations") which are equally probable *a priori*. Thus when tossing a coin we have the events 1 and 0 and $R_0 = 2$. In the case of a die we have 6 different outcomes, therefore $R_0 = 6$. Therefore the outcome of tossing a coin or throwing a die is interpreted as receiving a message and only one out of R_0 outcomes is actually realized. Apparently the greater R_0, the greater is the uncertainty before the message is received and the larger will be the amount of information after the message is received. Thus we may interpret the whole procedure in the following manner: In the initial situation we have no information I_0, i.e., $I_0 = 0$ with R_0 equally probable outcomes.

In the final situation we have an information $I_1 \neq 0$ with $R_1 = 1$, i.e., a single outcome. We now want to introduce a measure for the amount of information I which apparently must be connected with R_0. To get an idea how the connection between R_0 and I must appear we require that I is additive for independent events. Thus when we have two such sets with R_{01} or R_{02} outcomes so that the total number of outcomes is

$$R_0 = R_{01}R_{02}, \tag{3.1}$$

we require

$$I(R_{01}R_{02}) = I(R_{01}) + I(R_{02}). \tag{3.2}$$

This relation can be fulfilled by choosing

$$I = K \ln R_0 \tag{3.3}$$

where K is a constant. It can even be shown that (3.3) is the only solution to (3.2). The constant K is still arbitrary and can be fixed by some definition. Usually the following definition is used. We consider a so-called "binary" system which has only two symbols (or letters). These may be the head and the tail of a coin, or answers yes and no, or numbers 0 and 1 in a binomial system. When we form all possible "words" (or sequences) of length n, we find $R = 2^n$ realizations. We now want to identify I with n in such a binary system. We therefore require

$$I \equiv K \ln R \equiv Kn \ln 2 = n \tag{3.4}$$

which is fulfilled by

$$K = 1/\ln 2 = \log_2 e. \tag{3.5}$$

With this choice of K, another form of (3.4) reads

$$I = \log_2 R. \tag{3.4a}$$

Since a single position in a sequence of symbols (signs) in a binary system is called "bit", the information I is now directly given in bits. Thus if $R = 8 = 2^3$ we find $I = 3$ bits and generally for $R = 2^n$, $I = n$ bits. The definition of information for (3.3) can be easily generalized to the case when we have initially R_0 equally probable cases and finally R_1 equally probable cases. In this case the information is

$$I = K \ln R_0 - K \ln R_1 \tag{3.6}$$

which reduces to the earlier definition (3.3), if $R_1 = 1$. A simple example for this is given by a die. Let us define a game in which the even numbers mean gain and the odd numbers mean loss. Then $R_0 = 6$ and $R_1 = 3$. In this case the information content is the same as that of a coin with originally just two possibilities. We now derive a more convenient expression for the information. To this end we first consider the following example of a simplified[1] Morse alphabet with dash and dot. We consider a word of length G which contains N_1 dashes and N_2 dots, with

$$N_1 + N_2 = N. \tag{3.7}$$

We ask for the information which is obtained by the receipt of such a word. In the spirit of information theory we must calculate the total number of words which can be constructed out of these two symbols for *fixed* N_1, N_2. The consideration is quite similar to that of Section 2.11 page 33. According to the ways we can distribute the dashes and dots over the N positions, there are

$$R = \frac{N!}{N_1!\, N_2!} \tag{3.8}$$

[1] In the realistic Morse alphabet, the intermission is a third symbol.

possibilities. Or, in other words, R is the number of messages which can be transmitted by N_1 dashes and N_2 dots. We now want to derive the information per symbol, i.e., $i = I/N$. Inserting (3.8) into (3.3) we obtain

$$I = K \ln R = K[\ln N! - \ln N_1! - \ln N_2!]. \tag{3.9}$$

Using Stirling's formula presented in (2.68) in the approximation

$$\ln Q! \approx Q(\ln Q - 1) \tag{3.10}$$

which is good for $Q > 100$, we readily find

$$I \approx K[N(\ln N - 1) - N_1(\ln N_1 - 1) - N_2(\ln N_2 - 1)], \tag{3.11}$$

and with use of (3.7) we find

$$i \equiv \frac{I}{N} \approx -K\left[\frac{N_1}{N} \ln \frac{N_1}{N} + \frac{N_2}{N} \ln \frac{N_2}{N}\right] \tag{3.12}$$

We now introduce a quantity which may be interpreted as the probability of finding the sign "dash" or "dot". The probability is identical to the frequency with which dash or dot are found

$$p_j = \frac{N_j}{N}, \quad j = 1, 2. \tag{3.13}$$

With this, our final formula takes the form

$$i = \frac{I}{N} = -K(p_1 \ln p_1 + p_2 \ln p_2). \tag{3.14}$$

This expression can be easily generalized if we have not simply two symbols but several, such as letters in the alphabet. Then we obtain in quite similar manner as just an expression for the information per symbol which is given by

$$i = -K \textstyle\sum_j p_j \ln p_j. \tag{3.15}$$

p_j is the relative frequency of the occurrence of the symbols. From this interpretation it is evident that i may be used in the context of transmission of information, etc.

Before continuing we should say a word about information used in the sense here. It should be noted that "useful" or "useless" or "meaningful" or "meaningless" are not contained in the theory; e.g., in the Morse alphabet defined above quite a number of words might be meaningless. Information in the sense used here rather refers to scarcity of an event. Though this seems to restrict the theory considerably, this theory will turn out to be extremely useful.

The expression for the information can be viewed in two completely different ways. On the one hand we may assume that the p_i's are given by their numerical values, and then we may write down a number for I by use of formula (3.3). Of still greater importance, however, is a second interpretation; namely, to consider I as a *function* of the p_i's; that means if we change the values of the p_i's, the value of I changes correspondingly. To make this interpretation clear we anticipate an application which we will treat later on in much greater detail. Consider a gas of atoms moving freely in a box. It is then of interest to know about the spatial distribution of the gas atoms. Note that this problem is actually identical to that of Section 2.11 but we treat it here under a new viewpoint. We again divide the container into M cells of equal size and denote the number of particles in cell k by N_k. The total number of particles be N. The relative frequency of a particle to be found in cell k is then given by

$$\frac{N_k}{N} = p_k, \; k = 1, 2, \ldots, M. \tag{3.16}$$

p_k may be considered as distribution function of the particles over the cells k. Because the cells have equal size and do not differ in their physical properties, we expect that the particles will be found with equal probability in each cell, i.e.,

$$p_k = 1/M. \tag{3.17}$$

We now want to derive this result (3.17) from the properties of information. Indeed the information may be as follows: Before we make a measurement or obtain a message, there are R possibilities or, in other words, $K \ln R$ is a measure of our ignorance. Another way of looking at this is the following: R gives us the number of realizations which are possible in principle.

Now let us look at an ensemble of M containers, each with N gas atoms. We assume that in each container the particles are distributed according to different distribution functions p_k, i.e.,

$$p_k^{(1)}, p_k^{(2)}, p_k^{(3)}, \ldots$$

Accordingly, we obtain different numbers of realizations, i.e., different informations. For example, if $N_1 = N$, $N_2 = N_3 = \cdots = 0$, we have $p_1^{(1)} = 1$, $p_2^{(1)} = p_3^{(1)} = \cdots = 0$ and thus $I^{(1)} = 0$. On the other hand, if $N_1 = N_2 = N_3 = \cdots = N/M$, we have $p_1^{(2)} = 1/M$, $p_2^{(2)} = 1/M, \ldots,$ so that $I^{(2)} = -M \log_2 (1/M) = M \log_2 M$, which is a very large number if the number of boxes is large.

Thus when we consider any container with gas atoms, the probability that it is one with the second distribution function is much greater than one with the first distribution function. That means there is an overwhelming probability of finding that probability distribution p_k realized which has the greatest number of possibilities R and thus the greatest information. Hence we are led to require that

$$-\sum p_i \ln p_i = Extr! \tag{3.18}$$

is an extremum under the constraint that the total sum of the probabilities p_i equals unity

$$\textstyle\sum_{i=1}^{M} p_i = 1. \tag{3.19}$$

This principle will turn out to be fundamental for applications to realistic systems in physics, chemistry, and biology and we shall come back to it later.

The problem (3.18) with (3.19) can be solved using the so-called Lagrangian multipliers. This method consists in multiplying (3.19) by a still unknown parameter λ and adding it to the lhs of (3.18) now requiring that the total expression becomes an extremum. Here we are now allowed to vary the p_i's independently of each other, not taking into account the constraint (3.19). Varying the left-hand side of

$$-\textstyle\sum p_i \ln p_i + \lambda \sum_i p_i = \mathit{Extr}! \tag{3.20}$$

means taking the derivative of it with respect to p_i which leads to

$$-\ln p_i - 1 + \lambda = 0. \tag{3.21}$$

Eq. (3.21) allows for the solution

$$p_i = \exp(\lambda - 1) \tag{3.22}$$

which is independent of the index i, i.e., the p_i's are constant. Inserting them into (3.19) we may readily determine λ so that

$$M \exp(\lambda - 1) = 1, \tag{3.23}$$

or, in other words, we find

$$p_i = \frac{1}{M} \tag{3.24}$$

in agreement to (3.17) as expected.

Exercises on 3.1

1) Show by means of (3.8) and the binomial theorem that $R = 2^N$ if N_1, N_2 may be choosen arbitrarily (within (3.7)).
2) Given a container with five balls in it. These balls may be excited to equidistant energy levels. The difference in energy between two levels is denoted by Δ. Now put the energy $E = 5\Delta$ into the system. Which is the state with the largest number of realizations, if a state is characterized by the number of "excited" balls? What is the total number of realizations? Compare the probabilities for the single state and the corresponding number of realizations. Generalize the formula for the total number of states to 6 balls which get an energy $E = N\Delta$.

3) Convince yourself that the number of realizations of N_1 dashes and N_2 dots over N positions could be equally well derived by the distribution of N particles (or balls) over 2 boxes, with the particle (ball) numbers N_1, N_2 in boxes 1, 2 fixed.
Hint: Identify the (numbered) balls with the (numbered) positions and index 1 or 2 of the boxes with "dash" or "dot". Generalize this analogy to one between M symbols over N positions and N balls in M boxes.
4) On an island five different bird populations are found with the following relative abundance: 80%, 10%, 5%, 3%, 2%. What is the information entropy of this population?
Hint: Use (3.15).

3.2* Information Gain: An Illustrative Derivation

We consider a distribution of balls among boxes, which are labeled by 1, 2, . . . , We now ask in how many ways we can distribute N balls among these boxes so that finally we have N_1, N_2 etc. balls in the corresponding boxes. This task has been treated in Section 2.11 for two boxes 1, 2. (compare (3.8)). Similarly we now obtain as number of configurations

$$Z_1 = \frac{N!}{\prod_k N_k!}. \tag{3.25}$$

Taking the logarithm of Z_1 to the basis of two, we obtain the corresponding information. Now consider the same situation but with balls having two different colors, black and white (cf. Fig. 3.1). Now these black balls form a subset of all balls. The number of black balls be N', their number in box k be N'_k. We now calculate the number of configurations which we find if we distinguish between black and white balls. In each box we must now subdivide N_k into N'_k and $N_k - N'_k$. Thus we find

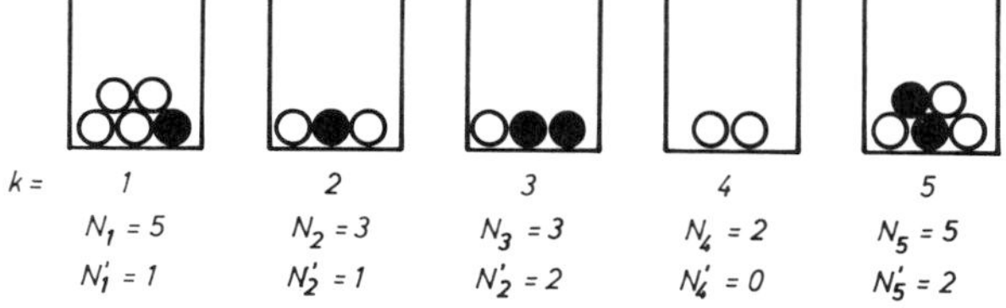

Fig. 3.1. Distribution of white and black balls over boxes

$$Z_2 = \frac{N'!}{\prod N'_k!} \cdot \frac{(N - N')!}{\prod (N_k - N'_k)!} \quad (\text{where } \prod = \prod_k) \tag{3.26}$$

realizations. Now we consider the ratio between the numbers of realizations

$$Z = \frac{Z_1}{Z_2} \tag{3.27}$$

or, using (3.25), (3.26)

$$Z = \frac{N!}{\prod N_k!} \cdot \frac{\prod N_k'! \prod (N_k - N_k')!}{N'! \quad (N - N')!}. \tag{3.28}$$

Using Stirling's formula (2.68), we may put after some analysis

$$\frac{N!}{(N - N')!} \approx N^{N'} \tag{3.29}$$

and

$$\frac{N_k!}{(N_k - N_k')!} \approx N_k^{N_k'} \tag{3.30}$$

and thus

$$Z \approx N^{N'} \frac{1}{N'!} \prod N_k'! \frac{1}{\prod N_k^{N_k'}}. \tag{3.31}$$

Using again Stirling's formula, we find

$$\ln Z = N' \ln N - N' \ln N' + \textstyle\sum_k N_k'(\ln N_k' - \ln N_k) \tag{3.32}$$

and using

$$N' = \textstyle\sum_k N_k' \tag{3.33}$$

we obtain

$$\ln Z = \textstyle\sum_k N_k' \ln \frac{N_k'}{N_k} - \sum_k N_k' \ln \frac{N'}{N} \tag{3.34}$$

which can be written in short in the form

$$\ln Z = N' \textstyle\sum_k \frac{N_k'}{N'} \ln \frac{N_k'/N'}{N_k/N}. \tag{3.35}$$

As above we now introduce the relative frequencies, or probabilities,

$$N_k/N = p_k \tag{3.36}$$

and

$$N_k'/N' = p_k'. \tag{3.37}$$

If we still divide both sides by N' (and then multiply by the constant K (3.5)), we obtain our final formula for the *information gain* in the form

$$K(p', p) = \frac{K}{N'} \ln Z = K \sum_k p'_k \ln \frac{p'_k}{p_k} \tag{3.38}$$

where

$$\sum p_k = 1, \tag{3.39}$$

and

$$\sum p'_k = 1. \tag{3.40}$$

The information gain $K(p', p)$ has the following important property, which we will use in later chapters:

$$K(p', p) \geq 0. \tag{3.41}$$

The equality sign holds if and only if

$$p' \equiv p, \text{ i.e. } p'_k = p_k \text{ for all } k\text{'s.}$$

3.3 Information Entropy and Constraints

In this and the following two sections we have especially applications of the information concept to *physics* in our minds so that we shall follow the convention to denote the information by S identifying the constant K in (3.3) with Boltzmann's constant k_B. For reasons which will appear later, S will be called information entropy. Because chemical and biological systems can be viewed as physical systems, our considerations apply equally well to these other systems. Still more important, the general formalism of this chapter is also applicable to other sciences, such as information processing, etc. We start from the basic expression

$$S = -k_B \sum_i p_i \ln p_i. \tag{3.42}$$

The indices i may be considered as describing individual features of the particles or subsystems. Let us explain this in some detail. The index i may describe, for instance, the position of a gas particle or it may describe its velocity or both properties. In our previous examples the index i referred to boxes filled with balls. In the initial chapters on probability and random variables, the index i represented the values that a random variable may acquire. In this paragraph we assume for simplicity that the index i is discrete.

A central task to be solved in this book consists in finding ways to determine the p_i's. (compare for example the gas molecules in a container where one wants to know their location). The problem we are confronted with in many disciplines is to make *unbiased* estimates leading to p_i's which are in agreement with all the possible knowledge available about the system. Consider an ideal gas in one dimension. What we could measure, for instance, is the center of gravity. In this case we would have as constraint an expression of the form

$$\sum_i p_i q_i = M \tag{3.43}$$

where q_i measures the position of the cell i. M is a fixed quantity equal Q/N, where Q is the coordinate of the center of gravity, and N the particle number. There are, of course, very many sets of p_i's which fulfill the relation (3.43). Thus we could choose a set $\{p_i\}$ rather arbitrarily, i.e., we would favor one set against another one. Similar to common life, this is a biased action. How may it be unbiased? When we look again at the example of the gas atoms, then we can invoke the principle stated in Section 3.1. With an overwhelming probability we will find those distributions realized for which (3.42) is a maximum. However, due to (3.43) *not all* distributions can be taken into account. Instead we have to seek the maximum of (3.42) under the constraint (3.43). This principle can be generalized if we have a set of constraints. Let for example the variable i distinguish between different velocities. Then we may have the constraint that the total kinetic energy E_{kin}^{tot} of the particles is fixed. Denoting the kinetic energy of a particle with mass m and velocity v_i by $f_i[f_i = (m/2)v_i^2]$ the mean kinetic energy per particle is given by

$$\sum_i p_i f_i = E_{kin}. \tag{3.43a}$$

In general the single system i may be characterized by quantities $f_i^{(k)}$, $k = 1, 2, \ldots, M$ (position, kinetic energy or other typical features). If these features are additive, and the corresponding sums are kept fixed at values f_k the constraints take the form

$$\sum_i p_i f_i^{(k)} = f_k. \tag{3.44}$$

We further add as usual the constraint that the probability distribution is normalized

$$\sum_i p_i = 1. \tag{3.45}$$

The problem of finding the extremum of (3.42) under the constraints (3.44) and (3.45) can be solved by using the method of Lagrange parameters λ_k, $k = 1, 2, \ldots, M$ (cf. 3.1). We multiply the lhs of (3.44) by λ_k and the lhs of (3.45) by $(\lambda - 1)$ and take the sum of the resulting expressions. We then subtract this sum from$(1/k_B)S$. The factor $1/k_B$ amounts to a certain normalization of λ, λ_k. We then have to vary the total sum with respect to the p_i's

$$\delta\left[\frac{1}{k_B} S - (\lambda - 1)\sum_i p_i - \sum_k \lambda_k \sum_i p_i f_i^{(k)}\right] = 0. \tag{3.46}$$

Differentiating with respect to p_i and putting the resulting expression equal to zero, we obtain

$$-\ln p_i - 1 - (\lambda - 1) - \sum_k \lambda_k f_i^{(k)} = 0 \tag{3.47}$$

which can be readily solved for p_i yielding

$$p_i = \exp\{-\lambda - \sum_k \lambda_k f_i^{(k)}\}. \tag{3.48}$$

Inserting (3.48) into (3.45) yields

$$e^{-\lambda} \sum_i \exp\{-\sum_k \lambda_k f_i^{(k)}\} = 1. \tag{3.49}$$

It is now convenient to abbreviate the sum over i, $\sum_i$ in (3.49) by

$$\sum_i \exp\{-\sum_k \lambda_k f_i^{(k)}\} = Z(\lambda_1, \ldots, \lambda_M) \tag{3.50}$$

which we shall call the partition function. Inserting (3.50) into (3.49) yields

$$e^{\lambda} = Z \tag{3.51}$$

or

$$\lambda = \ln Z, \tag{3.52}$$

which allows us to determine λ once the λ_k's are determined. To find equations for the λ_k's, we insert (3.48) into the equations of the constraints (3.44) which lead immediately to

$$\langle f_i^{(k)} \rangle = \sum_i p_i f_i^{(k)} = e^{-\lambda} \sum_i \exp\{-\sum_l \lambda_l f_i^{(l)}\} \cdot f_i^{(k)}. \tag{3.53}$$

Eq. (3.53) has a rather similar structure as (3.50). The difference between these two expressions arises because in (3.53) each exponential function is still multiplied by $f_i^{(k)}$. However, we may easily derive the sum occurring in (3.53) from (3.50) by differentiating (3.50) with respect to λ_k. Expressing the first factor in (3.53) by Z according to (3.51) we thus obtain

$$\langle f_i^{(k)} \rangle = \frac{1}{Z}\left(-\frac{\partial}{\partial \lambda_k}\right) \underbrace{\sum_i \exp\{-\sum_l \lambda_l f_i^{(l)}\}}_{Z} \tag{3.54}$$

or in still shorter form

$$f_k \equiv \langle f_i^{(k)} \rangle = -\frac{\partial \ln Z}{\partial \lambda_k} \tag{3.55}$$

Because the lhs are prescribed (compare (3.44)) and Z is given by (3.50) which is a function of the λ_k's in a special form, (3.55) is a concise form for a set of equations for the λ_k's.

We further quote a formula which will become useful later on. Inserting (3.48) into (3.42) yields

$$\frac{1}{k_B} S_{max} = \lambda \sum_i p_i + \sum_k \lambda_k \sum_i p_i f_i^{(k)} \tag{3.56}$$

which can be written by use of (3.44) and (3.45) as

$$\frac{1}{k_B} S_{max} = \lambda + \sum_k \lambda_k f_k. \tag{3.57}$$

The maximum of the information entropy may thus be represented by the mean values f_k and the Lagrange parameters λ_k. Those readers who are acquainted with the Lagrange equations of the first kind in mechanics will remember that the Lagrange parameters have a physical meaning, in that case, of forces. In a similar way we shall see later on that the Lagrange parameters λ_k have physical (or chemical or biological, etc.) interpretations. By the derivation of the above formulas (i.e., (3.48), (3.52) with (3.42), (3.55) and (3.57)) our original task to find the p's and S_{max} is completed.

We now derive some further useful relations. We first investigate how the information S_{max} is changed if we change the functions $f_i^{(k)}$ and f_k in (3.44). Because S depends, according to (3.57), not only on the f's but also on λ and λ_k's which are functions of the f's, we must exercise some care in taking the derivatives with respect to the f's. We therefore first calculate the change of λ (3.52)

$$\delta\lambda \equiv \delta \ln Z = \frac{1}{Z}\delta Z.$$

Inserting (3.50) for Z yields

$$\delta\lambda = e^{-\lambda} \sum_i \sum_k \{-\delta\lambda_k f_i^{(k)} - \lambda_k \delta f_i^{(k)}\} \exp\{-\sum_l \lambda_l f_i^{(l)}\}$$

which, by definition of p_i (3.48) transforms to

$$\delta\lambda = -\sum_k [\delta\lambda_k \sum_i p_i f_i^{(k)} + \lambda_k \sum_i p_i \delta f_i^{(k)}].$$

Eq. (3.53) and an analogous definition of $\langle \delta f_i^{(k)} \rangle$ allow us to write the last line as

$$= -\sum_k [\delta\lambda_k \langle f_i^{(k)} \rangle + \lambda_k \langle \delta f_i^{(k)} \rangle]. \tag{3.58}$$

Inserting this into δS_{max} (of (3.57)) we find that the variation of the λ_k's drops out and we are left with

$$\delta S_{max} = k_B \sum_k \lambda_k [\delta \langle f_i^{(k)} \rangle - \langle \delta f_i^{(k)} \rangle]. \tag{3.59}$$

We write this in the form

$$\delta S_{\max} = k_B \sum_k \lambda_k \delta Q_k \tag{3.60}$$

where we define a "generalized heat" by means of

$$\delta Q_k = \delta\langle f_i^{(k)}\rangle - \langle \delta f_i^{(k)}\rangle. \tag{3.61}$$

The notation "generalized heat" will become clearer below when contact with thermodynamics is made. In analogy to (3.55) a simple expression for the variance of $f_i^{(k)}$ (cf. (2.33)) may be derived:

$$\langle f_i^{(k)2}\rangle - \langle f_i^{(k)}\rangle^2 = \frac{\partial^2 \ln Z}{\partial \lambda_k^2}. \tag{3.62}$$

In many practical applications, $f_i^{(k)}$ depends on a further quantity α (or a set of such quantities $\alpha_1, \alpha_2, \ldots$). Then we want to express the change of the mean value (3.44), when α is changed. Taking the derivative of $f_{i,\alpha}^{(k)}$ with respect to α and taking the average value, we find

$$\left\langle \frac{\partial f_{i,\alpha}^{(k)}}{\partial \alpha} \right\rangle = \sum_i p_i \frac{\partial f_{i,\alpha}^{(k)}}{\partial \alpha}. \tag{3.63}$$

Using the p_i's in the form (3.48) and using (3.51), (3.63) may be written in the form

$$\frac{1}{Z}\sum_i \frac{\partial f_{i,\alpha}^{(k)}}{\partial \alpha} \exp\left(-\sum_j \lambda_j f_{i,\alpha}^{(j)}\right) \tag{3.64}$$

which may be easily expressed as a derivative of Z with respect to α

$$(3.64) = -\frac{1}{Z}\frac{1}{\lambda_k}\frac{\partial Z}{\partial \alpha}. \tag{3.65}$$

Thus we are led to the final formula

$$-\frac{1}{\lambda_k}\frac{\partial \ln Z}{\partial \alpha} = \left\langle \frac{\partial f_{i,\alpha}^{(k)}}{\partial \alpha} \right\rangle. \tag{3.66}$$

If there are several parameters α_l present, this formula can be readily generalized by providing on the left and right hand side the α with respect to which we differentiated by this index l.

As we have seen several times, the quantity Z (3.50), or its logarithm, is very useful (see e.g., (3.55), (3.62), (3.66)). We want to convince ourselves that $\ln Z \equiv \lambda$ (cf. (3.52)) may be directly determined by a variational principle. A glance at (3.46) reveals that (3.46) can also be interpreted in the following way: Seek the extremum of

$$\frac{1}{k_B} S - \sum_k \lambda_k \sum_i p_i f_i^{(k)} \tag{3.67}$$

under the only constraint

$$\sum p_i = 1! \tag{3.68}$$

Now, by virtue of (3.44), (3.57) and (3.52) the extremum of (3.67) is indeed identical with ln Z. Note that the spirit of the variational principle for ln Z is different from that for S. In the former case, we had to seek the maximum of S under the constraints (3.44), (3.45) with f_k fixed and λ_k unknown. Here now, only one constraint, (3.68), applies and the λ_k's are assumed as *given* quantities. How such a switching from one set of fixed quantities to another one can be done will become more evident by the following example from physics, which will elucidate also many further aspects of the foregoing.

Exercise on 3.3

To get a feeling for how much is achieved by the extremal principle, answer the following questions:

1) Determine all solutions of $p_1 + p_2 = 1$ (E1)

   with $0 \leq p_i \leq 1$,

   all solutions of $p_1 + p_2 + p_3 = 1$ (E2)

   all solutions of $\sum p_i = 1$ (E3)

2) In addition to (E1), (E2) one further constraint (3.44) is given. Determine now all solutions p_i.

   Hint: Interpret (E1) – (E3) as equation in the $p_1 - p_2$ plane, in the $p_1 - p_2 - p_3$ space, in the n-dimensional space. What is the geometrical interpretation of additional constraints? What does a specific solution $p_1^{(0)}, \ldots, p_n^{(0)}$ mean in the $(p_1, p_2, \ldots, p_n)$-space?

3.4 An Example from Physics: Thermodynamics

To visualize the meaning of the index i, let us identify it with the velocity of a particle. In a more advanced theory p_i is the occupation probability of a quantum state i of a many-particle system. Further, we identify $f_{i,\alpha}^{(k)}$ with energy E and the parameter α with the volume. Thus we put

$$f_{i,\alpha}^{(k)} = E_i(V); \quad k = 1, \tag{3.69}$$

and have the identifications

$$f_1 \leftrightarrow U \equiv \langle E_i \rangle; \alpha \leftrightarrow V; \lambda_1 = \beta. \tag{3.70}$$

We have, in particular, called $\lambda_1 = \beta$. With this, we may write a number of the previous formulas in a way which can be immediately identified with relations well known in thermodynamics and statistical mechanics. Instead of (3.48) we find

$$p_i = e^{-\lambda - \beta E_i(V)} \tag{3.71}$$

which is the famous *Boltzmann Distribution Function*.

Eq. (3.57) acquires the form

$$\frac{1}{k_B} S_{max} = \ln Z + \beta U \tag{3.72}$$

or, after a slight rearrangement of this equation

$$U - \frac{1}{k_B \beta} S_{max} = -\frac{1}{\beta} \ln Z. \tag{3.73}$$

This equation is well known in thermodynamics and statistical physics. The first term may be interpreted as the internal energy U, $1/\beta$ as the absolute temperature T multiplied by Boltzmann's constant k_B. S_{max} is the entropy. The rhs represents the free energy, $\mathcal{F}$, so that (3.73) reads in thermodynamic notation

$$U - TS = \mathcal{F}. \tag{3.74}$$

By comparison we find

$$\mathcal{F} = -k_B T \ln Z, \tag{3.75}$$

and $S = S_{max}$. Therefore we will henceforth drop the suffix "max". (3.50) reads now

$$Z = \sum_i \exp(-\beta E_i) \tag{3.76}$$

and is nothing but the usual partition function. A number of further identities of thermodynamics can easily be checked by applying our above formulas.

The only problem requiring some thought is to identify *independent* and *dependent* variables. Let us begin with the information entropy, S_{max}. In (3.57) it appears as a function of λ, λ_μ's and f_k's. However, λ and λ_μ's are themselves determined by equations which contain the f_k's and $f_i^{(k)}$'s as *given* quantities (cf. (3.50), (3.52), (3.53)). Therefore, the independent variables are f_k, $f_i^{(k)}$, and the dependent variables are λ, λ_k, and thus, by virtue of (3.57), S_{max}. In practice the $f_i^{(k)}$'s are fixed functions of i (e.g., the energy of state "i"), but still depending on parameters α (e.g., the volume, cf. (3.69)). Thus the truly independent variables in our approach are f_k's (as above) and the α's. In conclusion we thus find $S = S(f_k, \alpha)$. In our example, $f_1 = E \equiv U$, $\alpha = V$, and therefore

$$S = S(U,V). \tag{3.77}$$

Now let us apply the general relation (3.59) to our specific model. If we vary only the internal energy, U, but leave V unchanged, then

$$\delta\langle f_i^{(1)}\rangle \equiv \delta f_1 \equiv \delta U \neq 0, \tag{3.78}$$

and

$$\delta f_{i,\alpha}^{(1)} \equiv \delta E_i(V) = (\delta E_i(V)/\delta V)\delta V = 0, \tag{3.79}$$

and therefore

$$\delta S = k_B \lambda_1 \delta U$$

or

$$\frac{\delta S}{\delta U} = k_B \lambda_1 (\equiv k_B \beta). \tag{3.80}$$

According to thermodynamics, the lhs of (3.80) defines the inverse of the absolute temperature

$$\delta S/\delta U = 1/T. \tag{3.81}$$

This yields $\beta = 1/(k_B T)$ as anticipated above. On the other hand, varying V but leaving U fixed, i.e.,

$$\delta\langle f_i^{(1)}\rangle = 0, \tag{3.82}$$

but

$$\langle \delta f_i^{(1)}\rangle = \langle \delta E_i(V)/\delta V\rangle \cdot \delta V \neq 0 \tag{3.83}$$

yields in (3.59)

$$\delta S = k_B(-\lambda_1)\langle \delta E_i(V)/\delta V\rangle \delta V$$

or

$$\frac{\delta S}{\delta V} = -\frac{1}{T}\langle \delta E_i(V)/\delta V\rangle. \tag{3.84}$$

Since thermodynamics teaches us that

$$\frac{\delta S}{\delta V} = \frac{P}{T} \tag{3.85}$$

where P is the pressure, we obtain by comparison with (3.84)

$$\langle \delta E_i(V)/\delta V \rangle = -P. \tag{3.86}$$

Inserting (3.81) and (3.85) into (3.59) yields

$$\delta S = \frac{1}{T}\delta U + \frac{1}{T}P\delta V. \tag{3.87}$$

In thermodynamics the rhs is equal to dQ/T where dQ is heat. This explains the notation "generalized heat" used above after (3.61). These considerations may be generalized to different kinds of particles whose average numbers $\overline{N}_k$, $k = 1, \ldots m$ are prescribed quantities. We therefore identify f_1 with E, but $f_{k'+1}$ with $\overline{N}_{k'}$, $k' = 1, \ldots, m$ (Note the shift of index!). Since each kind of particle, l, may be present with different numbers N_l we generalize the index i to $i, N_1, \ldots, N_m$ and put

$$f_i^{(k+1)} \rightarrow f_{i,N_1,\cdots,N_m}^{(k+1)} = N_k.$$

To be in accordance with thermodynamics, we put

$$\lambda_{k+1} = -\frac{1}{k_{\mathrm{B}}T} \cdot \mu_k, \tag{3.88}$$

μ_k is called chemical potential.

Equation (3.57) with (3.52) acquires (after multiplying both sides by $k_{\mathrm{B}}T$) the form

$$TS = \underbrace{k_{\mathrm{B}}T \ln Z}_{-\mathscr{F}} + U - \mu_1\overline{N}_1 - \mu_2\overline{N}_2 - \cdots - \mu_m\overline{N}_m. \tag{3.89}$$

Eq. (3.59) permits us to identify

$$\frac{\delta S}{\delta \overline{N}_k} = -k_{\mathrm{B}}\lambda_{k+1} = \frac{1}{T}\mu_k. \tag{3.90}$$

The partition function reads

$$Z = \sum_{N_1 N_2 \cdots N_m} \sum_i \exp\left\{-\frac{1}{k_{\mathrm{B}}T}[E_i(V) - \mu_1 N_1 - \cdots - \mu_m N_m]\right\}. \tag{3.91}$$

While the above considerations are most useful for irreversible thermodynamics (see the following Section 3.5), in thermodynamics the role played by independent and dependent variables is, to some extent, exchanged. It is not our task to treat these transformations which give rise to the different thermodynamic potentials (Gibbs, Helmholtz, etc). We just mention one important case: Instead of U, V (and $N_1, \ldots, N_m$) as independent variables, one may introduce V and $T = (\partial S/\partial U)^{-1}$ (and $N_1, \ldots, N_n$) as new independent variables. As an example we treat the U-V case (putting formally $\mu_1, \mu_2, \ldots = 0$). According to (3.75) the free energy, $\mathscr{F}$, is there directly given as function of T. The differentiation $\partial\mathscr{F}/\partial T$ yields

$$-\frac{\partial\mathscr{F}}{\partial T} = k_B \ln Z + \frac{1}{T}\frac{1}{Z}\sum_i E_i \exp(-\beta E_i).$$

The second term on the rhs is just U, so that

$$-\frac{\partial\mathscr{F}}{\partial T} = k_B \ln Z + \frac{1}{T}U. \tag{3.92}$$

Comparing this relation with (3.73), where $1/\beta = k_B T$, yields the important relation

$$-\frac{\partial\mathscr{F}}{\partial T} = S \tag{3.93}$$

where we have dropped the suffix "max."

3.5* An Approach to Irreversible Thermodynamics

The considerations of the preceding chapter allow us to simply introduce and make understandable basic concepts of irreversible thermodynamics. We consider a system which is composed of two subsystems, e.g., a gas of atoms whose volume is divided into two subvolumes. We assume that in both subsystems probability distributions p_i and p_i' are given. We then define the entropies in the corresponding subsystems by

$$S = -k_B \sum_i p_i \ln p_i \tag{3.94a}$$

and

$$S' = -k_B \sum_i p_i' \ln p_i'. \tag{3.94b}$$

Similarly, in both subsystems constraints are given by

$$\sum_i p_i f_i^{(k)} = f_k \tag{3.95a}$$

and

$$\sum_i p_i' f_i^{(k)} = f_k'. \tag{3.95b}$$

We require that the sums of the f's in both subsystems are prescribed constants

$$f_k + f_k' = f_k^0, \tag{3.96}$$

(e.g., the total number of particles, the total energy, etc.). According to (3.57) the entropies may be represented as

$$\frac{1}{k_B} S = \lambda + \sum_k \lambda_k f_k, \tag{3.97a}$$

and

$$\frac{1}{k_B} S' = \lambda' + \sum_k \lambda'_k f'_k. \tag{3.97b}$$

(Here and in the following the suffix "max" is dropped). We further introduce the sum of the entropies in the total system

$$S + S' = S^0. \tag{3.98}$$

To make contact with irreversible thermodynamics we introduce "extensive variables" X. Variables which are additive with respect to the volume are called extensive variables. Thus if we divide the system in two volumes so that $V_1 + V_2 = V$, then an extensive variable has the property $X_{V_1} + X_{V_2} = X_V$. Examples are provided by the number of particles, by the energy (if the interaction is of short range), etc. We distinguish the different physical quantities (particle numbers, energies, etc.) by a superscript k and the values such quantities acquire in the state i by the subscript i (or a set of them). We thus write $X_i^{(k)}$.

Example: Consider the particle number and the energy as extensive variables. Then we choose $k = 1$ to refer to the number of particles, and $k = 2$ to refer to the energy. Thus

$$X_i^{(1)} = N_i,$$

and

$$X_i^{(2)} = E_i.$$

Evidently, we have to identify $X_i^{(k)}$ with $f_i^{(k)}$,

$$X_i^{(k)} \equiv f_i^{(k)}. \tag{3.99}$$

For the rhs of (3.95) we introduce correspondingly the notation X_k. In the first part of this chapter we confine our attention to the case where the extensive variables X_k may vary, but the $X_i^{(k)}$ are fixed. We differentiate (3.98) with respect to X_k. Using (3.97) and

$$X_k + X'_k = X_k^0 \tag{3.100}$$

(compare (3.96)) we find

$$\frac{\partial S^0}{\partial X_k} = \frac{\partial S}{\partial X_k} - \frac{\partial S'}{\partial X'_k} = k_B(\lambda_k - \lambda'_k). \tag{3.101}$$

Consider an example: take $X_k = U$ (internal energy) then $\lambda_k = \beta = 1/(k_B T)$ (cf. Sec. 3.4). T is again the absolute temperature.

Since we want to consider *processes* we now admit that the entropy S depends on time. More precisely, we consider two subsystems with entropies S and S' which are initially kept under different conditions, e.g., having different temperatures. Then we bring these two systems together so that for example the energy between these two systems can be exchanged. It is typical for the considerations of irreversible thermodynamics that no detailed assumptions about the transfer mechanisms are made. For instance one completely ignores the detailed collision mechanisms of molecules in gases or liquids. One rather treats the mechanisms in a very global manner which assumes local thermal equilibrium. This comes from the fact that the entropies S, S' are determined just as for an equilibrium system with given constraints. Due to the transfer of energy or other physical quantities the probability distribution p_i will change. Thus for example in a gas the molecules will acquire different kinetic energies when the gas is heated and so the p_i change. Because there is a steady transfer of energy or other quantities, the probability distribution p_i changes steadily as a function of time. When the p_i's change, of course, the values of the f_k's (compare (3.95a) or (3.95b)) are also functions of time. We now extend the formalism of finding the entropy maximum to the time-dependent case, i.e., we imagine that the physical quantities f_k are given (and changing) and the p_i's must be determined by maximizing the entropy

$$S = -k_B \sum_i p_i(t) \ln p_i(t) \tag{3.102}$$

under the constraints

$$\sum_i p_i(t) f_i^{(k)} = f_k(t). \tag{3.103}$$

In (3.103) we have admitted that the f_k's are functions of time, e.g., the energy U is a function of time. Taking now the derivative of (3.102) with respect to time and having in mind that S depends via the constraints (3.103) on $f_k \equiv X_k$ we obtain

$$\frac{dS}{dt} = \sum_k \frac{\partial S}{\partial X_k} \frac{dX_k}{dt}. \tag{3.104}$$

As we have seen before (compare (3.101))

$$\frac{1}{k_B} \frac{\partial S}{\partial X_k} = \lambda_k \tag{3.105}$$

introduces the Lagrange multipliers which define the so-called intensive variables. The second factor in (3.104) gives us the temporal change of the extensive quantities X_k, e.g., the temporal change of the internal energy. Since there exist conservation laws, e.g., for the energy, the decrease or increase of energy in one system must be caused by an energy flux between different systems. This leads us to define the flux by the equation

$$J_k = \frac{dX_k}{dt}. \tag{3.106}$$

Replacing S by S^0 in (3.104), and using (3.101) and (3.106), we obtain

$$\frac{dS^0}{dt} = \sum_k J_k(\lambda_k - \lambda'_k)k_B. \tag{3.107a}$$

The difference $(\lambda_k - \lambda'_k)k_B$ is called a *generalized thermodynamic force* (or *affinity*) and we thus put

$$F_k = (\lambda_k - \lambda'_k)k_B. \tag{3.107b}$$

The motivation for this nomenclature will become evident by the examples treated below. With (3.107b) we may cast (3.107a) into the form

$$\frac{dS^0}{dt} = \sum_k F_k J_k \tag{3.108}$$

which expresses the temporal change of the entropy S^0 of the total system $(1 + 2)$ in terms of forces and fluxes. If $F_k = 0$, the system is in equilibrium; if $F_k \neq 0$, an irreversible process occurs.

Examples: Let us consider two systems separated by a diathermal wall and take $X_k = U$ (U: internal energy). Because according to thermodynamics

$$\frac{\partial S}{\partial U} = \frac{1}{T}, \tag{3.109}$$

we find

$$F_k = \frac{1}{T} - \frac{1}{T'}. \tag{3.110}$$

We know that a difference in temperature causes a flux of heat. Thus the generalized forces occur as the cause of fluxes. For a second example consider X_k as the mole number. Then one derives

$$F_k = [\mu'/T' - \mu/T]. \tag{3.111}$$

A difference in chemical potential, μ, gives rise to a change of mole numbers.

In the above examples, 1 and 2, corresponding fluxes occur, namely in 1) a heat flux, and in 2) a particle flux. We now treat the case that the *extensive variables X* *are parameters of f.* In this case we have in addition to (3.99)

$$X_k = \alpha_k, \tag{3.112}$$

$$f_i^{(k)} = f_{i,\alpha_1,\dots,\alpha_j,\dots}^{(k)} \tag{3.113}$$

(Explicit examples from physics will be given below). Then according to (3.59) we have quite generally

$$\frac{\partial S}{\partial \alpha_j} = \sum_k k_B \lambda_k \left\{ \frac{\partial f_k}{\partial \alpha_j} - \left\langle \frac{\partial f_{i,\alpha_1\dots}^{(k)}}{\partial \alpha_j} \right\rangle \right\}. \tag{3.114}$$

We confine our subsequent considerations to the case where f_k does not depend on α, i.e., we treat f_k and α as independent variables. (Example: In

$$f_{i,\alpha} \equiv E_i(V) \tag{3.115}$$

the volume is an independent variable, and so is the internal energy $U = \langle E_i(V) \rangle$.) Then (3.114) reduces to

$$\frac{1}{k_B} \frac{\partial S}{\partial \alpha_j} = \sum_k \lambda_k \{ -\langle \partial f_{i,\alpha_1,\dots}^{(k)} / \partial \alpha_j \rangle \}. \tag{3.116}$$

For two systems, characterized by (3.98) and (3.100), we have now the additional conditions

$$\alpha_j(t) + \alpha_j'(t) = \alpha_j^{(0)}. \quad (1 \le j \le m) \tag{3.117}$$

(Example: α = volume)
The change of the entropy $S^{(0)}$ of the total system with respect to time, $dS^{(0)}/dt$ may be found with help of (3.107a), (3.116) and (3.117),

$$\frac{1}{k_B} \frac{dS^{(0)}}{dt} = \sum_k (\lambda_k - \lambda_k') \frac{dX_k}{dt} - \sum_{j,k} \left[\lambda_k \left\langle \frac{\partial X_i^{(k)}}{\partial \alpha_j} \right\rangle - \lambda_k' \left\langle \frac{\partial X_i'^{(k)}}{\partial \alpha_j'} \right\rangle \right] \cdot \frac{d\alpha_j}{dt}. \tag{3.118}$$

Using the definitions (3.106) and (3.107b) together with

$$\frac{d\alpha_j}{dt} = \tilde{J}_j, \tag{3.119}$$

and

$$\tilde{F}_j = k_B \sum_k \left[\lambda_k' \left\langle \frac{\partial X_i'^{(k)}}{\partial \alpha_j} \right\rangle - \lambda_k \left\langle \frac{\partial X_i^{(k)}}{\partial \alpha_j} \right\rangle \right] \tag{3.120}$$

we cast (3.118) into the form

$$\frac{dS^{(0)}}{dt} = \sum_{k=1}^n F_k J_k + \sum_{j=1}^m \tilde{F}_j \tilde{J}_j. \tag{3.121}$$

Putting

$$\left.\begin{aligned}\hat{F}_l &= F_l;\ \hat{J}_l = J_l \qquad\quad \text{if} \quad 1 \le l \le n \\ \hat{F}_l &= \tilde{F}_{l-n};\ \hat{J}_l = \tilde{J}_{l-n} \quad \text{if} \quad n+1 \le l \le n+m\end{aligned}\right\} \tag{3.122}$$

we arrive at an equation of the same structure as (3.107) but now taking into account the parameters α,

$$\frac{dS^{(0)}}{dt} = \sum_l \hat{F}_l \hat{J}_l. \tag{3.123}$$

The importance of this generalization will become clear from the following example which we already encountered in the foregoing chapter ($\alpha = V$), ($i \to i,\, N$)

$$\begin{aligned} f_{i,\alpha}^{(1)} &= E_i(V)\,, \quad f_1 \equiv X_1 = U;\ \lambda_1 = \frac{1}{k_B T}, \\ f_{i,\alpha}^{(2)} &= N, \qquad\quad f_2 \equiv X_2 = \bar{N};\ \lambda_2 = \frac{-\mu}{k_B T}. \end{aligned} \tag{3.124}$$

Inserting (3.124) into (3.118) we arrive at

$$\frac{dS^{(0)}}{dt} = \left(\frac{1}{T} - \frac{1}{T'}\right)\frac{dU}{dt} + \left(\frac{\mu'}{T'} - \frac{\mu}{T}\right)\frac{d\bar{N}}{dt} + \left(\frac{P}{T} - \frac{P'}{T'}\right)\frac{dV}{dt}. \tag{3.125}$$

The interpretation of (3.125) is obvious from the example given in (3.110, 111). The first two terms are included in (3.107), the other term is a consequence of the generalized (3.123).

The relation (3.125) is a well-known result in irreversible thermodynamics. It is, however, evident that the whole formalism is applicable to quite different disciplines, provided one may invoke there an analogue of local equilibrium. Since irreversible thermodynamics mainly treats *continuous systems* we now demonstrate how the above considerations can be generalized to such systems. We again start from the expression of the entropy

$$S = -k_B \sum_i p_i \ln p_i. \tag{3.126}$$

We assume that S and the probability distribution p_i depend on space and time.

The basic idea is this. We divide the total continuous system into subsystems (or the total volume in subvolumes) and assume that each subvolume is so large that we can still apply the methods of thermodynamics but so small that the probability distribution, or, the value of the extensive variable f may be considered as constant. We again assume that the p_i's are determined instantaneously and locally by the constraints (3.103) where $f_k(t)$ has now to be replaced by

$$f_k(x, t). \tag{3.127}$$

We leave it as an exercise to the reader to extend our considerations to the case where the $f^{(k)}$ depends on parameters α,

$$f^{(k)}_{i,\alpha(x,t)}. \tag{3.128}$$

We thus consider the equations

$$\textstyle\sum_i p_i(x,t) f_i^{(k)}(x,t) = f_k(x,t) \tag{3.129}$$

as determining the p_i's. We divide the entropy S by a unit volume V_0 and consider in the following the change of this entropy density

$$s = \frac{S}{V_0}. \tag{3.130}$$

We again identify the extensive variables with the f's

$$X_k \equiv f_k(x,t), \tag{3.131}$$

which are now functions of space and time. We further replace

$$\frac{X_k}{V_0} \to X_k, \quad \text{and} \quad \frac{f_k}{V_0} \to f_k. \tag{3.132}$$

It is a simple matter to repeat all the considerations of Section 3.3 in the case in which the f's are space- and time-dependent. One then shows immediately that one may write the entropy density again in the form

$$\frac{1}{k_B} s = \lambda_0 + \textstyle\sum_k \lambda_k(x,t) f_k(x,t), \tag{3.133}$$

which completely corresponds to the form (3.97).

It is our purpose to derive a continuity equation for the entropy density and to derive explicit expressions for the local temporal change of the entropy and the entropy flux. We first consider the local temporal change of the entropy by taking the derivative of (3.133) which gives us (as we have seen already in formula (3.59))

$$\frac{1}{k_B}\frac{\partial s}{\partial t} = \textstyle\sum_k \lambda_k \left[\frac{\partial}{\partial t} f_k - \left\langle \frac{\partial f^{(k)}}{\partial t} \right\rangle\right]. \tag{3.134}$$

We concentrate our consideration on the case in which $\langle \partial f^{(k)}/\partial t \rangle$ vanishes. $(\partial/\partial t) f_k$ gives us the local temporal change of the extensive variables, e.g., the energy, the momentum etc. Since we may assume that these variables X_k obey continuity equations of the form

$$\dot{X}_k + \nabla \boldsymbol{J}_k = 0, \tag{3.135}$$

where $\boldsymbol{J}_k$ are the corresponding fluxes, we replace in (3.134) $\dot{X}_k$ by $-\nabla \boldsymbol{J}_k$. Thus

(3.134) reads

$$\frac{1}{k_B}\frac{\partial s}{\partial t} = -\sum_k \lambda_k \nabla \boldsymbol{J}_k. \tag{3.136}$$

Using the identity $\lambda_k \nabla \boldsymbol{J}_k = \nabla(\lambda_k \boldsymbol{J}_k) - \boldsymbol{J}_k \nabla \lambda_k$ and rearranging terms in (3.136), we arrive at our final relation

$$\frac{1}{k_B}\frac{\partial s}{\partial t} + \nabla \sum_k \lambda_k \boldsymbol{J}_k = \sum_k (\nabla \lambda_k) \boldsymbol{J}_k \tag{3.137}$$

whose lhs has the typical form of a continuity equation. This leads us to the idea that we may consider the quantity

$$k_B \sum_k \lambda_k \boldsymbol{J}_k$$

as an adequate expression for the entropy flux. On the other hand the rhs may be interpreted as the local production rate of entropy so that we write

$$\frac{ds}{dt} = k_B \sum_k (\nabla \lambda_k) \boldsymbol{J}_k. \tag{3.138}$$

Thus (3.137) may be interpreted as follows: The local production of entropy leads to a temporal entropy change (first term on the lhs) and an entropy flux (second term). The importance of this equations rests on the following: The quantities $\boldsymbol{J}_k$ and $\nabla \lambda_k$ can be again identified with macroscopically measurable quantities. Thus for example if $\boldsymbol{J}_k$ is the energy flux, $k_B \nabla \lambda_k = \nabla(1/T)$ is the thermodynamic force. An important comment should be added. In the realm of irreversible thermodynamics several additional hypotheses are made to find equations of motion for the fluxes. The usual assumptions are 1) the system is Markovian, i.e., the fluxes at a given instant depend only on the affinities at that instant, and 2) one considers only linear processes, in which the fluxes are proportional to the forces. That means one assumes relations of the form

$$J_k = \sum_j L_{jk} F_j. \tag{3.139}$$

The coefficients L_{jk} are called *Onsager coefficients*. Consider as an example the heat flow $\boldsymbol{J}$. Phenomenologically it is related to temperature, T, by a gradient

$$\boldsymbol{J} = -\kappa \nabla T, \tag{3.140}$$

or, written in a somewhat different way by

$$\boldsymbol{J} = \kappa T^2 \nabla \frac{1}{T}. \tag{3.141}$$

By comparison with (3.139) and (3.138) we may identify $\nabla(1/T)$ with the affinity and κT^2 with the kinetic coefficient L_{11}. Further examples are provided by Ohm's law in the case of electric conduction or by Fick's law in the case of diffusion.

In the foregoing we have given a rough sketch of some basic relations. There are some important extensions. While we assumed an instantaneous relation between $p_i(t)$ and $f_k(t)$ (cf. (3.129)), more recently a retarded relation is assumed by Zubarev and others. These approaches still are subject to the limitations we shall discuss in Section 4.11.

Exercise on 3.5

1) Since $S^{(0)}$, S and S' in (3.98) can be expressed by probabilities $p_i, \ldots$ according to (3.42) it is interesting to investigate the implications of the additivity relation (3.98) with respect to the p's. To this end, prove the theorem: Given are the entropies (we omit the common factor k_B)

$$S^{(0)} = -\sum_{i,j} p(i,j) \ln p(i,j) \tag{E.1}$$

$$S = -\sum_i p_i \ln p_i \tag{E.2}$$

$$S' = -\sum_j p'_j \ln p'_j, \tag{E.3}$$

where

$$p_i = \sum_j p(i,j) \tag{E.4}$$

$$p'_j = \sum_i p(i,j) \tag{E.5}$$

Show: the entropies are additive, i.e.,

$$S^{(0)} = S + S' \tag{E.6}$$

holds, if and only if

$$p(i,j) = p_i p'_j \tag{E.7}$$

(i.e., the joint probability $p(i,j)$ is a product or, in other words, the events $X = i$, $Y = j$ are statistically independent).

Hint: 1) to prove "if", insert (E.7) into (E.1), and use $\sum_i p_i = 1$, $\sum_j p'_j = 1$,
2) to prove "only if", start from (E.6), insert on the lhs (E.1), and on the rhs (E.2) and (E.3), which eventually leads to

$$\sum_{ij} p(i,j) \ln (p(i,j)/(p_i p'_j)) = 0.$$

Now use the property (3.41) of $K(p,p')$, replacing in (3.38) the index k by the double index i, j.

2) Generalize (3.125) to several numbers, N_k, of particles.

3.6 Entropy—Curse of Statistical Mechanics?

In the foregoing chapters we have seen that the essential relations of thermodynamics emerged in a rather natural way from the concept of information. Those readers who are familiar with derivations in thermodynamics will admit that the present approach has an internal elegance. However, there are still deep questions and problems behind information and entropy. What we had intended originally was to make an unbiased guess about the p_i's. As we have seen, these p_i's drop out from the thermodynamic relations so that it actually does not matter how the p_i's look explicitly. It even does not matter how the random variables, over whose values i we sum up, have been chosen. This is quite in accordance with the so-called subjectivistic approach to information theory in which we just admit that we have only limited knowledge available, in our case presented by the constraints.

There is, however another school, called the objectivistic. According to it, it should be, at least in principle, possible to determine the p_i's directly. We then have to check whether our above approach agrees with the direct determination of the p_i's. As example let us consider physics. First of all we may establish a necessary condition. In statistical mechanics one assumes that the p_i's obey the so-called Liouville's equation which is a completely deterministic equation (compare the exercises on Section 6.3). This equation allows for some constants of motion (energy, momentum, etc.) which do not change with time. The constraints of Section 3.4 are compatible with the approach of statistical mechanics, provided the f_k's are such constants of motion. In this case one can prove that the p_i's fulfill indeed Liouville's equation (compare the exercises on Section 6.3).

Let us consider now a time-dependent problem. Here it is one of the basic postulates in thermodynamics that the entropy increases in a closed system. On the other hand, if the p_i's obey Liouville's equation and the initial state of the system is known with certainty, then it can be shown rigorously that S is time independent. This can be made plausible as follows: Let us identify the subscripts i with a coordinate $\boldsymbol{q}$ or with indices of cells in some space (actually it is a high-dimensional space of positions and momenta). Because p_i obeys a deterministic equation, we know always that $p_i = 1$ if that cell is occupied at a time t, and $p_i = 0$ if the cell is unoccupied. Thus the information remains equal to zero because there is no uncertainty in the whole course of time. To meet the requirement of an increasing entropy, two approaches have been suggested:

1) One averages the probability distribution over small volumes[2] replacing p_i by $1/\Delta v \int p_i dV = \bar{P}_i$ and forming $S = -k_B \sum_i \bar{P}_i \ln \bar{P}_i$ (or rather an integral over i). This S is called "coarse-grained entropy". By this averaging we take into account that we have no complete knowledge about the initial state. One can show that the initial distribution spreads more and more so that an uncertainty with respect to the actual distribution arises resulting in an increase of entropy. The basic idea of coarse graining can be visualized as follows (cf. page 1). Consider a drop of ink poured into water. If we follow the individual ink particles, their paths are com-

[2] Meant are volumes in *phase-space*.

pletely determined. However, if we average over volumes, then the whole state gets more and more diffuse. This approach has certain drawbacks because it appears that the increase of entropy depends on the averaging procedure.

2) In the second approach we assume that it is impossible to determine the time-dependence of p_i by the completely deterministic equations of mechanics. Indeed it is impossible to neglect the interaction of a system with its surroundings. The system is continuously subjected to fluctuations from the outer world. These can be, if no other contacts are possible, vacuum fluctuations of the electromagnetic field, fluctuations of the gravitational field, etc. In practice, however, such interactions are much more pronounced and can be taken into account by "heatbaths", or "reservoirs" (cf. Chap. 6). This is especially so if we treat *open systems* where we often need decay constants and transition rates explicitly. Thus we shall adopt in our book the attitude that the p_i's are always, at least to some extent, spreading which automatically leads to an increase of entropy (in a closed system), or more generally, to a decrease of information gain (cf. Section 3.2) in an open (or closed) system (for details cf. Exercises (1) and (2) of Section 5.3).

Lack of space does not allow us to discuss this problem in more detail but we think the reader will learn still more by looking at explicit examples. Those will be exhibited later in Chapter 4 in context with the master equation, in Chapter 6 in context with the Fokker-Planck equation and Langevin equation, and in Chapters 8 to 10 by explicit examples. Furthermore the impact of fluctuations on the entropy problem will be discussed in Section 4.10.

4. Chance

How Far a Drunken Man Can Walk

While in Chapter 2 we dealt with a fixed probability measure, we now study stochastic processes in which the probability measure changes with time. We first treat models of Brownian movement as example for a completely stochastic motion. We then show how further and further constraints, for example in the frame of a master equation, render the stochastic process a more and more deterministic process.

This Chapter 4, and Chapter 5, are of equal importance for what follows. Since Chapter 4 is somewhat more difficult to read, students may also first read 5 and then 4. On the other hand, Chapter 4 continues directly the line of thought of Chapters 2 and 3. In both cases, chapters with an asterisk in the heading may be omitted during a first reading.

4.1 A Model of Brownian Movement

This chapter will serve three purposes. 1) We give a most simple example of a stochastic process; 2) we show how a probability can be derived explicitly in this case; 3) we show what happens if the effects of many independent events pile up. The binomial distribution which we have derived in Section 2.11 allows us to treat Brownian movement in a nice fashion. We first explain what Brownian movement means: When a small particle is immersed in a liquid, one observes under a microscope a zig-zag motion of this particle. In our model we assume that a particle moves along a one-dimensional chain where it may hop from one point to one of its two neighboring points *with equal probability*. We then ask what is the probability that after n steps the particle reaches a certain distance, x, from the point where it had started. By calling a move to the right, success, and a move to the left, failure, the final position, x, of the particle is reached after, say, s successes and $(n - s)$ failures. If the distance for an elementary step is a, the distance, x, reached reads

$$x = a(s - (n - s)) = a(2s - n) \tag{4.1}$$

from which we may express the number of successes by

$$s = \frac{n}{2} + \frac{x}{2a}. \tag{4.2}$$

Denoting the transition time per elementary step by τ, we have for the time t after n steps

$$t = n\tau. \tag{4.3}$$

The probability of finding the configuration with n trials out of which s are successful has been derived in Section 2.11. It reads

$$P(s, n) = \binom{n}{s}\left(\frac{1}{2}\right)^n \tag{4.4}$$

since $p = q = 1/2$. Replacing s and n by the directly measurable quantities x (distance), and t (time) via (4.2) and (4.3), we obtain instead of P (4.4)

$$\hat{P}(x, t) \equiv P(s, n) = \binom{t/\tau}{t/(2\tau) + x/(2a)}\left(\frac{1}{2}\right)^{t/\tau} \tag{4.4a}$$

Apparently, this formula is not very handy and, indeed, difficult to evaluate exactly. However, in many cases of practical interest we make our measurements in times t which contain many elementary steps so that n may be considered as a large number. Similarly we also assume that the final position x requires many steps s so that also s may be considered as large. How to simplify (4.4a) in this limiting case requires a few elementary thoughts in elementary calculus. (Impatient readers may quickly pass over to the final result (4.29) which we need in later chapters). To evaluate (4.4) for large s and n we write (4.4) in the form

$$\frac{n!}{s!\,(n - s)!}\left(\frac{1}{2}\right)^n \tag{4.5}$$

and apply Stirling's formula (compare Section 2.14) to the factorials. We use Stirling's formula in the form

$$n! = \mathrm{e}^{-n}n^n\sqrt{2\pi n} \tag{4.6}$$

and correspondingly for $s!$ and $(n - s)!$ Inserting (4.6) and the corresponding relations into (4.4), we obtain

$$\hat{P}(x, t) \equiv P(s, n) = A\cdot B\cdot C\cdot D \tag{4.7}$$

where

$$A = \frac{\mathrm{e}^{-n}}{\mathrm{e}^{-s}\mathrm{e}^{-(n-s)}} = 1, \tag{4.8}$$

$$B = \exp[n \ln n - s \ln s - (n - s) \ln (n - s)], \tag{4.9}$$

$$\begin{aligned} C &= n^{1/2}(2\pi s(n - s))^{-1/2}, \\ F &= \exp(-n \ln 2). \end{aligned} \tag{4.10}$$

A stems from the powers of e in (4.6) which in (4.8) cancel out. B stems from the second factor in (4.6). The factor C stems from the last factor in (4.6), while F stems from the last factor in (4.5). We first show how to simplify (4.9). To this end we replace s by means of (4.2), put

$$\frac{x}{2a} = \delta, \tag{4.11}$$

and use the following elementary formulas for the logarithm

$$\ln (n/2 \pm \delta) = \ln n - \ln 2 + \ln \left(1 \pm \frac{2\delta}{n}\right), \tag{4.12}$$

and

$$\ln (1 + \alpha) = \alpha - \frac{\alpha^2}{2} + \cdots \tag{4.13}$$

Inserting (4.12) with (4.13) into (4.9) and keeping only terms up to second order in δ, we obtain for $F \cdot B$

$$B \cdot F = \exp\{-2\delta^2/n\} = \exp\left\{-\frac{x^2\tau}{2a^2t}\right\}. \tag{4.14}$$

We introduce the abbreviation

$$D = \frac{a^2}{\tau} \tag{4.15}$$

where D is called the diffusion constant. In the following we require that when we let a and τ go to zero the diffusion constant D remains a finite quantity. Thus (4.14) takes the form

$$B \cdot F = \exp\left\{-\frac{x^2}{2Dt}\right\}. \tag{4.16}$$

We now discuss the factor C in (4.7). We shall see later that this factor is closely related to the normalization constant. Inserting (4.3) and (4.2) into C, we find

$$C = (2\pi)^{-1/2}\left(\frac{n}{\left(\frac{n}{2} - \frac{x}{2a}\right)\left(\frac{n}{2} + \frac{x}{2a}\right)}\right)^{1/2} \tag{4.17}$$

or, after elementary algebra

$$C = (2\pi)^{-1/2}2\left\{\left(t^2 - x^2\tau\cdot\underbrace{\frac{\tau}{a^2}}_{=\frac{1}{D}}\right)\left(\frac{1}{t\tau}\right)\right\}^{-1/2} \tag{4.18}$$

Because we have in mind to let τ and a go to zero *but letting* D fixed we find that

$$\tau D \to 0$$

so that the term containing x^2 in (4.18) can be dropped and we are left with

$$C = (2\pi)^{-1/2} 2\tau^{1/2} t^{-1/2}. \tag{4.19}$$

Inserting the intermediate results A (4.8), $B \cdot F$ (4.16) and C (4.19) into (4.7), we obtain

$$\hat{P}(x, t) = (2\pi)^{-1/2} \cdot 2\tau^{1/2} t^{-1/2} \exp\left\{-\frac{x^2}{2Dt}\right\}. \tag{4.20}$$

The occurrence of the factor $\tau^{1/2}$ is, of course, rather disturbing because we want to let $\tau \to 0$. This problem can be resolved, however, when we bear in mind that it does not make sense to introduce a continuous variable x[1] and seek a *probability* to find a *specific value* x. This probability must necessarily tend to zero. We therefore have to ask what is the probability to find the particle after time t in an interval between x_1 and x_2 or, in the original coordinates, between s_1 and s_2. We therefore have to evaluate (compare Section 2.5)

$$\sum_{s_1}^{s_2} P(s, n) = P(s_1 \leq s \leq s_2). \tag{4.21}$$

If $P(s, n)$ is a slowly varying function in the interval $s_1, \ldots, s_2$, we may replace the sum by an integral

$$\sum_{s_1}^{s_2} P(s, n) = \int_{s_1}^{s_2} P(s, n)\, ds. \tag{4.22}$$

We now pass over to the new coordinates x, t. We note that on account of (4.2)

$$ds = \frac{1}{2a}\, dx. \tag{4.23}$$

Using furthermore (4.7) we find instead of (4.22)

$$\sum_{s_1}^{s_2} P(s, n) = \int_{x_1}^{x_2} \hat{P}(x, t) \frac{1}{2a}\, dx. \tag{4.24}$$

As we shall see in a minute, it is useful to abbreviate

$$\frac{1}{2a} \hat{P}(x, t) = f(x, t) \tag{4.25}$$

[1] Note that through (4.15) and (4.1) x becomes a continuous variable as τ tends to zero!

so that (4.22) reads finally

$$\sum_{s_1}^{s_2} P(s, n) = \int_{x_1}^{x_2} f(x, t)\, dx. \tag{4.26}$$

We observe that when inserting the probability distribution (4.7), which is proportional to $\tau^{1/2}$, into (4.21) the factor $\tau^{1/2}$ occurs simultaneously with $1/a$. We therefore introduce in accordance with (4.15)

$$\frac{\tau^{1/2}}{a} = D^{-1/2} \tag{4.27}$$

so that (4.21) acquires the final form

$$\int_{x_1}^{x_2} \exp\left\{-\frac{x^2}{2Dt}\right\}(2\pi)^{-1/2}t^{-1/2}D^{-1/2}\, dx. \tag{4.28}$$

When we compare the final form (4.28) with (2.21), we may identify

$$f(x, t) = (2\pi Dt)^{-1/2} \exp\left\{-\frac{x^2}{2Dt}\right\} \tag{4.29}$$

as the probability density (Fig. 4.1). In the language of physicists, $f(x, t)\, dx$ gives us the probability of finding the particle after time t in the interval $x \ldots x + dx$. The above steps (implying (4.21) – (4.29)) are often abbreviated as follows: Assuming again a slow, continuous variation of $P(s, n)$ and of f, the sum on the lhs of (4.26) is written as

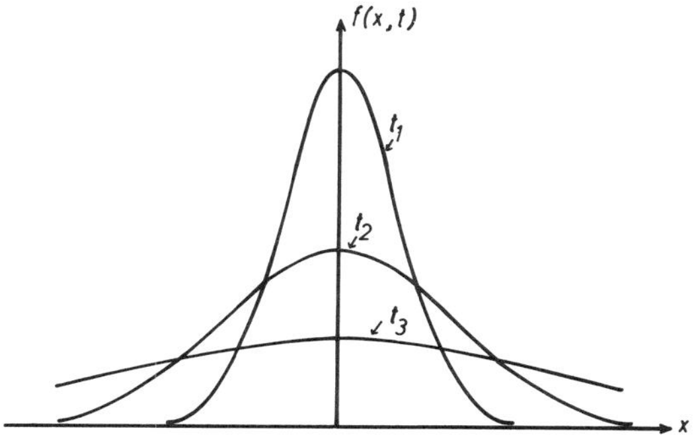

Fig. 4.1. The distribution function (4.29) as a function of the space point x for three different times

$$P(s, n)\Delta s \quad (\Delta s \equiv s_2 - s_1)$$

whereas the rhs of (4.26) is written as

$$f(x, t)\Delta x \quad (\Delta x = x_2 - x_1).$$

Passing over to infinitesimal intervals, (4.26) is given the form

$$P(s, n)\,ds = f(x, t)\,dx. \tag{4.30}$$

Explicit examples can be found in a number of disciplines, e.g., in physics. The result (4.29) is of fundamental importance. It shows that when many independent events (namely going backward or forward by one step) pile up the probability of the resulting quantity (namely the distance x) is the given by (4.29), i.e., the normal distribution. A nice example for (4.4) or (4.29) is provided by the gambling machine of Fig. 4.2. Another important result: In the limit of continuously varying variables, the original probability takes a very simple form.

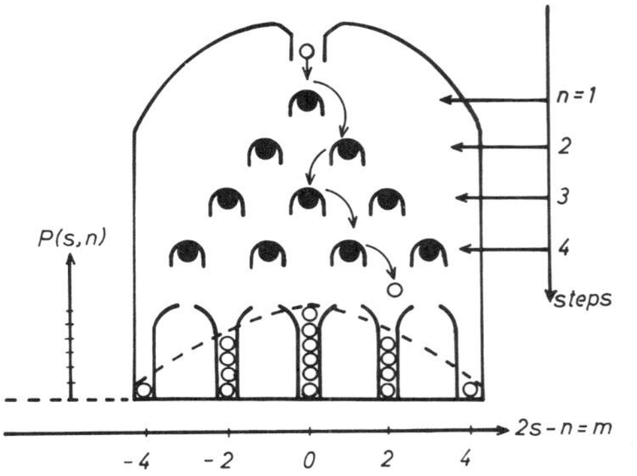

Fig. 4.2. Random walk of a ball in gambling machine yields the binomial distribution (4.4). Here $n = 4$. Shown is *average number* of balls in final boxes

Let us now discuss *restricted Brownian movement*. We again treat a hopping process, but insert between each two points a diaphragm so that the particle can only hop in one direction, say, to the right. Consequently,

$$p = 1 \quad \text{and} \quad q = 0. \tag{4.31}$$

In this case, the number, s, of successes coincides with the number, n, of hops

$$s = n. \tag{4.32}$$

Thus the probability measure is

$$P(s, n) = \delta_{s,n}. \tag{4.33}$$

We again introduce a coordinate x by sa and time by $t = n\tau$, and let $a \to 0$, $\tau \to 0$, so that x and t become continuous variables. Using (4.30) we obtain a probability density

$$f(x, t) = \delta(x - vt), \quad v = \frac{a}{\tau}, \tag{4.34}$$

because the Kronecker δ becomes Dirac's δ-function in the continuum. Note that $\lim_{\tau\to 0}$ and $\lim_{a\to 0}$ must be taken such that $v = a/\tau$ remains finite.

Exercise on 4.1

Calculate the first two moments for (4.29) and (4.34) and compare their time dependence. How can this be used to check whether nerve conduction is a diffusion or a unidirectional process?

4.2 The Random Walk Model and Its Master Equation

In the preceding Section 4.1 we have already encountered an example of a stochastic process going on in time; namely, the motion of a particle which is randomly hopping backward or forward. It is our goal to derive an equation for the probability measure, or, more precisely, we wish to establish an equation for the probability that after $n + 1$ pushes the particle is at the position m, $m = 0, \pm 1, \pm 2, \pm 3$. We denote this probability by $P(m; n + 1)$. Since the particle can hop at one elementary act only over a distance "1", it must have been after n pushes at $m - 1$ or $m + 1$ that it now reaches exactly the position m. There it had arrived with probabilities $P(m - 1; n)$ or $P(m + 1; n)$. Thus $P(m; n + 1)$ consists of two parts stemming from two possibilities for the particle to jump. The particle moves from $m - 1$ to m with the transition probability

$$w(m, m - 1) = \tfrac{1}{2}\,(= p). \tag{4.35}$$

The probability, $P(m; n + 1)$, of finding the particle at m after $n + 1$ pushes when it has been at time n, at $m - 1$ is given by the product $w(m, m - 1)\cdot P(m - 1; n)$. A completely analogous expression can be derived when the particle comes from $m + 1$. Since the particle must come either from $m - 1$ or $m + 1$ and these are independent events, $P(m; n + 1)$ can be written down in the form

$$P(m; n + 1) = w(m, m - 1)P(m - 1; n) + w(m, m + 1)P(m + 1; n) \tag{4.36}$$

where

$$w(m, m + 1) = w(m, m - 1) = 1/2. \tag{4.37}$$

Note that we did not use (4.37) explicitly so that (4.36) is also valid for the general case, provided

$$p + q \equiv w(m, m + 1) + w(m, m - 1) = 1. \tag{4.38}$$

Our present example allows us to explain some concepts in a simple manner. In Section 2.6 we introduced the joint probability which can now be generalized to time dependent processes. Let us consider as an example the probability $P(m, n + 1;$

m', n) to find the particle at step n at point m' and at step $n + 1$ at m. This probability consists of two parts as we have already discussed above. The particle must be at step n at point m' which is described by the probability $P(m'; n)$. Then it must pass from m' to m which is governed by the transition probability $w(m, m')$. Thus the joint probability can generally be written in the form

$$P(m, n + 1; m', n) = w(m, m')P(m'; n) \tag{4.39}$$

where w is apparently identical with the conditional probability (compare Section 2.8). Since the particle is pushed at each time away from its original position,

$$m \neq m', \tag{4.40}$$

and because it can jump only over an elementary distance "1"

$$|m - m'| = 1, \tag{4.41}$$

it follows that $w(m, m') = 0$ unless $m' = m \pm 1$ (in our example). We return to (4.36) which we want to transform further. We put

$$w(m, m \pm 1)/\tau = \tilde{w}(m, m \pm 1),$$

and shall refer to $\tilde{w}(m, m \pm 1)$ as transition probability per second (or per unit time). We subtract $P(m; n)$ from both sides of (4.36) and divide both sides by τ. Using furthermore (4.38) the following equation results

$$\frac{P(m; n + 1) - P(m; n)}{\tau} = \tilde{w}(m, m - 1)P(m - 1; n) + \tilde{w}(m, m + 1)P(m + 1; n)$$
$$- (\tilde{w}(m + 1, m) + \tilde{w}(m - 1, m))P(m; n). \tag{4.42}$$

In our special example the $\tilde{w}$'s are both equal to $1/(2\tau)$. We relate the discrete variable n with the time variable t by writing $t = n\tau$ and accordingly introduce now a probability measure $\tilde{P}$ by putting

$$\tilde{P}(m, t) = P(m; n) \equiv P(m; t/\tau). \tag{4.43}$$

$\tilde{P}$ in (4.43) is an abbreviation for the function $P(m, t/\tau)$. We now approximate the difference on the lhs of (4.42) by the time derivative

$$\frac{P(m; n + 1) - P(m; n)}{\tau} \approx \frac{d\tilde{P}}{dt} \tag{4.44}$$

so that the original equation (4.36) takes the form

$$\frac{d\tilde{P}(m, t)}{dt} = \tilde{w}(m, m - 1)\tilde{P}(m - 1, t) + \tilde{w}(m, m + 1)\tilde{P}(m + 1, t)$$
$$-(\tilde{w}(m + 1, m) + \tilde{w}(m - 1, m))P(m, t). \tag{4.45}$$

This equation is known in the literature as *master equation*.

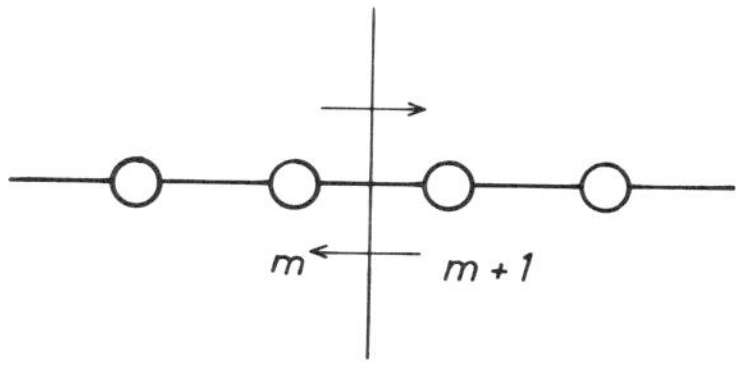

Fig. 4.3. How to visualize detailed balance

Let us now consider the stationary state in which the probability P is time-independent. How can this be reached in spite of the hopping process going on all the time? Consider a line between m and $m + 1$. Then P does not change, if (per unit time) the same number of transitions occur in the right and in the left direction. This requirement is called the *principle of detailed balance* (Fig. 4.3) and can be written mathematically in the form

$$\tilde{w}(m, m')P(m'; n) = \tilde{w}(m', m)P(m; n). \tag{4.46}$$

In Section 4.1 we found it still more advantageous to replace m by a continuous variable, x, by putting $m = x/a$. When x is finite and a tends to 0, m must necessarily become a very large number. In this case the unity "1" may be considered as a small quantity compared to m. Therefore we replace, in a formal way, 1 by ε. We treat the special case

$$\tilde{w}(m, m + 1) = \tilde{w}(m, m - 1) = \frac{1}{2\tau}.$$

In this case the right hand side of (4.42) reads

$$\frac{1}{2\tau}(\tilde{P}(m - \varepsilon, t) + \tilde{P}(m + \varepsilon, t) - 2\tilde{P}(m, t)) \tag{4.47}$$

which we expand into a Taylor series up to second order in ε

$$(4.47) \approx \frac{1}{2\tau}\{\tilde{P}(m, t) + \tilde{P}(m, t) - 2\tilde{P}(m, t) + [\tilde{P}'(m, t) - \tilde{P}'(m, t)]\varepsilon$$
$$+ \tfrac{1}{2}[\tilde{P}''(m, t) + \tilde{P}''(m, t)]\varepsilon^2\}, \tag{4.48}$$

where the primes denote differentiation of $\tilde{P}$ with respect to m. Because the first and second terms $\propto \varepsilon^0$ or ε cancel, the first nonvanishing term is that of the second derivative of $\tilde{P}$. Inserting this result into (4.42) yields

$$\frac{d\tilde{P}}{dt} = \frac{1}{2\tau}\frac{d^2\tilde{P}}{dm^2}\varepsilon^2. \tag{4.49}$$

It is now appropriate to introduce a new function f by

$$\tilde{P}(m, t)\Delta m = f(x, t)\Delta x. \tag{4.50}$$

(Why have we introduced Δm and Δx here? (cf. Section 4.1).) Using furthermore

$$\frac{d}{dm} = \frac{a \cdot d}{dx}, \quad \varepsilon = 1, \tag{4.51}$$

we find the fundamental equation

$$\frac{df(x, t)}{dt} = \frac{1}{2} D \frac{d^2 f(x, t)}{dx^2}, \quad D = \frac{a^2}{\tau} \tag{4.52}$$

which we shall refer to as *diffusion equation*. As we will see much later (cf. Section 6.3), this equation is a very simple example of the so-called Fokker-Planck equation. The solution of (4.52) is, of course, much easier to find than that of the master equation. The ε-expansion relies on a *scaling* property. Indeed, letting $\varepsilon \to 0$ can be identified with letting the length scale a go to zero. We may therefore repeat the above steps in a more formal manner, introducing the length scale a as expansion parameters. We put (since $\Delta x/\Delta m = a$)

$$\tilde{P}(m, t) = af(x, t), \tag{4.53a}$$

and, concurrently, since $m + 1$ implies $x + a$:

$$\tilde{P}(m \pm 1, t) = af(x \pm a, t). \tag{4.53b}$$

Inserting (4.53a) and (4.53b) into (4.42) or (4.47) yields, after expanding f up to second order in a, again (4.52). The above procedure is only valid under the condition that for fixed a, the function f varies very slowly over the distance $\Delta x \mp a$. Thus the whole procedure implies a self-consistency requirement.

Exercises on 4.2

1) Show that $P(s, n)$ (4.4) fulfils (4.36) with (4.37) if s, n are correspondingly related to m, n. The initial condition is $P(m, 0) = \delta_{m,0}$ (i.e. $= 1$ for $m = 0$, and $= 0$ for $m \neq 0$).
2) Verify that (4.29) satisfies (4.52) with the initial condition $f(x, 0) = \delta(x)$ (δ: Dirac's δ-function).
3) Derive the generalization of equation (4.52) if $w(m, m + 1) \neq w(m, m - 1)$.
4) *Excitation transfer between two molecules by a hopping process.* Let us denote the two molecules by the index $j = 0, 1$ and let us denote the probability to find the molecule j in an excited state at time t by $P(j, t)$. We denote the transition rate per unit time by $\tilde{w}$. The master equation for this process reads

$$\begin{aligned} \dot{P}(0, t) &= \tilde{w}P(1, t) - \tilde{w}P(0, t) \\ P(1, t) &= \tilde{w}P(0, t) - \tilde{w}P(1, t). \end{aligned} \tag{E.1}$$

Show that the equilibrium distribution is given by

$$P(j) = \tfrac{1}{2}, \tag{E.2}$$

determine the conditional probability

$$P(j, t \mid j', 0), \tag{E.3}$$

and the joint probability

$$P(j, t; j', t'). \tag{E.4}$$

Hint: Solve equations (E.1) by the hypothesis

$$P(j, t \mid j', 0) = \alpha_{1j} e^{-\lambda_1 t} + \alpha_{2j} e^{-\lambda_2 t} \tag{E.5}$$

using for $j' = 0, j' = 1$ the initial condition ($t = 0$)

$$P(0, 0) = 1, P(1, 0) = 0. \tag{E.6}$$

4.3* Joint Probability and Paths. Markov Processes. The Chapman-Kolmogorov Equation. Path Integrals

To explain the main ideas of this section, let us first consider the example of the two foregoing sections: A particle hopping randomly along a chain. When we follow up the different positions the particle occupies in a specific realization, we may draw the plot of Fig. 4.4a. At times $t_1, t_2, \ldots, t_n$ the particle is found at n specific positions $m_1, m_2, \ldots, m_n$. If we repeat the experiment, we shall find another realization, say that of Fig. 4.4b. We now ask for the probability of finding the particle at times $t_1, t_2, \ldots$ at the corresponding points $m_1, m_2, \ldots$ We denote this probability by

$$P_n(m_n, t_n; m_{n-1}, t_{n-1}; \ldots; m_1, t_1). \tag{4.54}$$

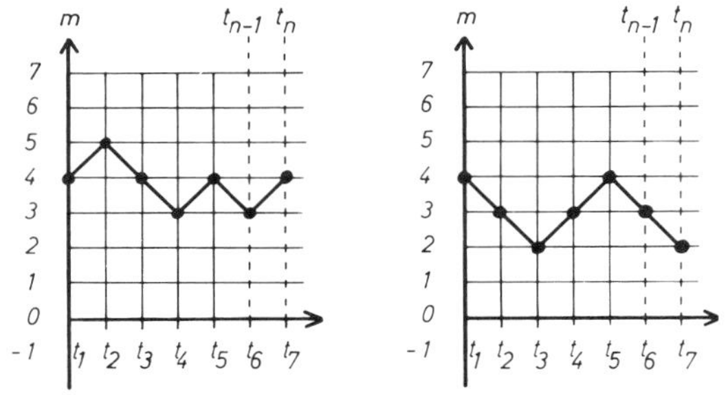

Fig. 4.4. Two paths of a Brownian particle $t_1 = 0$, $m_1 = 4, \ldots,$ $n = 7$

Apparently P_n is a joint probability in the sense of Section 2.6. Once we know the joint probability with respect to n different times, we can easily find other probability distributions containing a smaller number of arguments m_j, t_j. Thus for example if we are interested only in the joint probability at times $t_1, \ldots, t_{n-1}$ we have to sum up (4.54) over m_n

$$P_{n-1}(m_{n-1}, t_{n-1}; \ldots; m_1, t_1) = \sum_{m_n} P_n. \tag{4.55}$$

Another interesting probability is that for finding the particle at time t_n at the position m_n and at time t_1 at position m_1, irrespective of the positions the particle may have acquired at intermediate times. We find the probability by summing (4.54) over all intermediate positions:

$$P_2(m_n, t_n; m_1, t_1) = \sum_{m_2 m_3 \cdots m_{n-1}} P_n(m_n, t_n; m_{n-1}, t_{n-1}; \ldots; n_1, t_1). \tag{4.56}$$

Let us now consider an experiment (a realization of the random process) in which the particle was found at times $t_1, \ldots, t_{n-1}$ at the corresponding positions $m_1, \ldots, m_{n-1}$. We then ask what is the probability of finding it at time t_n at position m_n? We denote this conditional probability by

$$P(m_n, t_n \mid m_{n-1}, t_{n-1}; m_{n-2}, t_{n-2}; \ldots; m_1, t_1). \tag{4.57}$$

According to Section 2.8 this conditional probability is given by

$$P(m_n, t_n \mid m_{n-1}, t_{n-1}; m_{n-2}, t_{n-2}; \ldots; m_1, t_1) = \frac{P_n(m_n, t_n; m_{n-1}, t_{n-1}; \ldots; m_1, t_1)}{P_{n-1}(m_{n-1}, t_{n-1}; \ldots; m_1, t_1)}. \tag{4.58}$$

So far our considerations apply to any process. However, if we consider our special example of Sections 4.1 and 4.2, we readily see that the probability for the final position m_n depends only on the probability distribution at time t_{n-1} and not on any earlier time. In other words, the particle has lost its memory. In this case the conditional probability (4.57) depends only on the arguments at times t_n and t_{n-1} so that we may write

$$P(m_n, t_n \mid m_{n-1}, t_{n-1}; \ldots; m_1, t_1) = p_{t_n, t_{n-1}}(m_n, m_{n-1}). \tag{4.59}$$

If the conditional probability obeys the equation (4.59) the corresponding process is called a *Markov process*. In the following we want to derive some general relations for Markov processes.

The rhs of (4.59) is often referred to as transition probability. We now want to express the joint probability (4.54) by means of the transition probability (4.59). In a first step we multiply (4.58) by P_{n-1}

$$P_n = p_{t_n, t_{n-1}}(m_n, m_{n-1}) P_{n-1}. \tag{4.60}$$

Replacing in this equation n by $n - 1$ we find

$$P_{n-1} = p_{t_{n-1},t_{n-2}}(m_{n-1}, m_{n-2})P_{n-2}. \tag{4.61}$$

Reducing the index n again, and again we finally obtain

$$P_2 = p_{t_2,t_1}(m_2, m_1)P_1, \tag{4.62}$$

and

$$P_1 \equiv P_1(m_1, t_1). \tag{4.63}$$

Consecutively substituting P_{n-1} by P_{n-2}, P_{n-2} by P_{n-3}, etc., we arrive at

$$P_n(m_n, t_n; m_{n-1}, t_{n-1}; \ldots ; m_1, t_1)$$
$$= p_{t_n,t_{n-1}}(m_n, m_{n-1})p_{t_{n-1},t_{n-2}}(m_{n-1}, m_{n-2}) \ldots p_{t_2,t_1}(m_2, m_1)P_1(m_1, t_1). \tag{4.64}$$

Thus the joint probability of a Markov process can be obtained as a mere product of the transition probabilities. Thus the probability of finding for example a particle at positions m_j for a time sequence t_j can be found by simply following up the path of the particle and using the individual transition probabilities from one point to the next point.

To derive the Chapman-Kolmogorov equation we consider three arbitrary times subject to the condition

$$t_1 < t_2 < t_3. \tag{4.65}$$

We specialize equation (4.56) to this case:

$$P_2(m_3, t_3; m_1, t_1) = \sum_{m_2} P_3(m_3, t_3; m_2, t_2; m_1, t_1). \tag{4.66}$$

We now use again formula (4.62) which is valid for any arbitrary time sequence. In particular we find that for the lhs of (4.65)

$$p_{t_3,t_1}(m_3, m_1)P_1(m_1, t_1). \tag{4.67}$$

With help of (4.64) we write P_3 on the rhs of (4.66) in the form

$$P_3(m_3, t_3; m_2, t_2; m_1, t_1) = p_{t_3,t_2}(m_3, m_2)p_{t_2 t_1}(m_2, m_1)P_1(m_1, t_1). \tag{4.68}$$

Inserting (4.67) and (4.68), we find a formula in which on both sides $P_1(m_1, t_1)$ appears as a factor. Since this initial distribution can be chosen arbitrarily, that relation must be valid without this factor. This yields as final result the *Chapman-Kolmogorov equation*

$$p_{t_3,t_1}(m_3, m_1) = \sum_{m_2} p_{t_3,t_2}(m_3, m_2)p_{t_2,t_1}(m_2, m_1). \tag{4.69}$$

Note that (4.69) is not so innocent as it may look because it must hold for any time sequence between the initial time and the final time.

The relation (4.69) can be generalized in several ways. First of all we may replace m_j by the M-dimensional vector $\boldsymbol{m}_j$. Furthermore we may let $\boldsymbol{m}_j$ become a continuous variable $\boldsymbol{q}_j$ so that P becomes a distribution function (density). We leave it to the reader as an exercise to show that in this latter case (4.69) acquires the form

$$p_{t_3,t_1}(\boldsymbol{q}_3, \boldsymbol{q}_1) = \int \cdots \int p_{t_3,t_2}(\boldsymbol{q}_3, \boldsymbol{q}_2) p_{t_2,t_1}(\boldsymbol{q}_2, \boldsymbol{q}_1) d^M q_2. \tag{4.70}$$

Another form of the Chapman-Kolmogorov equation is obtained if we multiply (4.69) by $P(m_1, t_1)$ and sum up over m_1. Using

$$p_{t_3,t_1}(m_3, m_1) P_1(m_1, t_1) = P_2(m_3, t_3; m_1, t_1), \tag{4.71}$$

and changing the indices appropriately we obtain

$$P(m, t) = \sum_{m'} p_{t,t'}(m, m') P(m', t') \tag{4.72}$$

in the discrete case, and

$$P(\boldsymbol{q}, t) = \int \cdots \int p_{t,t'}(\boldsymbol{q}, \boldsymbol{q}') P(\boldsymbol{q}', t')\, d^M q' \tag{4.73}$$

for continuous variables. (Here and in the following we drop the index 1 of P).

We now proceed to consider infinitesimal time intervals, i.e., we put $t = t' + \tau$. We then form the following expression

$$\frac{1}{\tau}\{P(m, t + \tau) - P(m, t)\}, \tag{4.74}$$

and let τ go to zero or, in other words, we take the derivative of (4.72) with respect to the time t, which yields

$$\dot{P}(m, t) = \sum_{m'} \dot{p}_t(m, m') P(m', t) \tag{4.75}$$

where $\dot{p}_t(m, m') = \lim_{\tau \to 0} \tau^{-1}(p_{t+\tau,t}(m, m') - p_{t,t}(m, m'))$. The discussion of $\dot{p}_t$ requires some care. If $m' \neq m$, $\dot{p}_t$ is nothing but the number of transitions from m' to m per second for which we put

$$\dot{p}_t(m, m') = w(m, m') \tag{4.76}$$

or, in other words, w is the transition probability per second (or unit time). The discussion of $\dot{p}_t(m, m)$ with the same indices $m = m'$ is somewhat more difficult because it appears that no transition would occur. For a satisfactory discussion we must go back to the definition of p according to (4.59) and consider the difference of the conditional probabilities

$$P(m, t + \tau \mid m, t) - P(m, t \mid m, t). \tag{4.77}$$

This difference represents the change of probability of finding the particle at point m at a later time, $t + \tau$, if it had been at the same point at time t. This change of probability is caused by all processes in which the particle leaves its original place m. Thus, if we divide (4.77) by τ and let τ go to zero, (4.77) is equal to the sum over rates (4.76) for the transitions from m to all other states

$$\dot{p}_t(m, m) = -\textstyle\sum_{m'} w(m', m). \tag{4.78}$$

Inserting now (4.76) and (4.78) into (4.75) leaves us with the so-called *master equation*

$$\dot{P}(\boldsymbol{m}, t) = \textstyle\sum_{\boldsymbol{m}'} w(\boldsymbol{m}, \boldsymbol{m}')P(\boldsymbol{m}', t) - P(\boldsymbol{m}, t) \sum_{\boldsymbol{m}'} w(\boldsymbol{m}', \boldsymbol{m}). \tag{4.79}$$

In passing from (4.75) to this equation we have generalized it by replacing m by the vector $\boldsymbol{m}$.

4.3.1 Example of a Joint Probability: The Path Integral as Solution of the Diffusion Equation

Our somewhat abstract considerations about the joint probability at different times can be illustrated by an example which we have already met several times in our book. It is the random motion of a particle in one dimension. We adopt a continuous space coordinate q. The corresponding diffusion equation was given in Section 4.2 by (4.52). We write it here in a slightly different form by calling the function $f(q)$ of (4.52) now $p_{tt'}(q, q')$:

$$\left(\frac{\partial}{\partial t} - \frac{D}{2}\frac{\partial^2}{\partial q^2}\right)p_{tt'}(q, q') = 0, \quad t > t'. \tag{4.80}$$

The reason for this new notation is the following: we subject the solution of (4.80) to the initial condition:

$$p_{tt'}(q, q') = \delta(q - q') \quad \text{for} \quad t = t'. \tag{4.81}$$

That is we assume that a time t equal to the initial time t' the particle is with certainty at the space point q'. Therefore $p_{tt'}(q, q')\,dq$ is the conditional probability to find the particle at time t in the interval $q \cdots q + dq$ provided it has been at time t' at point q' or, in other words, $p_{tt'}(q, q')$ is the transition probability introduced above. For what follows we make the replacements

$$q' \to q_j \tag{4.82}$$

and

$$q \to q_{j+1} \tag{4.83}$$

where j will be an index defined by a time sequence t_j. Fortunately, we already know the solution to (4.80) with (4.81) explicitly. Indeed in Section 4.1 we derived the conditional probability to find a particle undergoing a random walk after a time t at point q provided it has been at the initial time at $q' = 0$. By a shift of the coordinate system, $q \to q - q' \equiv q_{j+1} - q_j$ and putting $t_{j+1} - t_j = \tau$ that former solution (4.29) reads

$$p_{tt'}(q_{j+1}, q_j) = (2\pi D\tau)^{-1/2} \exp\left\{-\frac{1}{2D\tau}(q_{j+1} - q_j)^2\right\}. \tag{4.84}$$

By inserting (4.84) into (4.80) one readily verifies that (4.84) fulfils this equation. We further note that for $\tau \to 0$ (4.84) becomes a δ-function (see Fig. 2.10). It is now a simple matter to find an explicit expression for the joint probability (4.54). To this end we need only to insert (4.84) (with $j = 1, 2, \ldots$) into the general formula (4.64) which yields

$$P(q_n, t_n; q_{n-1}, t_{n-1}; \ldots; q_1, t_1)$$
$$= (2\pi D\tau)^{-(n-1)/2} \exp\left\{-\frac{1}{2D\tau} \sum_{j=1}^{n-1} (q_{j+1} - q_j)^2\right\} P(q_1, t_1). \tag{4.85}$$

This can be interpreted as follows: (4.85) describes the probability that the particle moves along the path $q_1 q_2 q_3 \ldots$. We now proceed to a continuous time scale by replacing

$$\tau \to dt \tag{4.86}$$

and

$$\frac{1}{\tau}(q_{j+1} - q_j) = \frac{dq}{dt} \tag{4.87}$$

which allows us to write the exponential function in (4.85) in the form

$$\exp\left\{-\frac{1}{2D}\int_{t'}^{t} \left(\frac{dq}{dt}\right)^2 dt\right\}. \tag{4.88}$$

(4.88) together with a normalization factor is the simplest form of a probability distribution for a path. If we are interested only in the probability of finding the final coordinate $q_n = q$ at time $t_n = t$ irrespective of the special path chosen, we have to integrate over all intermediate coordinates $q_{n-1}, \ldots, q_1$ (see (4.56)). This probability is thus given by

$$P(q_n, t) = \lim_{\tau \to 0, n \to \infty} (2\pi D\tau)^{-(n-1)/2}$$
$$\times \int_{-\infty}^{+\infty} \int_{-\infty}^{+\infty} dq_1 \cdots dq_{n-1} \exp\left\{-\frac{1}{2D\tau} \sum_{j=1}^{n-1} (q_{j+1} - q_j)^2\right\} P(q_1, t_1) \tag{4.89}$$

which is called a *path integral.* Such path integrals play a more and more important role in statistical physics. We will come back to them later in our book.

Exercise on 4.3

Determine

$$P_3(m_3, t_3; m_2, t_2; m_1, t_1), \quad t_2 = t_1 + \tau, t_3 = t + 2\tau,$$

for the random walk model of Section 4.1, for the initial condition $P(m_1, t_1) = 1$ for $m_1 \equiv 0$; $= 0$ otherwise.

4.4* How to Use Joint Probabilities. Moments. Characteristic Function. Gaussian Processes

In the preceding section we have become acquainted with joint probabilities for time-dependent processes

$$P(m_n, t_n; m_{n-1}, t_{n-1}; \ldots ; m_1, t_1), \tag{4.90}$$

where we now drop the index n of P_n. Using (4.90) we can define moments by generalization of concepts introduced in Section 2.7

$$\langle m_n^{\nu_n} \cdots m_1^{\nu_1} \rangle = \sum_{m_1 \cdots m_n} m_n^{\nu_n} \cdots m_1^{\nu_1} P(m_n, t_n; m_{n-1}, t_{n-1}; \ldots ; m_1, t_1). \tag{4.91}$$

In this equation some of the ν's may be equal to zero. A particularly important case is provided by

$$\langle m_n^{\nu_n} \cdot m_1^{\nu_1} \rangle \tag{4.92}$$

where we have only the product of two powers of m at different times. When we perform the sum over all other m's according to the definition (4.91) we find the probability distribution which depends only on the indices 1 and n. Thus (4.92) becomes

$$\langle m_n^{\nu_n} m_1^{\nu_1} \rangle = \sum_{m_1} \sum_{m_n} m_n^{\nu_n} m_1^{\nu_1} P(m_n, t_n; m_1, t_1). \tag{4.93}$$

We mention a few other notations. Since the indices $j = 1 \ldots n$ refer to times we can also write (4.92) in the form

$$\langle m^{\nu_n}(t_n) m^{\nu_1}(t_1) \rangle. \tag{4.94}$$

This notation must not be interpreted that m is a given function of time. Rather, t_n and t_1 must be interpreted as indices. If we let the t_j's become a continuous time sequence we will drop the suffices j and use t or t' as index so that (4.94) is written in the form

$$\langle m^{\nu}(t) m^{\nu'}(t') \rangle . \tag{4.95}$$

In a completely analogous manner we can repeat (4.93) so that we obtain the relation

$$\langle m^{\nu}(t) m^{\nu'}(t') \rangle = \sum_{m(t), m(t')} m^{\nu}(t) m^{\nu'}(t') P(m(t), t; m(t'), t'). \tag{4.96}$$

Another way of writing is m_t instead of $m(t)$. As already mentioned before we can let m become a continuous variable q in which case P becomes a probability density. To mention as an example again the case of two times, (4.96) acquires the form

$$\langle q_t^{\nu} q_{t'}^{\nu'} \rangle = \iint q_t^{\nu} q_{t'}^{\nu'} P(q_t, t \mid q_{t'}, t') P(q_{t'}, t') \, dq_t \, dq_{t'}. \tag{4.97}$$

In analogy of Section 2.10 we introduce the *characteristic function* by

$$\langle \exp \{ i \textstyle\sum_{l=1}^{n} u_l q_l \} \rangle = \Phi(u_n, t_n; u_{n-1}, t_{n-1}; \ldots ; u_1, t_1) \tag{4.98}$$

in the case of a discrete time sequence, and by

$$\left\langle \exp \left\{ i \int_{t_0}^{t} dt' u_{t'} q_{t'} \right\} \right\rangle = \Phi(\{u_t\}) \tag{4.99}$$

in the case of a continuous time sequence. By taking ordinary or functional derivatives we can easily recover *moments* in the case of a single variable by

$$\frac{1}{i^{\nu}} \frac{\delta^{\nu}}{\delta u_{t_1}^{\nu}} \Phi(\{u_t\})|_{u=0} = \langle q_{t_1}^{\nu} \rangle, \tag{4.100}$$

and in the case of variables at different times by

$$(-i)^{\nu_1 + \cdots + \nu_n} \frac{\delta^{\nu_1 + \cdots + \nu_n}}{\delta u_{t_1}^{\nu_1} \delta u_{t_2}^{\nu_2} \cdots \delta u_{t_n}^{\nu_n}} \Phi|_{u=0} = \langle q_{t_n}^{\nu_n} \cdots q_{t_1}^{\nu_1} \rangle. \tag{4.101}$$

We furthermore define *cumulants* by

$$\Phi(u_n, t_n; \ldots, u_1, t_1) = \exp \left\{ \textstyle\sum_{s=1}^{\infty} \frac{i^s}{s!} \sum_{\alpha_1, \cdots \alpha_s = 1}^{n} k_s(t_{\alpha_1}, \ldots, t_{\alpha_s}) u_{\alpha_1} \cdots u_{\alpha_s} \right\}. \tag{4.102}$$

We call a process *Gaussian* if all cumulants except the first two vanish, i.e., if

$$k_3 = k_4 = \cdots = 0 \tag{4.103}$$

holds. In the case of a Gaussian process, the characteristic function thus can be written in the form

$$\Phi(u_n, t_n; \ldots ; u_1, t_1) = \exp \{i \textstyle\sum_{\alpha=1}^{n} k_1(t_\alpha)u_\alpha - \frac{1}{2} \sum_{\alpha,\beta=1}^{n} k_2(t_\alpha, t_\beta)u_\alpha u_\beta\}. \quad (4.104)$$

According to (4.101) all moments can be expressed by the first two cumulants or, because k_1 and k_2 can be expressed by the first and second moment, all higher moments can be expressed by the first two moments.

The great usefulness of the joint probability and of correlation functions for example of the form (4.97) rests in the fact that these quantities allow us to discuss time dependent correlations. Consider for example the correlation function

$$\langle q_t q_{t'} \rangle, \quad t > t', \quad (4.105)$$

and let us assume that the mean values at the two times t and t'

$$\langle q_t \rangle = 0, \quad \langle q_{t'} \rangle = 0 \quad (4.106)$$

vanish. If there is no correlation between q's at different times we can split the joint probability into a product of probabilities at the two times. As a consequence (4.105) vanishes. On the other hand if there are correlation effects then the joint probability cannot be split into a product and (4.105) does not vanish in general. We will exploit this formalism later on in greater detail to check how fast fluctuations decrease, or, how long a coherent motion persists. If the mean values of q do not vanish, one can replace (4.105) by

$$\langle (q_t - \langle q_t \rangle)(q_{t'} - \langle q_{t'} \rangle) \rangle \quad (4.107)$$

to check correlations.

Exercise on 4.4

Calculate the following moments (or correlation functions) for the diffusion process under the initial conditions

1) $P(q, t = 0) = \delta(q)$,

2) $P(q, t = 0) = (\beta/\pi)^{1/2} \exp(-\beta q^2)$,

$\langle q(t) \rangle, \langle q^2(t) \rangle, \langle q(t)q(t') \rangle, \langle q^2(t)q^2(t') \rangle$

Hint: $\int_{-\infty}^{+\infty} q^\nu e^{-\alpha q^2} dq = 0, \quad \nu\text{: odd},$

$$\sqrt{\alpha/\pi} \int_{-\infty}^{+\infty} q^2 e^{-\alpha q^2} dq = \frac{1}{2\alpha},$$

$$\sqrt{\alpha/\pi} \int_{-\infty}^{+\infty} q^4 e^{-\alpha q^2} dq = \tfrac{3}{4}\alpha^{-2}.$$

Try in $\langle \cdots \rangle$ a replacement of q_t by $q_t + q_{t'}$!

4.5 The Master Equation

This section can be read without knowledge of the preceding ones. The reader should be acquainted, however, with the example of Section 4.2. The master equation we shall derive in this chapter is one of the most important means to determine the probability distribution of a process. In Section 4.2 we have already encountered the example of a particle which is randomly pushed back or forth. Its motion was described by a stochastic equation governing the change of the probability distribution as a function of time. We will now consider the general case in which the system is described by discrete variables which can be lumped together to a vector $\boldsymbol{m}$. To visualize the process think of a particle moving in three dimensions on a lattice.

The probability of finding the system at point $\boldsymbol{m}$ at a time t increases due to transitions from other points $\boldsymbol{m}'$ to the point under consideration. It decreases due to transitions leaving this point, i.e., we have the general relation

$$\dot{P}(\boldsymbol{m}, t) = \text{rate in} - \text{rate out}. \tag{4.108}$$

Since the "rate in" consists of all transitions from initial points $\boldsymbol{m}'$ to $\boldsymbol{m}$, it is composed of the sum over the initial points. Each term of it is given by the probability to find the particle at point $\boldsymbol{m}'$, multiplied by the transition probability per unit time to pass from $\boldsymbol{m}'$ to $\boldsymbol{m}$. Thus we obtain

$$\text{rate in} = \sum_{\boldsymbol{m}'} w(\boldsymbol{m}, \boldsymbol{m}')P(\boldsymbol{m}', t). \tag{4.109}$$

In a similar way we find for the outgoing transitions the relation

$$\text{rate out} = P(\boldsymbol{m}, t)\cdot\sum_{\boldsymbol{m}'} w(\boldsymbol{m}', \boldsymbol{m}). \tag{4.110}$$

Putting (4.109) and (4.110) into (4.108) we obtain

$$\dot{P}(\boldsymbol{m}, t) = \sum_{\boldsymbol{m}'} w(\boldsymbol{m}, \boldsymbol{m}')P(\boldsymbol{m}', t) - P(\boldsymbol{m}, t)\sum_{\boldsymbol{m}'} w(\boldsymbol{m}', \boldsymbol{m}) \tag{4.111}$$

which is called the *master equation* (Fig. 4.5). The crux to derive a master equation is not so much writing down the expressions (4.109) and (4.110), which are rather obvious, but to determine the transition rates w explicitly. This can be done in two ways. Either we can write down the w's by means of plausibility arguments. This has been done in the example of Section 4.2. Further important examples will be given later, applied to chemistry and sociology. Another way, however, is to derive the w's from first principles where mainly quantum statistical methods have to be used.

Exercise on 4.5

Why has one to assume a Markov process to write down (4.109) and (4.110)?

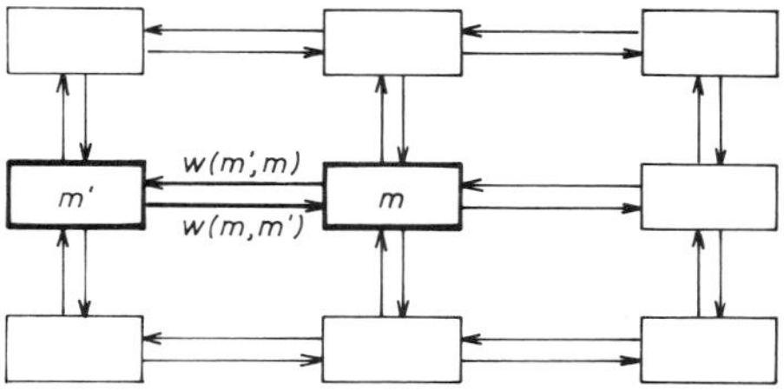

Fig. 4.5. Example of a network of the master equation

4.6 Exact Stationary Solution of the Master Equation for Systems in Detailed Balance

In this section we prove the following: If the master equation has a unique stationary solution ($\dot{P} = 0$) and fulfils the principle of detailed balance, this solution can be obtained explicitly by mere summations or, in the continuous case, by quadratures. The *principle of detailed balance* requires that there are as many transitions per second from state $\boldsymbol{m}$ to state $\boldsymbol{n}$ as from $\boldsymbol{n}$ to $\boldsymbol{m}$ by the inverse process. Or, in mathematical form:

$$w(\boldsymbol{n}, \boldsymbol{m})P(\boldsymbol{m}) = w(\boldsymbol{m}, \boldsymbol{n})P(\boldsymbol{n}). \tag{4.112}$$

In physics, the principle of detailed balance can be (in most cases) derived for systems in thermal equilibrium, using microreversibility. In physical systems far from thermal equilibrium or in nonphysical systems, it holds only in special case (for a counter example cf. exercise 1).

Equation (4.112) represents a set of homogeneous equations which can be solved only if certain conditions are fulfilled by the w's. Such conditions can be derived, for instance, by symmetry considerations or in the case that w can be replaced by differential operators. We are not concerned, however, with this question here, but want to show how (4.112) leads to an explicit solution. In the following we assume that $P(\boldsymbol{n}) \neq 0$. Then (4.112) can be written as

$$\frac{P(\boldsymbol{m})}{P(\boldsymbol{n})} = \frac{w(\boldsymbol{m}, \boldsymbol{n})}{w(\boldsymbol{n}, \boldsymbol{m})}. \tag{4.113}$$

Writing $\boldsymbol{m} = \boldsymbol{n}_{j+1}$, $\boldsymbol{n} = \boldsymbol{n}_j$, we pass from $\boldsymbol{n}_0$ to $\boldsymbol{n}_N$ by a chain of intermediate states. Because there exists a unique solution, at least one chain must exist. We then find

$$\frac{P(\boldsymbol{n}_N)}{P(\boldsymbol{n}_0)} = \prod_{j=0}^{N-1} \frac{w(\boldsymbol{n}_{j+1}, \boldsymbol{n}_j)}{w(\boldsymbol{n}_j, \boldsymbol{n}_{j+1})}. \tag{4.114}$$

Putting

$$P(\boldsymbol{m}) = \mathcal{N} \exp \Phi(\boldsymbol{m}) \tag{4.115}$$

where $\mathcal{N}$ is the normalization factor, (4.114) may be written as

$$\Phi(\boldsymbol{n}_N) - \Phi(\boldsymbol{n}_0) = \sum_{j=0}^{N-1} \ln \{w(\boldsymbol{n}_{j+1}, \boldsymbol{n}_j)/w(\boldsymbol{n}_j, \boldsymbol{n}_{j+1})\}. \tag{4.116}$$

Because the solution was assumed to be unique, $\Phi(\boldsymbol{n}_N)$ is independent of the path chosen. Taking a suitable limit one may apply (4.116) to continuous variables.

As an *example* let us consider a linear chain with nearest neighbor transitions. Since detailed balance holds (cf. exercise 2) we may apply (4.114). Abbreviating transition probabilities by

$$w(m, m-1) = w_+(m) \tag{4.117}$$

$$w(m, m+1) = w_-(m) \tag{4.118}$$

we find

$$P(m) = P(0) \prod_{m'=0}^{m-1} \frac{w(m'+1, m')}{w(m', m'+1)} = P(0) \prod_{m'=0}^{m-1} \frac{w_+(m'+1)}{w_-(m')}. \tag{4.119}$$

In many practical applications, w_+ and w_- are "smooth" functions of m, since m is generally large compared to unity in the regions of interest. Plotting $P(m)$ gives then a smooth curve showing extrema (compare Fig. 4.6). We establish *conditions for extrema*. An extremum (or stationary value) occurs if

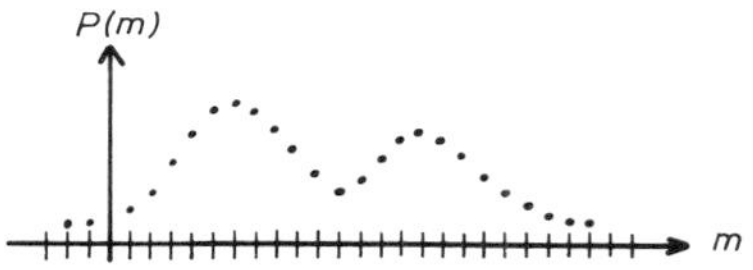

Fig. 4.6. Example of $P(m)$ showing maxima and minima

$$P(m_0) = P(m_0 + 1). \tag{4.120}$$

Since we obtain $P(m_0 + 1)$ from $P(m_0)$ by multiplying

$$P(m_0) \text{ by } \frac{w_+(m_0 + 1)}{w_-(m_0)},$$

$$\text{(4.120) implies } \frac{w_+(m_0 + 1)}{w_-(m_0)} = 1. \tag{4.121}$$

$P(m_0)$ is a maximum, if

$$\left.\begin{array}{ll} P(m) < P(m_0) & \text{for } \quad m < m_0 \\ & \text{and for } \quad m > m_0 \end{array}\right\} \tag{4.122}$$

Equivalently, $P(m_0)$ is a maximum if (and only if)

$$\begin{aligned}&\frac{w_+(m+1)}{w_-(m)} > 1 \quad \text{for} \quad m < m_0,\\&\frac{w_+(m+1)}{w_-(m)} < 1 \quad \text{for} \quad m > m_0.\end{aligned} \tag{4.123}$$

In both cases, (4.122) and (4.123), the numbers m belong to a finite surrounding of m_0.

Exercises on 4.6

1) Verify that the process depicted in Fig. 4.7 does not allow for detailed balance.

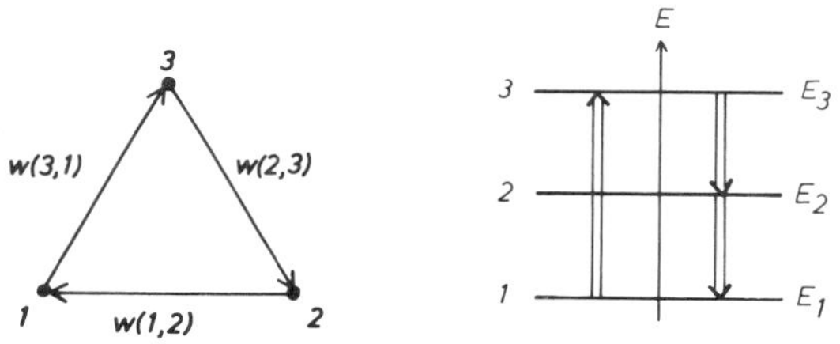

Fig. 4.7. Circular transitions violating the principle of detailed balance, e.g., in a three-level atom (right) 1 → 3: pump from external source, 3 → 2, 2 → 1 recombination of electron

2) Show: in a linear chain m with nearest neighbor transitions $m \to m \pm 1$ the principle of detailed balance always holds.
3) What are the conditions that $P(m_0)$ is a minimum?
4) Generalize the extremal condition to several dimensions, i.e., $m \to \boldsymbol{m} = (m_1, m_2, \ldots, m_N)$, and nearest neighbor transitions.
5) Determine extrema and determine $P(m)$ explicitly for

a) $w(m, m \pm 1) = w,$

$w(m, m + n) = 0, n \neq \pm 1$

Note: normalize P only in a finite region $-M \leq m \leq M$, and put $P(m) = 0$ otherwise.

b) $w(m, m+1) = \dfrac{m+1}{N} w_0 \quad \text{for} \quad 0 \leq m \leq N-1$

$w(m, m-1) = \dfrac{N-m+1}{N} w_0 \quad \text{for} \quad 1 \leq m \leq N$

$w(m, m') = 0 \quad$ otherwise.

c) $w(m+1, m) = w_+(m+1) = \alpha(m+2) \quad m \geq 0$

$w(m, m+1) = w_-(m) = \beta(m+1)(m+2).$

Show that $P(m)$ is the Poisson distribution $\pi_{k,\mu}(\equiv a_k)$ (2.57) with $m \leftrightarrow k$ and $\mu \leftrightarrow \alpha/\beta$.
Hint: Determine $P(0)$ by means of the normalization condition

$$\sum_{m=0}^{\infty} P(m) = 1$$

d) Plot $P(m)$ in the cases a) – c).

4.7* The Master Equation with Detailed Balance. Symmetrization, Eigenvalues and Eigenstates

We write the master equation (4.111) in the form

$$\sum_n L_{mn} P(n) = \dot{P}(m) \tag{4.124}$$

where we have used the abbreviation

$$L_{m,n} = w(m, n) - \delta_{m,n} \sum_l w(l, n). \tag{4.125}$$

The master equation represents a set of linear differential equations of first order. To transform this equation into an ordinary algebraic equation we put

$$P(m) = e^{-\lambda t} \varphi_m \tag{4.126}$$

where φ_m is time independent. Inserting (4.126) into (4.124) yields

$$\sum_n L_{mn} \varphi_n^{(\alpha)} = -\lambda_\alpha \varphi_m^{(\alpha)}. \tag{4.127}$$

The additional index α arises because these algebraic equations allow for a set of eigenvalues λ and eigenstates φ_n which we distinguish by an index α. Since in general the matrix L_{mn} is not symmetric the eigenvectors of the adjoint problem

$$\sum_m \chi_m^{(\alpha)} L_{mn} = -\lambda_\alpha \chi_n^{(\alpha)} \tag{4.128}$$

are different from the eigenvectors of (4.127). However, according to well-known results of linear algebra, φ and χ form a biorthonormal set so that

$$(\chi^{(\alpha)}, \varphi^{(\beta)}) \equiv \sum_n \chi_n^{(\alpha)} \varphi_n^{(\beta)} = \delta_{\alpha,\beta}. \tag{4.129}$$

In (4.129) the lhs is an abbreviation for the sum over n. With help of the eigenvectors φ and χ, L_{mn} can be written in the form

$$L_{mn} = -\sum_\alpha \lambda_\alpha \varphi_m^{(\alpha)} \chi_n^{(\alpha)}. \tag{4.130}$$

We now show that the matrix occurring in (4.127) can be symmetrized. We first define the symmetrized matrix by

$$L^s_{m,n} = w(m, n)\frac{P^{1/2}(n)}{P^{1/2}(m)} \tag{4.131}$$

for $m \neq n$. For $m = n$ we adopt the original form (4.125). $P(n)$ is the stationary solution of the master equation. It is assumed that the detailed balance condition

$$w(m, n)P(n) = w(n, m)P(m) \tag{4.132}$$

holds. To prove that (4.131) represents a symmetric matrix, L^s, we exchange in (4.131) the indices n, m

$$L^s_{n,m} = w(n, m)\frac{P^{1/2}(m)}{P^{1/2}(n)}. \tag{4.133}$$

By use of (4.132) we find

$$(4.133) = w(m, n)\frac{P(n)}{P(m)}\cdot\frac{P^{1/2}(m)}{P^{1/2}(n)} \tag{4.134}$$

which immediately yields

$$L^s_{m,n} \tag{4.135}$$

so that the symmetry is proven. To show what this symmetrization means for (4.127), we put

$$\varphi^{(\alpha)}_n = P^{1/2}_{(n)}\tilde{\varphi}^{(\alpha)}_n \tag{4.136}$$

which yields

$$\sum\nolimits_n L_{mn}P^{1/2}_{(n)}\tilde{\varphi}^{(\alpha)}_n = -\lambda_\alpha P^{1/2}_{(m)}\tilde{\varphi}^{(\alpha)}_m. \tag{4.137}$$

Dividing this equation by $P^{1/2}_{(m)}$ we find the symmetrized equation

$$\sum\nolimits_n L^s_{mn}\cdot\tilde{\varphi}^{(\alpha)}_n = -\lambda_\alpha\tilde{\varphi}^{(\alpha)}_m. \tag{4.138}$$

We proceed with (4.128) in an analogous manner. We put

$$\chi^{(\alpha)}_n = P^{-1/2}_{(n)}\tilde{\chi}^{(\alpha)}_n, \tag{4.139}$$

insert it into (4.128) and multiply by $P^{1/2}(n)$ which yields

$$\sum\nolimits_m \tilde{\chi}^{(\alpha)}_m L^s_{mn} = -\lambda_\alpha\tilde{\chi}^{(\alpha)}_n. \tag{4.140}$$

The $\tilde{\chi}$'s may be now identified with the $\tilde{\varphi}$'s because the matrix L^s_{mn} is symmetric.

This fact together with (4.136) and (4.139) yields the relation

$$\varphi_n^{(\alpha)} = P(n)\chi_n^{(\alpha)}. \tag{4.141}$$

Because the matrix L^s is symmetric, the eigenvalues λ can be determined by the following variational principles as can be shown by a well-known theorem of linear algebra. The following expression must be an extremum:

$$-\lambda = \text{Extr.}\left\{\frac{(\tilde{\chi}L^s\tilde{\chi})}{(\tilde{\chi}\tilde{\chi})}\right\} = \text{Extr.}\left\{\frac{(\chi L\varphi)}{(\chi\varphi)}\right\}. \tag{4.142}$$

$\tilde{\chi}_n$ has to be chosen so that it is orthogonal to all lower eigenfunctions. Furthermore, one immediately establishes that if we chose $\chi_n^{(0)} = 1$, then on account of

$$\sum_m L_{m,n} = 0 \tag{4.143}$$

the eigenvalue $\lambda = 0$ associated with the stationary solution results. We now show that all eigenvalues λ are nonnegative. To this end we derive the numerator in (4.142)

$$\sum_{m,n} \chi_m L_{m,n} P(n)\chi_n \tag{4.144}$$

(compare (4.139), (4.141), and (4.131)) in a way which demonstrates that this expression is nonpositive. We multiply $w(m, n)P(n) \geq 0$ by $-1/2(\chi_m - \chi_n)^2 \leq 0$ so that we obtain

$$-\sum_{m,n} \tfrac{1}{2}(\chi_m - \chi_n)^2 w(m, n)P(n) \leq 0. \tag{4.145}$$

The evaluation of the square bracket yields

$$\underset{(1)}{-\tfrac{1}{2}\sum_{m,n} \chi_m^2 w(m, n)P(n)} \underset{(2)}{- \tfrac{1}{2}\sum_{m,n} \chi_n^2 w(m, n)P(n)} + \underset{(3)}{\sum_{m,n} \chi_m\chi_n w(m, n)P(n)}. \tag{4.146}$$

In the second sum we exchange the indices m, n and apply the condition of detailed balance (4.132). We then find that the second sum equals the first sum, thus that (4.146) agrees with (4.144). Thus the variational principle (4.142) can be given the form

$$\lambda = \text{Extr}\left\{\frac{\tfrac{1}{2}\sum_{m,n} (\chi_m - \chi_n)^2 w(m, n)P(n)}{\sum_n \chi_n^2 P(n)}\right\} \geq 0! \tag{4.147}$$

from which it is evident that λ is nonnegative. Furthermore, it is evident that if we chose $\chi = \text{const.}$, we obtain the eigenvalue $\lambda = 0$.

4.8* Kirchhoff's Method of Solution of the Master Equation

We first present a simple counter example to the principle of detailed balance. Consider a system with three states 1, 2, 3, between which only transition rates $w(1, 2)$, $w(2, 3)$ and $w(3, 1)$ are nonvanishing. (Such an example is provided by a three-level atom which is pumped from its first level to its third level, from where it decays to the second, and subsequently to the first level, cf. Fig. 4.7). On physical grounds it is obvious that $P(1)$ and $P(2) \neq 0$ but due to $w(2, 1) = 0$, the equation

$$w(2, 1)P(1) = w(1, 2)P(2) \tag{4.148}$$

required by detailed balance cannot be fulfilled. Thus other methods for a solution of the master equation are necessary. We confine our treatment to the stationary solution in which case the master equation, (4.111), reduces to a linear algebraic equation. One method of solution is provided by the methods of linear algebra. However, that is a rather tedious procedure and does not use the properties inherent in the special form of the master equation. We rather present a more elegant method developed by Kirchhoff, originally for electrical networks. To find the stationary solution $P(n)$ of the master equation[2].

$$\sum_{n=1}^{N} w(m, n)P(n) - P(m) \sum_{n=1}^{N} w(n, m) = 0, \tag{4.149}$$

and subject to the normalization condition

$$\sum_{n=1}^{N} P(n) = 1 \tag{4.150}$$

we use a little bit of graph theory.

We define a graph (or, in other words a figure) which is associated with (4.149). This graph G contains all vertices and edges for which $w(m, n) \neq 0$. Examples of graphs with three or four vertices are provided by Fig. 4.8. For the following solution, we must consider certain parts of the graph G which are obtained from G by omitting certain edges. This subgraph, called maximal tree $T(G)$, is defined as follows:

1) $T(G)$ covers subgraph so that
   a) all edges of $T(G)$ are edges of G,
   b) $T(G)$ contains all vertices of G.
2) $T(G)$ is connected.
3) $T(G)$ contains no circuits (cyclic sequence of edges).

This definition, which seems rather abstract, can best be understood by looking at examples. The reader will immediately realize that one has to drop a certain minimum number of edges of G. (Compare Figs. 4.9 and 4.10). Thus in order to

[2] When we use n instead of the vector, this is not a restriction because one may always rearrange a *discrete* set in the form of a sequence of single numbers.

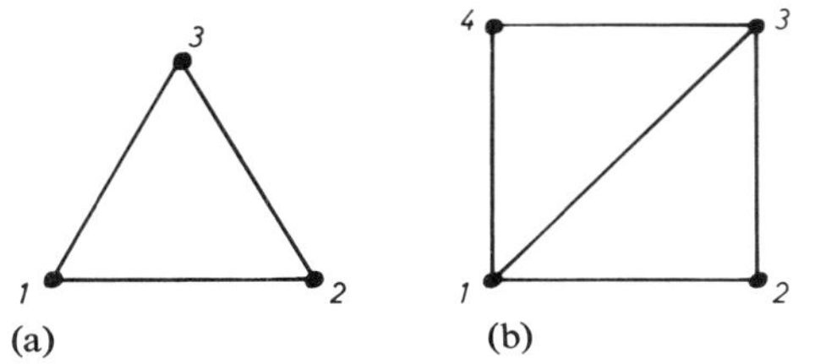

Fig. 4.8a and b. Examples of graphs with 3 or 4 vertices

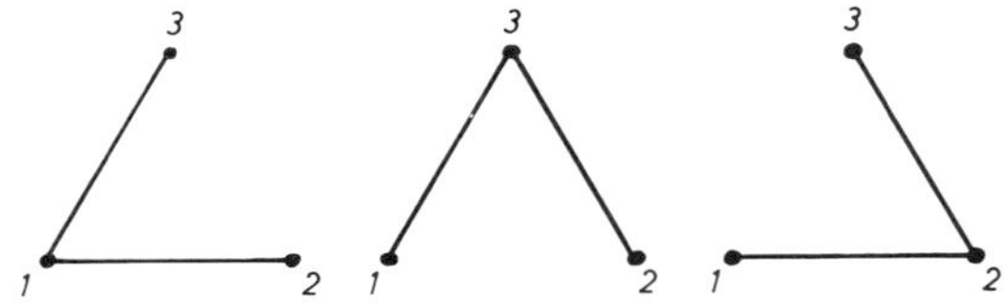

Fig. 4.9. The maximal trees $T(G)$ belonging to the graph of Fig. 4.8a

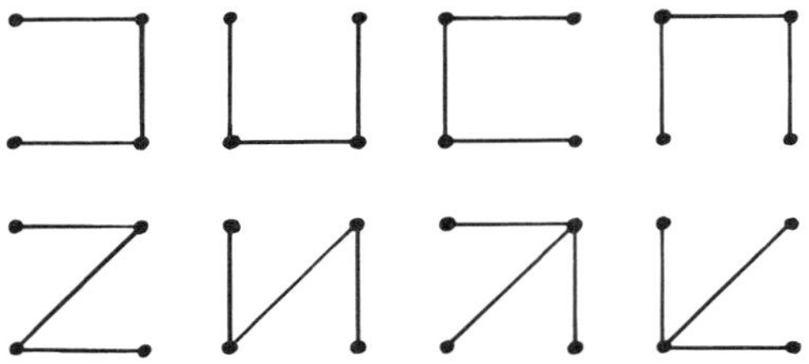

Fig. 4.10. The maximal trees $T(G)$ belonging to the graph G of Fig. 4.8b

obtain the maximal trees of Fig. 4.8b one has to drop in Fig. 4.8b either one side *and* the diagonal or two sides.

We now define a directed maximal tree with index n, $T_n(G)$. It is obtained from $T(G)$ by directing all edges of $T(G)$ towards the vertex with index n. The directed maximal trees belonging to $n = 1$, Fig. 4.9, are then given by Fig. 4.11. After these preliminaries we can give a recipe how to construct the stationary solution $P(n)$. To this end we ascribe to each directed maximal tree a numerical value called A: A $(T_n(G))$: This value is obtained as product of all transition rates $w(n, m)$ whose edges occur in $T_n(G)$ in the corresponding direction. In the example of Fig. 4.11 we thus obtain the following different directed maximal trees

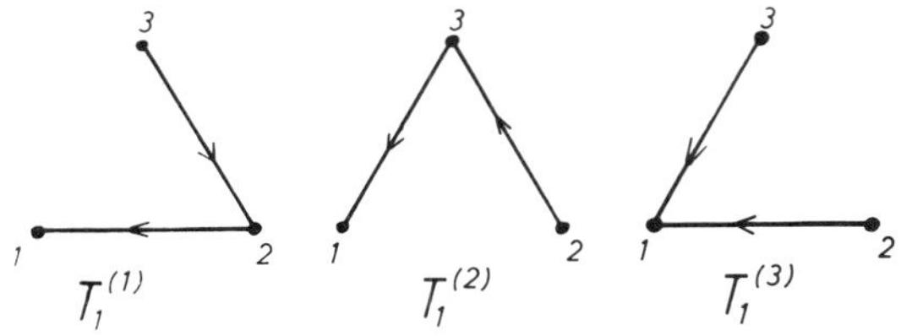

Fig. 4.11. The directed maximal trees T_1 belonging to Fig. 4.9 and $n = 1$

$$T_1^{(1)}: w(1, 2)w(2, 3) = A(T_1^{(1)}) \tag{4.151}$$

$$T_1^{(2)}: w(1, 3)w(3, 2) = A(T_1^{(2)}) \tag{4.152}$$

$$T_1^{(3)}: w(1, 2)w(1, 3) = A(T_1^{(3)}) \tag{4.153}$$

(It is best to read all arguments from right to left). Note that in our example Fig. 4.11 $w(3, 2) = w(1, 3) = 0$. We now come to the last step. We define S_n as the sum over all maximal directed trees with the same index n, i.e.,

$$S_n = \sum_{\text{all } T_n(G)} A(T_n(G)). \tag{4.154}$$

In our triangle example we would have, for instance,

$$S_1 = w(1, 2)w(2, 3) + w(1, 3)w(3, 2) + w(1, 2)w(1, 3) \tag{4.155}$$
$$= w(1, 2)w(2, 3) \quad (\text{since } w(3, 2) = w(1, 3) = 0).$$

Kirchhoff's formula for the probability distribution P_n is then given by

$$P_n = \frac{S_n}{\sum_{l=1}^{N} S_l}. \tag{4.156}$$

In our standard example we obtain using (4.151) etc.

$$P_1 = \frac{w(1, 2)w(2, 3)}{w(1, 2)w(2, 3) + w(2, 3)w(3, 1) + w(3, 1)w(1, 2)}. \tag{4.157}$$

Though for higher numbers of vertices this procedure becomes rather tedious, at least in many practical cases it allows for a much deeper insight into the construction of the solution. Furthermore it permits decomposing the problem into several parts for example if the master equation contains some closed circles which are only connected by a single line.

Exercise on 4.8

Consider a chain which allows only for nearest neighbor transitions. Show that Kirchhoff's formula yields exactly the formula (4.119) for detailed balance.

4.9* Theorems about Solutions of the Master Equation

We present several theorems which are important for applications of the master equation. Since the proofs are purely mathematical, without giving us a deeper insight into the processes, we drop them. We assume the w's are *independent* of time.

1) There exists always at least one stationary solution $P(m)$, $\dot{P}(m) = 0$.
2) This stationary solution is unique, provided the graph G of the master equation is connected (i.e., any two pairs of points m, n can be connected by at least one sequence of lines (over other points)).
3) If at an initial time, $t = 0$,

$$0 \leq P(m, 0) \leq 1 \tag{4.158}$$

for all m, and

$$\sum_m P(m, 0) = 1, \tag{4.159}$$

then also for all later times, $t > 0$,

$$0 \leq P(m, t) \leq 1, \tag{4.160}$$

and

$$\sum_m P(m, t) = 1. \tag{4.161}$$

Thus the normalization property and the *positiveness* of probability are guaranteed for all times.

4) Inserting $P(m, t) = a_m \exp(-\lambda t)$ into the master equation yields a set of linear algebraic equations for a_m with eigenvalues λ. These eigenvalues λ have the following properties:
   a) The real part of λ is nonnegative Re $\lambda \geq 0$,
   b) if detailed balance holds, all λ's are purely real.
5) If the stationary solution is unique, then the time-dependent solution $P(m, t)$ for any initial distribution $P^0(m)$ (so that $P(m, 0) = P^0(m)$) tends for $t \to \infty$ to the stationary solution. For a proof by means of the information gain compare exercise 1 of Section 5.3.

4.10 The Meaning of Random Processes. Stationary State, Fluctuations, Recurrence Time

In the foregoing we investigated processes caused by random actions. In this section we want to discuss some rather general aspects, mainly utilizing a specific model. We consider what happens when we bring together two boxes filled with gas. Then the gas atoms of one box will diffuse into the other box and vice versa. The transitions may be considered as completely random because they are caused by the many pushes that the gas atoms suffer. Another example is provided by chemical processes where a reaction can take place only if two corresponding molecules hit each other, which is again a random event. Therefore, it is not surprising that such random processes occur in many different disciplines and are of utmost importance for the understanding of ordering phenomena.

To illuminate the basic ideas we consider a very simple model the so-called Ehrenfest urn model. This model was originally devised to discuss the meaning of the so-called H-theorem in thermodynamics (cf. exercises 1), 2) of Section 5.3). Here, however, we treat this model to illustrate some typical effects inherent in random processes and also in establishing equilibrium. Let us consider two boxes (urns) A, B filled with N balls which are labelled by $1, 2, \ldots, N$. Let us start with an initial distribution in which there are N_1 balls in box A, and N_2 balls in box B. Let us assume now that we have a mechanism which randomly selects one of the numbers $1, \ldots, N$, with the same probability $1/N$, and repeat this selection process regularly in time intervals τ. If the number is selected, one removes the corresponding ball from its present box and puts it into the other one. We are then interested in the change of the numbers N_1 and N_2 in the course of time. Since N_2 is given by $N - N_1$, the only variable we must consider is N_1. We denote the probability of finding N_1 after s steps at time t (i.e., $t = s\tau$) by $P(N_1, s)$.

We first establish an equation which gives us the change of the probability distribution as a function of time and then discuss several important features. The probability distribution, P, is changed in either of two ways. Either one ball is added to box A or removed from it. Thus the total probability $P(N_1, s)$ is a sum of the probabilities corresponding to these two events. In the first case, adding a ball A, we must start from a situation in which there are $N_1 - 1$ balls in A. We denote the probability that one ball is added by $w(N_1, N_1 - 1)$. If a ball is removed, we must start from a situation in which there are $N_1 + 1$ balls in A at "time" $s - 1$. We denote the transition rate corresponding to the removal of a ball by $w(N_1, N_1 + 1)$. Thus we find the relation

$$P(N_1, s) = w(N_1, N_1 - 1)P(N_1 - 1, s - 1) + w(N_1, N_1 + 1)P(N_1 + 1, s - 1). \tag{4.162}$$

Since the probability for picking a definite number is $1/N$ and there are $N_2 + 1 = N - N_1 + 1$ balls in urn B the transition rate $w(N_1, N_1 - 1)$ is given by

$$w(N_1, N_1 - 1) = \frac{N_2 + 1}{N} = \frac{N - N_1 + 1}{N}. \tag{4.163}$$

Correspondingly the transition rate for picking a ball in urn A is given by

$$w(N_1, N_1 + 1) = \frac{N_1 + 1}{N}. \tag{4.164}$$

Thus (4.162) acquires the form

$$P(N_1, s) = \frac{N - N_1 + 1}{N} P(N_1 - 1, s - 1) + \frac{N_1 + 1}{N} P(N_1 + 1, s - 1). \tag{4.165}$$

If any initial distribution of balls over the urns is given we may ask to which final distribution we will come. According to Section, 4.9 there is a unique final stationary solution to which any initial distribution tends. This solution is given by

$$\lim_{s\to\infty} P(N_1, s) \equiv P(N_1) = \frac{N!}{N_1!(N - N_1)!}\alpha \equiv \frac{N!}{N_1!\,N_2!}\alpha \tag{4.166}$$

where the normalization constant α is given by the condition

$$\textstyle\sum_{N_1} P(N_1, s) = 1, \tag{4.167}$$

and can easily be determined to be

$$\alpha = 2^{-N}. \tag{4.168}$$

We leave it to the reader as an exercise to check the correctness of (4.166) by inserting it into (4.165). (Note that P does not depend on s any more). We already met this distribution function much earlier, namely starting from quite different considerations in Section 3.1 we considered the number of configurations with which we can realize a macrostate by N_1 balls in A, and N_2 balls in B for different microstates in which the labels of the balls are different but the total numbers N_1, N_2 remain constant. This model allows us to draw several very important and rather general conclusions.

First we have to distinguish between a given single system and an *ensemble* of systems. When we consider what a given single system does in course of time, we find that the number N_1 (which is just a random variable), acquires certain values

$$N_1^{(1)}(s = 1), N_1^{(2)}(s = 2), \ldots, N_1^{(s)}, \ldots \tag{4.169}$$

Thus, in the language of probability theory, a single event consists of the sequence (4.169). Once this sequence is picked, there is nothing arbitrary left. On the other hand, when talking about probability, we treat a set of events, i.e., the sample set (sample space). In thermodynamics, we call this set "an ensemble" (including its changes in course of time). An individual system corresponds to a sample point.

In thermodynamics, but also in other disciplines, the following question is treated: If a system undergoes a random process for a very long time, does the temporal average coincide with the ensemble average? In our book we deal with the ensemble average if not otherwise noted. The ensemble average is defined as the mean value of any function of random variables with respect to the joint probability of Section 4.3.

In the following discussion the reader should always carefully distinguish between a single system being discussed, or the whole ensemble. With this warning in mind, let us discuss some of the main conclusions: The stationary solution is by no means completely sharp, i.e., we do not find $N/2$ balls in box A and $N/2$ balls in box B with probability unity. Due to the selection process, there is always a certain chance of finding a different number, $N_1 \neq N/2$, in box A. Thus fluctuations occur. Furthermore, if we had initially a given number N_1, we may show that the system may return to this particular number after a certain time. Indeed, to each individual process, say $N_1 \to N_1 + 1$, there exists a finite probability that the inverse process, $N_1 \to N_1 - 1$, occurs (Evidently this problem can be cast into a more rigorous mathematical formulation but this is not our concern here). Thus the total system does not approach a unique equilibrium state, $N_1 = N/2$. This seems to be in striking contrast to what we would expect on thermodynamic grounds, and this is the difficulty which has occupied physicists for a long time.

It is, however, not so difficult to reconcile these considerations with what we expect in thermodynamics, namely, the approach to equilibrium. Let us consider the case that N is a very large number, which is a typical assumption in thermodynamics (where one even assumes $N \to \infty$). Let us first discuss the stationary distribution function. If N is very large we may convince ourselves very quickly that it is very sharply peaked around $N_1 = N/2$. Or, in other words, the distribution function effectively becomes a δ-function. Again this sheds new light on the meaning of entropy. Since there are $N!/N_1!N_2!$ realizations of state N_1, the entropy is given by

$$S = k_B \ln \frac{N!}{N_1!\,N_2!} \tag{4.170}$$

or

$$S = k_B(-N_1 \ln N_1 - N_2 \ln N_2 + N \ln N) \tag{4.171}$$

or after dividing it by N (i.e. per ball)

$$S/N = -k_B(p_1 \ln p_1 + p_2 \ln p_2) \tag{4.172}$$

where

$$p_j = \frac{N_j}{N}. \tag{4.173}$$

Since the probability distribution $P(N_1)$ is strongly peaked, we shall find in all practical cases $N_1 = N/2$ realized so that when experimentally picking up any distribution we may expect that the entropy has acquired its maximum value where $p_1 = p_2 = 1/2$. On the other hand we must be aware that we can construct other initial states in which we have the condition that N_1 is equal a given number $N_0 \neq N/2$. If N_0 is given, we possess a maximal knowledge at an initial time. If we now let time run, P will move to a new distribution and the whole process has the character of an *irreversible process*. To substantiate this remark, let us consider the development of an initial distribution where all balls are in one box. There is a probability equal to one that one ball is removed. The probability is still close to unity if only a small number of balls has been removed. Thus the system tends very quickly away from its initial state. On the other hand, the probability for passing a ball from box A to box B becomes approximately equal to the probability for the inverse process if the boxes are equally filled. But these transition probabilities are by no means vanishing, i.e., there are still fluctuations of the particle numbers possible in each box. If we wait an extremely long time, t_r, there is a finite probability that the whole system comes back to its initial state. This recurrence time, t_r, has been calculated to be

$$t_r \sim \tau/P(N_1), \tag{4.174}$$

where t_r is defined as the mean time between the appearance of two identical macrostates.

In conclusion we may state the following: Even if N is large, N_1 is not fixed but may fluctuate. Large deviations from $N_1 = N/2$ are scarce, however. The most probable state is $N_1 = N/2$. If no information about the state is available or, in other words, if we have not prepared the system in a special manner, we must assume that the stationary state is present. The above considerations resolve the contradiction between an irreversible process which goes on in only one direction towards an "equilibrium state", and fluctuations which might drive back an individual system even to its initial state. Both may happen and it is just a question of probability depending on the preparation of the initial state as to which process actually takes place. The important role that fluctuations play in different kinds of systems will transpire in later chapters.

Exercises on 4.10

1) Why is $\sigma^2 = \langle N_1^2 \rangle - \langle N_1 \rangle^2$ a measure for the fluctuations of N_1? How large is σ^2?
2) Discuss Brownian movement by a single system and by an ensemble.

4.11* Master Equation and Limitations of Irreversible Thermodynamics

Let us assume that a system composed of subsystems can be described by a master equation. We wish to show that the approach used by irreversible thermodynamics which we explained in Section 3.5 implies a very restrictive assumption. The essential point can be demonstrated if we have only two subsystems. These subsystems need not necessarily be separated in space, for instance in the laser the two subsystems are the atoms and the light field. To make contact with earlier notation we denote the indices referring to one subsystem by i, those to the other subsystem by i'. The probability distribution then carries the indices ii'. The corresponding master equation (4.111) for ($P_{jj'} \equiv P(j, j', t)$, $(j, j' \leftrightarrow \boldsymbol{m})$, reads

$$\dot{P}_{ii'} = \sum_{jj'} w_{ii';jj'} P_{jj'} - P_{ii'} \sum_{jj'} w_{jj';ii'}. \tag{4.175}$$

We have encountered such indices already in Section 3.3 and 3.4. In particular we have seen that (Sect. 3.5) the entropy S is constructed from the probability distribution, p_i, whereas the entropy of system S' is determined by a second probability distribution $p_{i'}$. Furthermore we have assumed that the entropies are additive. This additivity implies that $P_{ii'}$ factorizes (cf. Exercise 1 on 3.5)

$$P_{ii'} = p_i p'_{i'}. \tag{4.176}$$

Inserting (4.176) into (4.175) shows after some inspection that (4.175) can be solved by (4.176) only under very special assumptions, namely, if

$$w_{ii';jj'} = \delta_{i'j'} w_{ij}^{(1)} + \delta_{ij} w_{i'j'}^{(2)} \tag{4.177}$$

(4.177) implies that there is no interaction between the two subsystems. The hypothesis (4.176) to solve (4.175) can thus only be understood as an approximation similar to the Hartree-Fock approximation in quantum theory (compare exercise below). Our example shows that irreversible thermodynamics is valid only if the correlations between the two subsystems are not essential and can be taken care of in a very global manner in the spirit of a self-consistent field approach.

Exercise on 4.11

Formulate a variational principle and establish the resulting equations for p_i, $p'_{i'}$ in the case that the master equation obeys detailed balance.
Hint: Make the hypothesis (4.176), transform the master equation to one with a self-adjoint operator (according to Sect. 4.7) and vary now the resulting expression (4.142) with respect to $p_i p'_{i'} = \tilde{\chi}_{ii'}$. For those who are acquainted with the Hartree-Fock procedure, convince yourself that the present procedure is identical with the Hartree-Fock self-consistent field procedure in quantum mechanics.

5. Necessity

Old Structures Give Way to New Structures

This chapter deals with completely deterministic processes. The question of stability of motion plays a central role. When certain parameters change, stable motion may become unstable and completely new types of motion (or structures) appear. Though many of the concepts are derived from mechanics, they apply to many disciplines.

5.1 Dynamic Processes

a) An Example: The Overdamped Anharmonic Oscillator

In practically all disciplines which can be treated quantitatively, we observe changes of certain quantities as a function of time. These changes of quantities result from certain causes. A good deal of the corresponding terminology has evolved from mechanics. Let us consider as an example the acceleration of a particle with mass m under the action of a force F_0. The velocity v of the particle changes in time according to Newton's equation

$$m\frac{dv}{dt} = F_0. \tag{5.1}$$

Let us further assume that the force F_0 may be decomposed into a "driving force" F and a friction force which we assume proportional to the velocity, v. Thus we replace F_0 by

$$F_0 \to F - \gamma v \tag{5.2}$$

and obtain as equation of motion

$$m\frac{dv}{dt} + \gamma v = F. \tag{5.3}$$

In many practical cases F is a function of the particle coordinate, q. In the case of a harmonic oscillator, Fig. 5.1, F is proportional to the elongation, q, from the equilibrium position. Denoting Hooke's constant by k, we have (compare Fig. 5.1b)

$$F(q) = -kq. \tag{5.4}$$

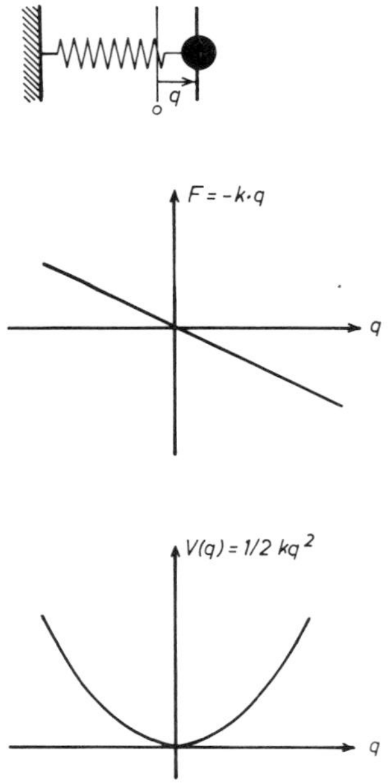

(a) (b) (c)

Fig. 5.1a–c. The harmonic oscillator. (a) Configuration with spring and point mass m. "O" indicates the equilibrium position. (b) Force as function of elongation q. (c) Potential (cf. (5.14), (5.15))

The minus sign results from the fact that the elastic force tends to bring the particle back to its equilibrium position. We express the velocity by the derivative of the coordinate with respect to time and use a dot above q to indicate this

$$v = \frac{dq}{dt} \equiv \dot{q}. \tag{5.5}$$

With (5.2) and (5.5), (5.1) acquires the form

$$m\ddot{q} + \gamma\dot{q} = F(q). \tag{5.6}$$

We shall use this equation to draw several important general conclusions which are valid for other systems and allow us to motivate the terminology. We mainly consider a special case in which m is very small and the damping constant γ very large, so that we may neglect the first term against the second term on the lhs of (5.6). In other words, we consider the so-called overdamped motion. We further note that by an appropriate time scale

$$t = \gamma t' \tag{5.7}$$

we may eliminate the damping constant γ. Eq. (5.6) then acquires the form

$$\dot{q} = F(q). \tag{5.8}$$

Equations of this form are met in many disciplines. We illustrate this by a few examples: In chemistry q may denote the density of a certain kind Q of molecules, which are created by the reaction of two other types of molecules A and B with concentrations a and b, respectively. The production rate of the density q is described by

$$\dot{q} = kab \tag{5.9}$$

where k is the reaction coefficient. An important class of chemical reactions will be encountered later. These are the so-called autocatalytic reactions in which one of the constituents, for example B, is identical with Q so that (5.9) reads

$$\dot{q} = kaq. \tag{5.10}$$

Equations of the type (5.10) occur in biology where they describe the multiplication of cells or bacteria, or in ecology, where q can be identified with the number of a given kind of animals. We shall come back to such examples in Section 5.4 and, in a much more general form, later on in our book in Chapters 8 through 11. For the moment we want to exploit the mechanical example. Here one introduces the notation of "work" which is defined by work = force times distance. Consider for example a body of weight G (stemming from the gravitational force that the earth exerts on the body). Thus we may identify G with the force F. When we lift this body to a height h ($\equiv$ distance q) we "do work"

$$W = G \cdot h = F \cdot q. \tag{5.11}$$

In general cases the force F depends on the position, q. Then (5.11) can be formulated only for an infinitesimally small distance, dq, and we have instead of (5.11)

$$dW = F(q)\, dq. \tag{5.12}$$

To obtain the total work over a finite distance, we have to sum up, or, rather to integrate

$$W = \int_{q_0}^{q_1} F(q)\, dq. \tag{5.13}$$

The negative of W is called the potential, V. Using (5.12) we find

$$F(q) = -\frac{dV}{dq}. \tag{5.14}$$

Let us consider the example of the harmonic oscillator with F given by (5.4). One readily establishes that the potential has the form

$$V(q) = \tfrac{1}{2}kq^2 \tag{5.15}$$

(besides an additive constant which we have put equal to zero). To interpret V, which is plotted for this example in Fig. 5.1c, we compare it with the work done when lifting a weight. This suggests interpreting the solid curve in Fig. 5.2 as the slope of a hill. When we bring the particle to a certain point on the slope, it will fall back down the slope and eventually come to rest at the bottom of the hill. Due to the horizontal slope at $q = 0$, $F(q)$, (5.14) vanishes and thus $\dot{q} = 0$. The particle is at an equilibrium point. Because the particle returns to this equilibrium point when we displace the particle along the slope, this position is *stable*.

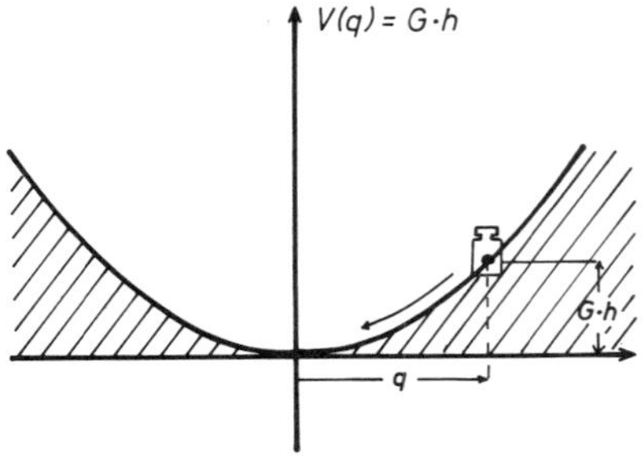

Fig. 5.2. Potential curve interpreted as slope of a hill

Now consider a slightly more complicated system, which will turn out later on to be of fundamental importance for self-organization (though this is at present not at all obvious). We consider the so-called anharmonic oscillator which contains a cubic term besides a linear term in its force F.

$$F(q) = -kq - k_1 q^3. \tag{5.16}$$

The equation of motion then reads

$$\dot{q} = -kq - k_1 q^3. \tag{5.17}$$

The potential is plotted in Fig. 5.3 for two different cases, namely, for $k > 0$ and

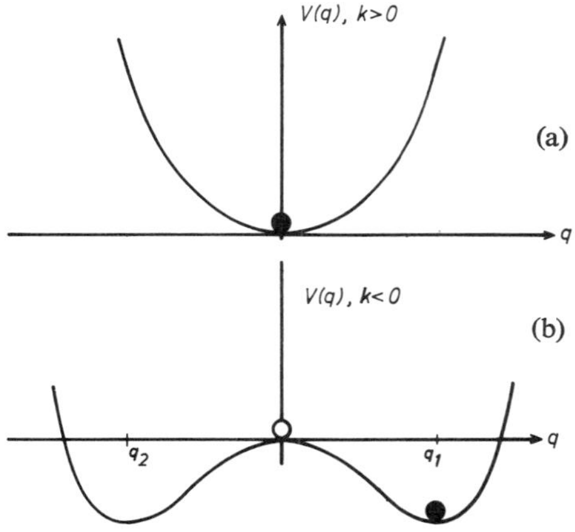

Fig. 5.3a and b. The potential of the force (5.16) for $k>0$ (a) and $k<0$ (b)

$k < 0$ ($k_1 > 0$). The equilibrium points are determined by

$$\dot{q} = 0. \tag{5.18}$$

From Fig. 5.3, it is immediately clear that we have two completely different situations corresponding to whether $k > 0$ or $k < 0$. This is fully substantiated by an algebraic discussion of (5.17) under the condition (5.18). The only solution in case

a) $k > 0, k_1 > 0$ is

$$q = 0, \text{ stable}, \tag{5.19}$$

whereas in case

b) $k < 0, k_1 > 0$,

we find three solutions, namely, $q = 0$ which is evidently unstable, and two stable solutions $q_{1,2}$ so that

$$q = 0 \text{ unstable}, \; q_{1,2} = \pm\sqrt{|k|/k_1} \text{ stable}. \tag{5.20}$$

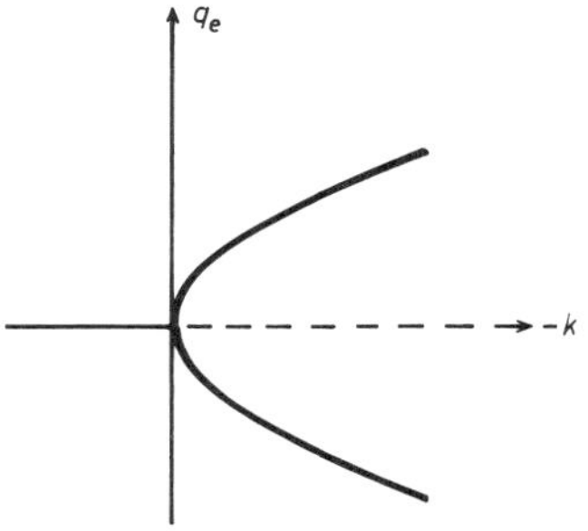

Fig. 5.4. The equilibrium coordinate q_e as function of k (cf. (5.19), and (5.20)). For $k>0$, $q_e=0$, but for $k<0$, $q_e=0$ becomes unstable (dashed line) and is replaced by two stable positions (solid fork)

There are now two physically stable equilibrium positions, (Fig. 5.4). In each of them the particle is at rest and stays there for ever.

By means of (5.17) we may now introduce the concept of symmetry. If we replace everywhere in (5.17) q by $-q$ we obtain

$$(-\dot{q}) = -k(-q) - k_1(-q^3) \tag{5.17a}$$

or, after division of both sides by -1, we obtain the old (5.17). Thus (5.17) remains unchanged (invariant) under the transformation

$$q \to -q \tag{5.17b}$$

or, in other words, (5.17) is symmetric with respect to the inversion $q \to -q$. Simultaneously the potential

$$V(q) = \tfrac{1}{2}kq^2 + \tfrac{1}{4}k_1q^4 \tag{5.21}$$

remains invariant under this transformation

$$V(q) \to V(-q) = V(q). \tag{5.17c}$$

Though the problem described by (5.17) is completely symmetric with respect to the inversion $q \to -q$, the symmetry is now broken by the actually realized solution. When we gradually change k from positive values to negative values, we come to $k = 0$ where the stable equilibrium position $q = 0$ becomes unstable. The whole phenomenon may be thus described as a *symmetry breaking instability*. This phenomenon can be still expressed in other words. When k passes from $k > 0$ to $k < 0$ the stable equilibrium positions are exchanged, i.e., we have the so-called *exchange of stability*. When we deform the potential curve from $k > 0$ to $k < 0$, it becomes flatter and flatter in the neighborhood of $q = 0$. Consequently the particle falls down the potential curve more and more slowly, a phenomenon called *critical slowing down*. For later purposes we shall call the coordinate q now r. When passing from $k > 0$ to $k < 0$ the stable position $r = 0$ is replaced by an unstable position at $r = 0$ and a stable one at $r = r_0$. Thus we have the scheme

$$\text{stable point} \begin{array}{l} \nearrow \text{ unstable point} \\ \searrow \text{ stable point} \end{array} \tag{5.22}$$

Since this scheme has the form of a fork, the whole phenomenon is called "bifurcation." Another example of bifurcation is provided by Fig. 5.7 below where the two points (stable at r_1, unstable at r_0) vanish when the potential curve is deformed.

b) *Limit Cycles*

We now proceed from our one-dimensional problem to a two-dimensional problem as follows: We imagine that we rotate the whole potential curve $V(r)$ around the V-axis which gives us Fig. 5.5. We consider a case in which the particle runs along

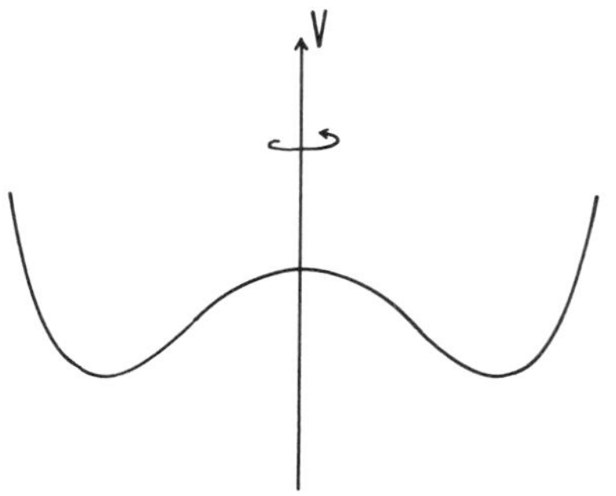

Fig. 5.5. Rotating symmetric potential

the bottom of the valley with a constant velocity in tangential direction. We may either use cartesian coordinates q_1 and q_2 or polar coordinates (radius r and angle

$\varphi)^1$. Since the angular velocity, $\dot{\varphi}$, is constant, the equations of motion have the form

$$\dot{r} = F(r), \qquad (5.23)$$
$$\dot{\varphi} = \omega.$$

We do not claim here that such equations can be derived for purely mechanical systems. We merely use the fact that the interpretation of V as a mechanical potential is extremely convenient for visualizing our results. Often the equations of motion are not given in polar coordinates but in cartesian coordinates. The relation between these two coordinate systems is

$$q_1 = r \cos \varphi,$$
$$q_2 = r \sin \varphi. \qquad (5.24)$$

Since the particle moves along the valley, its path is a circle. Having in mind the potential depicted in Fig. 5.5 and letting start the particle close to $q = 0$ we see that it spirals away to the circle of Fig. 5.6. The point $q = 0$ from which the particle spirals away is called an *unstable focus*. The circle which it ultimately approaches is called a *limit cycle*. Since our particle also ends up in this cycle, if it is started from the outer side, this limit cycle is *stable*. Of course, there may be other forms of the potential in radial direction, for example that of Fig. 5.7a now allowing for a

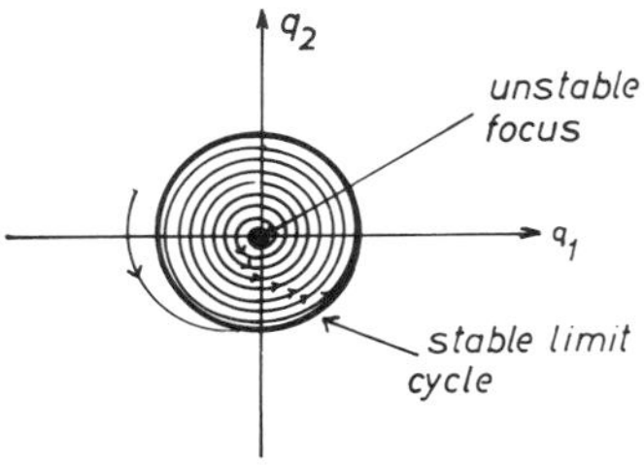

Fig. 5.6. Unstable focus and limit cycle

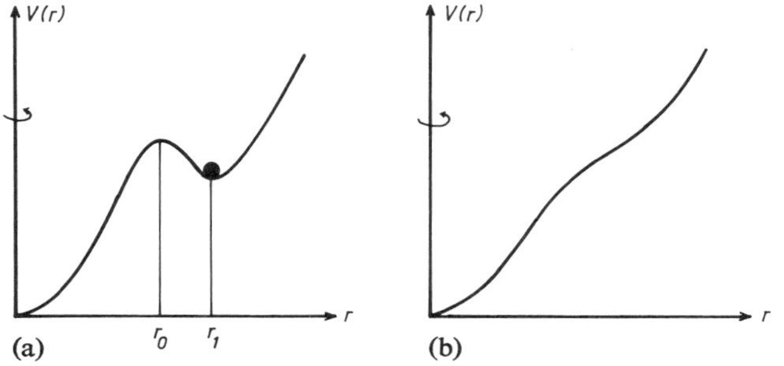

Fig. 5.7a and b. (a) An unstable limit cycle (r_0) and a stable limit cycle (r_1). (b) The limit cycles coalesce and disappear

[1] In mechanics, q_1 and q_2 may be sometimes identified with the coordinate q and the momentum p of a particle.

stable and an unstable limit cycle. When this potential is deformed, these two cycles coalesce and the limit cycles disappear. We then have the following bifurcation scheme

$$\left.\begin{array}{l}\text{stable limit cycle} \\ \text{unstable limit cycle}\end{array}\right\rangle \text{—no limit cycle} \tag{5.25}$$

c) *Soft and Hard Modes, Soft and Hard Excitations*

When we identify the coordinate q with the elongation of a pendulum, we may imagine that the present formalism is capable of describing clocks and watches. Similarly these equations can be applied to oscillations of radio tubes or to lasers (cf. Sect. 8.1). The great importance of limit cycles lies in the fact that now we can understand and mathematically treat self-sustained oscillations. As our above examples show clearly, the final curve (trajectory) is followed by the particle independently of initial conditions. Consider the example depicted by Fig. 5.5 or 5.6. If we start the system near to the unstable focus $\boldsymbol{q} = 0$, the oscillation starts on its own. We have the case of *self-excitation*. Since an infinitesimally small initial perturbation suffices to start the system, one calls this kind of excitation *soft self-excitation*. The watch or clock starts immediately. Fig. 5.7a is an example of the so-called *hard excitation*.

To bring the particle or coordinate from the equilibrium value $\boldsymbol{q} = 0\,(\equiv r = 0)$ to the stable limit cycle at $r = r_1$, the potential hill at $r = r_0$, i.e., a certain threshold value, must be surmounted. In the literature considerable confusion has arisen with the notation "soft" and "hard" modes and "soft" and "hard" excitations. In our notation soft and hard mode refers to $\omega = 0$ and $\omega \neq 0$, respectively. The form of the potential in the r-direction causes soft or hard excitations. When the system rotates, the bifurcation scheme (5.22) must be replaced by the scheme

$$\text{stable focus} \left\langle \begin{array}{l} \nearrow \text{stable limit cycle} \\ \searrow \text{unstable focus} \end{array}\right. \tag{5.26}$$

In view of (5.26) the bifurcation of limit cycles can be formulated as follows. Consider that $F(r)$ in (5.23) is a polynomial. To have circular motion requires $dr/dt = 0$, i.e., $F(r)$ must have real positive roots. A bifurcation or inverse bifurcation occurs if one double (or multiple) real root r_0 becomes complex for certain values of external parameters (in our above case k) so that the condition $r_0 = \text{real}$ cannot be fulfilled. While in the above examples we could easily find closed "trajectories" defining limit cycles, it is a major problem in other cases to decide whether given differential equations allow for stable or unstable limit cycles. Note that limit cycles need not be circles but can be *other closed trajectories* (cf. Sects. 5.2, 5.3). An important tool is the Poincaré-Bendixson theorem, which we will present in Section 5.2.

5.2* Critical Points and Trajectories in a Phase Plane. Once Again Limit Cycles

In this section we consider the coupled set of first-order differential equations

$$\dot{q}_1 = F_1(q_1, q_2), \tag{5.27}$$

$$\dot{q}_2 = F_2(q_1, q_2). \tag{5.28}$$

The usual equation of motion of a particle with mass m in mechanics

$$m\ddot{q} - F(q, \dot{q}) = 0 \tag{5.29}$$

is contained in our formalism. Because putting

$$m\dot{q} = p \; (p = \text{momentum}) \tag{5.30}$$

we may write (5.29) in the form

$$m\ddot{q} = \dot{p} = F_2(q, p) \equiv F(q, p), \tag{5.31}$$

and supplement it by (5.30) which has the form

$$\dot{q} = \frac{p}{m} \equiv F_1(q, p). \tag{5.32}$$

This is identical with (5.27), identifying $q_1 = q$, $q_2 = p$. We confine our analysis to so-called *autonomous systems* in which F_1, F_2 do not explicitly depend on time. Writing out the differentials on the left-hand side in (5.27) and (5.28) and dividing (5.28) by (5.27), one immediately establishes[2]

$$\frac{dq_2}{dq_1} = \frac{F_2(q_1, q_2)}{F_1(q_1, q_2)}. \tag{5.33}$$

The meaning of (5.27) and (5.28) becomes clearer when we write $\dot{q}_1$ in the form

$$\dot{q}_1 = \frac{dq_1}{dt} = \lim_{\Delta t \to 0} \frac{\Delta q_1}{\Delta t}, \tag{5.34}$$

where

$$\Delta q_1 = q_1(t + \tau) - q_1(t), \tag{5.35}$$

and

$$\Delta t = \tau. \tag{5.36}$$

[2] This "division" is done here in a formal way, but it can be given a rigorous mathematical basis, which, however, we shall not discuss here.

Thus we may write (5.27) in the form

$$q_1(t + \tau) = q_1(t) + \tau F_1(q_1(t), q_2(t)). \tag{5.37}$$

This form (and an equivalent one for q_2) lends itself to the following interpretation: If q_1 and q_2 are given at time t, then their values can be determined at a later time $t + \tau$ by means of the rhs of (5.37) and

$$q_2(t + \tau) = q_2(t) + \tau F_2(q_1(t), q_2(t)) \tag{5.38}$$

uniquely, which can also be shown quite rigorously. Thus when at an initial time q_1 and q_2 are given, we may proceed from one point to the next. Repeating this procedure, we find a unique trajectory in the q_1,q_2-plane (Fig. 5.8). We can let this trajectory begin at different initial values. Thus a given trajectory corresponds to an infinity of motions differing from each other by the "phase" (or initial value) $q_1(0)$, $q_2(0)$.

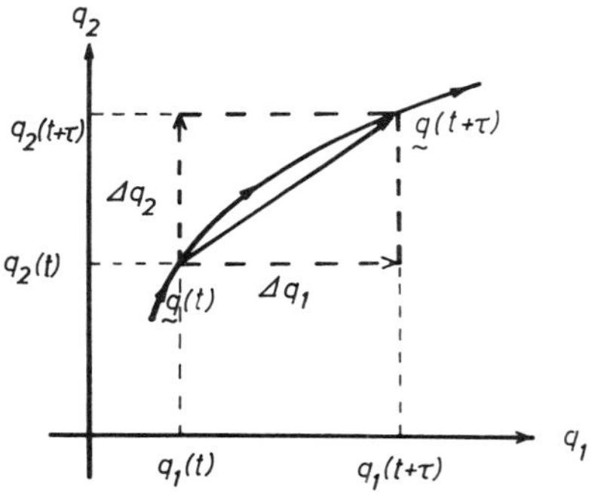

Fig. 5.8. Each trajectory can be approximated by a polygonial track which enables us to construct a trajectory. Here we have shown the approximation by means of secants. Another well known approach consists in proceeding from $\mathbf{q}(t)$ to $\mathbf{q}(t+\tau)$ by taking the tangents to the true trajectory using the rules of (5.37), (5.38)

Let us now choose other points q_1, q_2 not lying on this trajectory but close to it. Through these points other trajectories pass. We thus obtain a "field" of trajectories, which may be interpreted as streamlines of a fluid. An important point to discuss is the structure of these trajectories. First it is clear that trajectories can never cross each other because at the crossing point the trajectory must continue in a unique way which would not be so if they split up into two or more trajectories. The geometrical form of the trajectories can be determined by eliminating the time-dependence from (5.27), (5.28) leading to (5.33). This procedure breaks down, however, if simultaneously

$$F_1(q_1^0, q_2^0) = F_2(q_1^0, q_2^0) = 0 \tag{5.39}$$

for a couple q_1^0, q_2^0. In this case (5.33) results in an expression 0/0 which is meaningless. Such a point is called a *singular* (or *critical*) *point*. Its coordinates are determined by (5.39). Due to (5.27) and (5.28), (5.39) implies $\dot{q}_1 = \dot{q}_2 = 0$, i.e., the singular point is also an equilibrium point. To determine the nature of equilibrium (stable, unstable, neutral), we have to take into account trajectories close to the singular point. We call each singular point asymptotically stable if all trajectories starting sufficiently near it tend to it asymptotically for $t \to \infty$ (compare Fig. 5.9a). A singular point is asymptotically unstable if all trajectories which are sufficiently close to it tend asymptotically ($t \to -\infty$) away from it. Interpreting trajectories as streamlines, we call singular points which are asymptotically stable, "sinks", because the streamlines terminate at them. Correspondingly, asymptotically unstable singular points are called "sources".

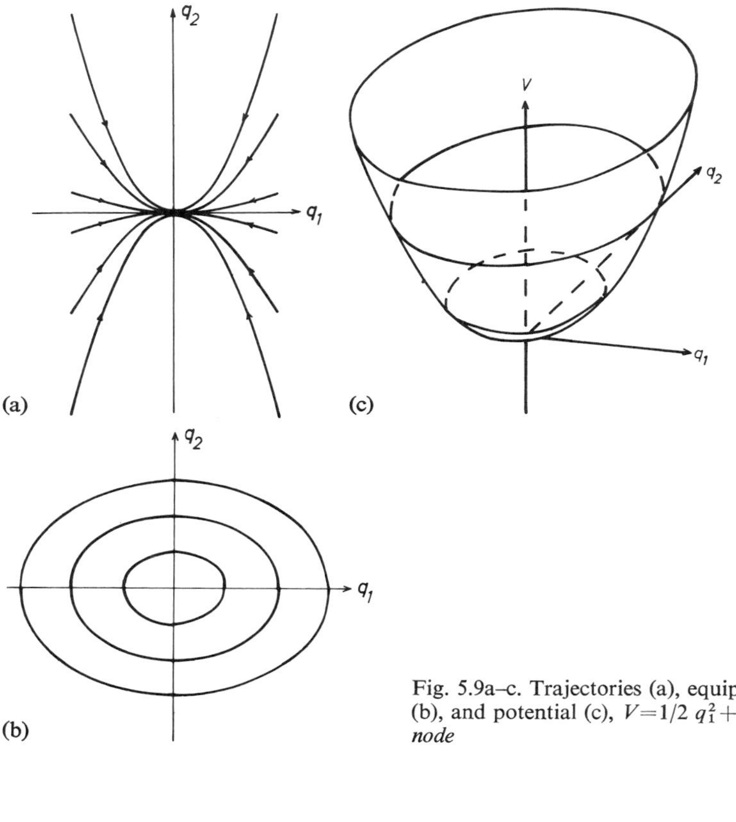

Fig. 5.9a–c. Trajectories (a), equipotential curves (b), and potential (c), $V=1/2\ q_1^2+a/2\ q_2^2$, $a>0$ of a *node*

The behavior of trajectories near critical points may be classified. To introduce these classes we first treat several special cases. We assume that the singular point lies in the origin of the coordinate system which can always be achieved by a shift of the coordinate system. We further assume that F_1 and F_2 can be expanded into a Taylor series and that at least F_1 or F_2 starts with a term linear in q_1, q_2. Neglecting higher powers in q_1, q_2, we have to discuss (5.27) and (5.28) in a form where F_1 and F_2 are linear in q_1 and q_2. We give a discussion of the different classes which may occur:

1) *Nodes and Saddle Points*

Here (5.27) and (5.28) are of the form

$$\begin{aligned} \dot{q}_1 &= q_1, \\ \dot{q}_2 &= aq_2, \end{aligned} \tag{5.40}$$

which allow for the solution

$$\begin{aligned} q_1^0 &= q_2^0 = 0, \\ q_1 &= c_1 e^t, \quad q_2 = c_2 e^{at}. \end{aligned} \tag{5.41}$$

Similarly the equations

$$\begin{aligned} \dot{q}_1 &= -q_1, \\ \dot{q}_2 &= -aq_2 \end{aligned} \tag{5.42}$$

have the solutions

$$\begin{aligned} q_1 &= c_1 e^{-t}, \\ q_2 &= c_2 e^{-at}. \end{aligned} \tag{5.43}$$

For $q_1 \neq 0$ (5.33) acquires the form

$$\frac{dq_2}{dq_1} = \frac{aq_2}{q_1} \tag{5.44}$$

with the solution

$$q_2 = Cq_1{}^a, \tag{5.45}$$

which can also be obtained from (5.41) and (5.43) by the elimination of time. For $a > 0$ we obtain parabolic integral curves depicted in Fig. 5.9a. In this case we have for the slope dq_2/dq_1

$$\frac{dq_2}{dq_1} = Caq_1^{a-1}. \tag{5.46}$$

We now distinguish between positive and negative exponents of q_1. If $a > 1$ we find $dq_2/dq_1 \to 0$ for $q_1 \to 0$. Every integral curve with exception of the q_2-axis approaches the singular point along the q_1 axis. If $a < 1$ one immediately establishes that the rôles of q_1 and q_2 are interchanged. Singular points which are surrounded by curves of the form of Fig. 5.9a are called *nodes*. For $a = 1$ the integral curves are half lines converging to or radiating from the singular point. In the case of the node every integral curve has a limiting direction at the singular point.

We now consider the case $a < 0$. Here we find the hyperbolic curves

$$q_2 q_1{}^{|a|} = C. \tag{5.47}$$

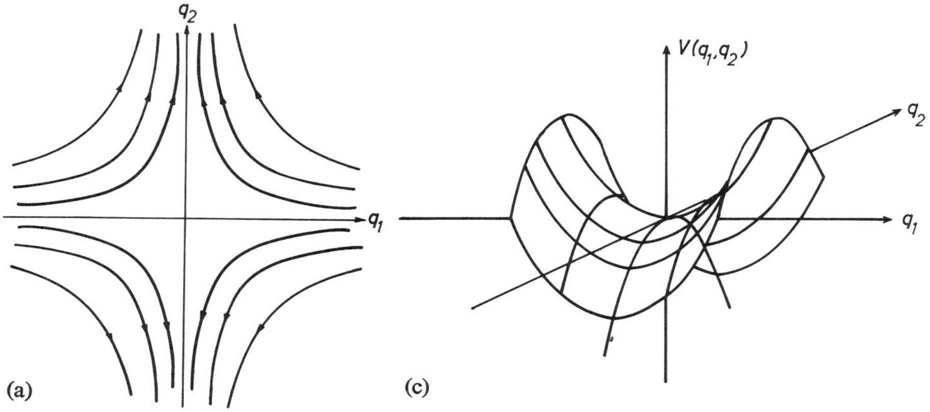

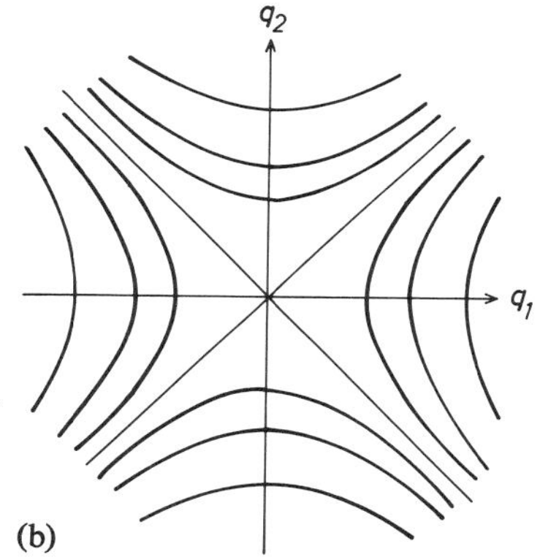

Fig. 5.10a–c. Trajectories (a), equipotential curves (b), and potential (c), $V=1/2\ q_1^2-|a|/2\ q_2^2$, $a<0$ of a *saddle point*

For $a = -1$ we have ordinary hyperbolas. The curves are depicted in Fig. 5.10. Only four trajectories tend to the singular point namely A_S, D_S for $t \to \infty$, D_S, C_S for $t \to -\infty$. The corresponding singular point is called *saddle point*.

2) *Focus and Center*

We now consider equations of the form

$$\frac{dq_1}{dt} = -aq_1 - q_2 \tag{5.48}$$

$$\frac{dq_2}{dt} = +q_1 - aq_2. \tag{5.49}$$

For $a > 0$, by use of polar coordinates (compare (5.24)), (5.48) and (5.49) acquire the form

$$\dot{r} = -ar \tag{5.50}$$

$$\dot{\varphi} = 1, \tag{5.51}$$

which have the solutions

$$r = C_1 e^{-at} \tag{5.52}$$

$$\varphi = t + C_2. \tag{5.53}$$

C_1, C_2 are integration constants. We have already encountered trajectories of this kind in Section 5.1. The trajectories are spirals approaching the singular point at the origin. The radius vector rotates anticlockwise with the frequency (5.51). This point is called a *stable focus* (Fig. 5.11). In the case $a < 0$ the motion departs from the focus: we have an *unstable focus*. For $a = 0$ a *center results* (Fig. 5.12).

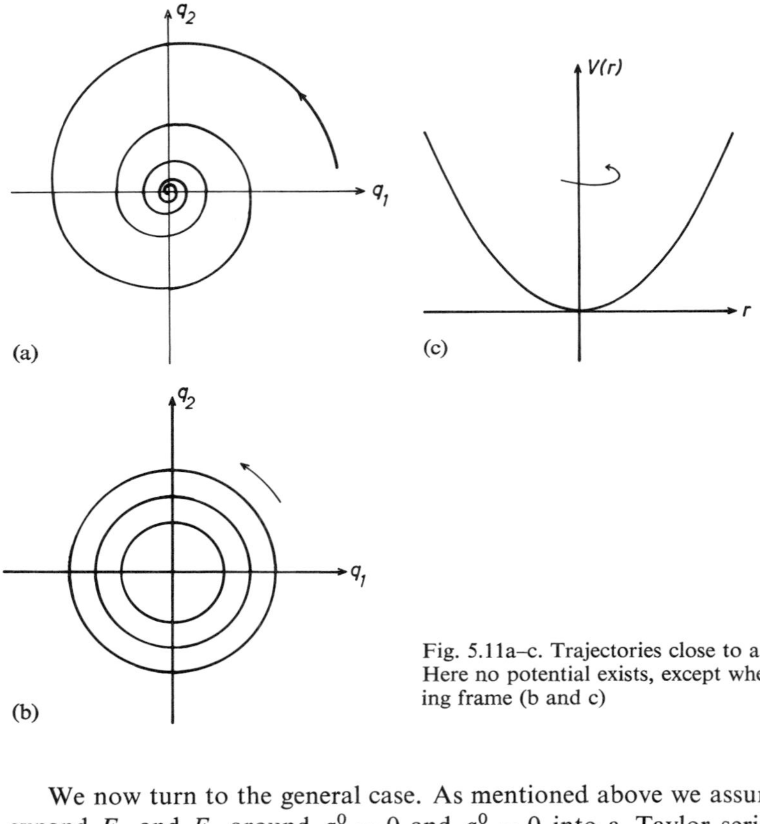

Fig. 5.11a–c. Trajectories close to a *stable focus*. Here no potential exists, except when using a rotating frame (b and c)

We now turn to the general case. As mentioned above we assume that we can expand F_1 and F_2 around $q_1^0 = 0$ and $q_2^0 = 0$ into a Taylor series and that we may keep as leading terms those linear in q_1 and q_2, i.e.,

$$\dot{q}_1 = a q_1 + b q_2, \tag{5.54}$$

$$\dot{q}_2 = c q_1 + d q_2. \tag{5.55}$$

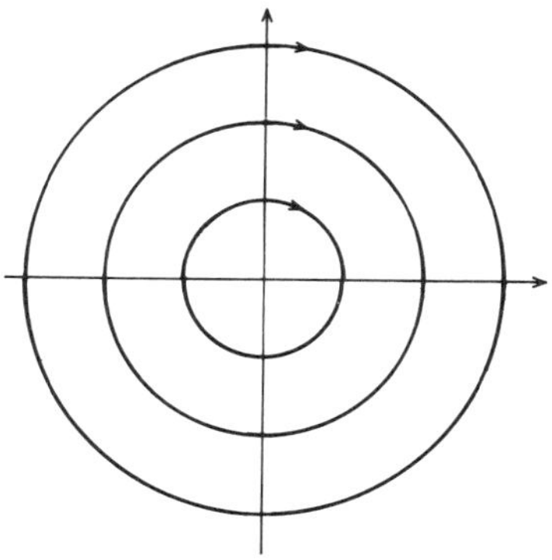

Fig. 5.12. A center. No potential exists, except when using a rotating frame. In it, V=const

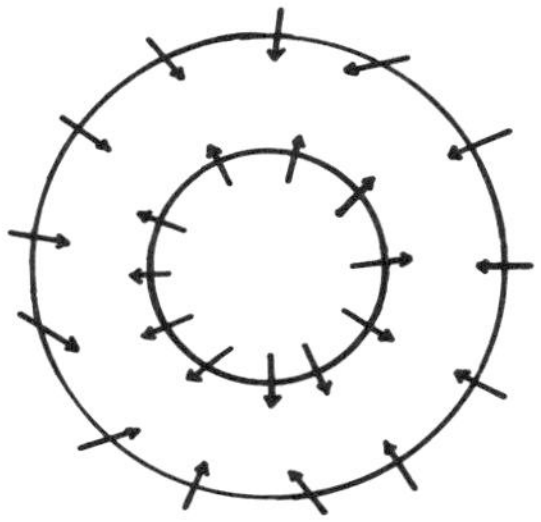

Fig. 5.13. Compare text

One then makes a linear transformation

$$\xi = \alpha q_1 + \beta q_2$$
$$\eta = \gamma q_1 + \delta q_2 \tag{5.56}$$

so that (5.54) and (5.55) are transformed into

$$\dot{\xi} = \lambda_1 \xi \tag{5.57}$$
$$\dot{\eta} = \lambda_2 \eta. \tag{5.58}$$

The further discussion may be performed as above, if λ_1, λ_2 are real. If $\lambda_1 (= \lambda_2^*)$ are complex, inserting $\xi = \eta^* = re^{i\varphi}$ into (5.57), (5.58) and separating the equations into their real and imaginary parts yields equations of the type (5.50), (5.51) (with $a = -\text{Re}\,\lambda_1$, and $+1$ replaced by $\text{Im}\,\lambda_1$).

Let us now discuss a further interesting problem, namely. *How to find limit cycles in the plane: The Poincaré-Bendixson theorem.* Let us pick a point $\boldsymbol{q}^0 = (q_1^0, q_2^0)$ which is assumed to be nonsingular, and take it as initial value for the solution of (5.27), (5.28). For later times, $t > t_0$, $\boldsymbol{q}(t)$ will move along that part of the trajectory which starts at $\boldsymbol{q}(t_0) = \boldsymbol{q}^0$. We call this part *half-trajectory*. The *Poincaré-Bendixson theorem* then states: If a half-trajectory remains in a finite domain without approaching singular points, then this half-trajectory is either a limit cycle or approaches such a cycle. (There are also other forms of this theorem available in the literature). We do not give a proof here, but discuss some ways of applying it invoking the analogy between trajectories and streamlines in hydrodynamics. Consider a finite domain called D which is surrounded by an outer and an inner curve (compare Fig. 5.13). In more mathematical terms: D is doubly connected. If all trajectories enter the domain and there are no singular points in it or on its boundaries, the conditions of the theorem are fulfilled[3]. When we let shrink the inner boundary to a point we see that the conditions of the Poincaré-Bendixson theorem are still fulfilled *if that point is a source.*

The Potential Case (n Variables)

We now consider n variables q_j which obey the equations

$$\dot{q}_j = F_j(q_1, \ldots, q_n). \tag{5.59}$$

The case in which the forces F_j can be derived from a potential V by means of

$$F_j = -\frac{\partial V(q_1 \cdots q_n)}{\partial q_j} \tag{5.60}$$

represents a straightforward generalization of the one-dimensional case considered in Section 5.1. It can be directly checked by means of the forces F_j if they can be derived from a potential. For this they have to satisfy the conditions

$$\frac{\partial F_j}{\partial q_k} = \frac{\partial F_k}{\partial q_j} \quad \text{for all} \quad j, k = 1, \ldots, n. \tag{5.61}$$

The system is in equilibrium if (5.60) vanishes. This can happen for certain points $q_1^0, \ldots, q_n^0$, but also for lines or for hypersurfaces. When we let V depend on parameters, quite different types of hypersurfaces of equilibrium may emerge from each other, leading to more complicated types of bifurcation. A theory of some types has been developed by Thom (cf. Sect. 5.5).

Exercise on 5.2

Extend the above considerations of the Poincaré-Bendixson theorem to the time-reversed case by inverting the directions of the streamlines.

5.3* Stability

The state of a system may change considerably if it loses its stability. Simple examples were given in Section 5.1 where a deformation of the potential caused an instability of the original state and led to a new equilibrium position of a "fictitious particle". In later chapters we will see that such changes will cause new structures on a macroscopic scale. For this reason it is desirable to look more closely into the question of stability and first to define stability more exactly. Our above examples referred to critical points where $\dot{q}_j = 0$. The idea of stability can be given a much more general form, however, which we want to discuss now.

---

[3] For the experts we remark that Fig. 5.5 without motion in *tangential direction* represents a "pathological case". The streamlines end perpendicularly on a circle. This circle represents a line of critical points but not a limit cycle.

To this end we consider the set of differential equations

$$\dot{q}_j = F_j(q_1, \ldots, q_n), \tag{5.62}$$

which determine the trajectories in $\boldsymbol{q}$ space. Consider a solution $q_j = u_j(t)$ of (5.62) which can be visualized as path of a particle in $\boldsymbol{q}$ space. This solution is uniquely determined by the original values at initial time $t = t_0$. In practical applications the initial values are subject to perturbations. Therefore, we ask what happens to the path of the "particle", if its initial condition is not exactly the one we have considered above. Intuitively it is clear that we shall call the trajectory $u_j(t)$ *stable* if other trajectories which are initially close to it remain close to it for later times. Or, in other words, if we prescribe a surrounding of $u_j(t)$, then $u_j(t)$ is stable, if all solutions which start initially in a sufficiently close surrounding of $u_j(0)$ remain in the prescribed surrounding, (Fig. 5.14). If we cannot find an initial surrounding

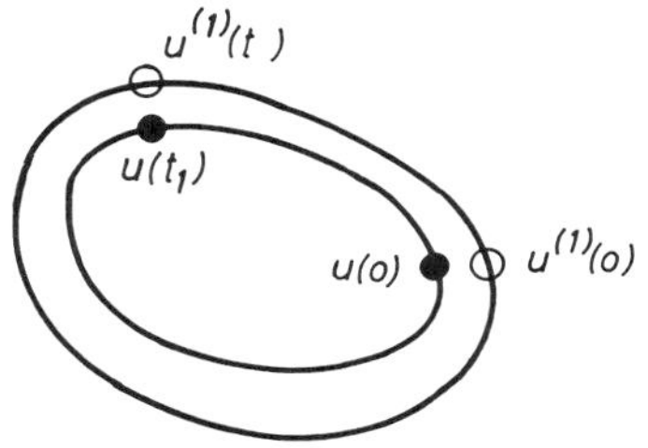

Fig. 5.14. Example of a stable trajectory u. $u^{(1)}$: neighboring trajectory

S_0 so that this criterion can be fulfilled, $u_j(t)$ is called unstable. Note that this definition of stability does not imply that the trajectories $v_j(t)$ which are initially close to $u_j(t)$ approach it so that the distance between u_j and v_j vanishes eventually. To take care of this situation we define asymptotic stability. The trajectory u_j is called *asymptotic stable* if for all $v_j(t)$ which fulfil the stability criterion the condition

$$|u_j(t) - v_j(t)| \to 0 \quad (t \to \infty) \tag{5.63}$$

holds in addition. (cf. Fig. 5.6, where the limit cycle is asymptotically stable). So far we have been following up the motion of the representative point (particle on its trajectory). If we do not look at the actual time dependence of $\boldsymbol{u}$ but only at the form of the trajectories then we may define *orbital stability*. We first give a qualitative picture and then cast the formulation into a mathematical frame.

This concept is a generalization of the stability definition introduced above. The stability consideration there started from neighboring points. Stability then meant that moving points always remain in a certain neighborhood in the course of time. In the case of orbital stability we do not confine ourselves to single points, but consider a whole trajectory C. Orbital stability now means, a trajectory in a sufficiently close neighborhood of C remains in a certain neighborhood in the course of time. This condition does not necessarily imply that two points which

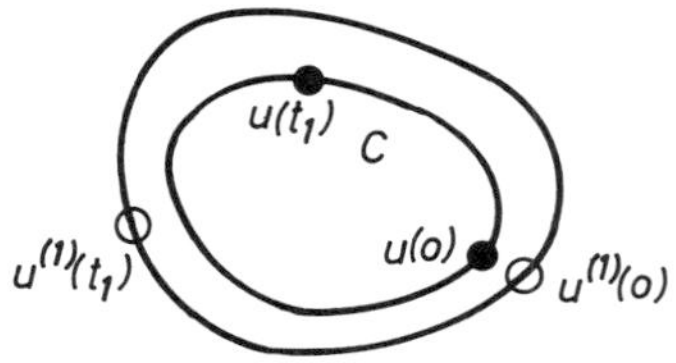

Fig. 5.15. Example of orbital stability

belong to two different trajectories and have been close together originally, do closely remain together at later times (cf. Fig. 5.15). In mathematical terms this may be stated as follows: C is orbitally stable if given $\varepsilon > 0$ there is $\eta > 0$ such that if R is a representative point of another trajectory which is within $\eta(C)$ at time τ then R remains within a distance ε for $t > \tau$. If no such η exists, C is unstable.

We further may define *asymptotic orbital stability* by the condition that C is orbitally stable, and in addition that the distance between R and C tends to 0 as t goes to ∞. To elucidate the difference between stability and orbital stability consider as an example the differential equations

$$\dot{\Theta} = r, \tag{5.64a}$$

$$\dot{r} = 0. \tag{5.64b}$$

According to (5.64b) the orbits are circles and according to (5.64a) the angular velocity increases with the distance. Thus two points on two neighboring circles which are initially close are at a later time far apart. Thus there is no stability. However, the orbits remain close and are orbitally stable.

One of the most fundamental problems now is how to check whether in a given problem the trajectories are stable or not. There are two main approaches to this problem, a local criterion and a global criterion.

1) *Local Criterion*

In this we start with a given trajectory $q_j(t) = u_j(t)$ whose asymptotic stability we want to investigate. To this end we start with the neighboring trajectory

$$q_j(t) = u_j(t) + \xi_j(t) \tag{5.65}$$

which differs from u_j by a small quantity ξ_j. Apparently we have asymptotic stability if $\xi_j \to 0$ for $t \to \infty$. Inserting (5.65) into (5.62) yields

$$\dot{u}_j(t) + \dot{\xi}_j(t) = F_j(u_1(t) + \xi_1(t), \ldots, u_n(t) + \xi_n(t)). \tag{5.66}$$

Since the ξ's are assumed to be small, we may expand the right-hand side of (5.66) into a Taylor series. Its lowest order term cancels against $\dot{u}_j$ on the lhs. In the next order we obtain

$$\dot{\xi}_j(t) = \sum_k \left.\frac{\partial F_j}{\partial \xi_k}\right|_0 \xi_k(t). \tag{5.67}$$

In general the solution of (5.67) is still a formidable problem because $dF/d\xi$ is still a function of time since the u_j's were functions of time. A number of cases can be treated. One is the case in which the u_j's are periodic functions. Another very simple case is provided if the trajectory u_j consists of a singular point so that $u_j =$ constant. In that case we may put

$$\frac{\partial F_j}{\partial q_k} = A_{jk}, \tag{5.68}$$

where A_{jk} are constants. Introducing the matrix

$$A = (A_{jk}) \tag{5.69}$$

the set of (5.67) acquires the form

$$\dot{\xi} = A\xi, \qquad \xi = \begin{pmatrix} \xi_1 \\ \vdots \\ \xi_n \end{pmatrix}. \tag{5.70}$$

These first-order linear differential equations of constant coefficients may be solved in a standard manner by the hypothesis

$$\xi = \xi_0 e^{\lambda t}. \tag{5.71}$$

Inserting (5.71) into (5.70), performing the differentiation with respect to time, and dividing the equation by $e^{\lambda t}$, we are left with a set of linear homogeneous algebraic equations for ξ_0. The solvability condition requires that the determinant vanishes, i.e.,

$$\begin{vmatrix} A_{11} - \lambda & A_{12} & \cdots & A_{1n} \\ \vdots & A_{22} - \lambda & & \\ \vdots & & \ddots & A_{nn} - \lambda \end{vmatrix} = 0. \tag{5.72}$$

The evaluation of this determinant leads to the characteristic equation

$$f(\lambda) \equiv C_0\lambda^n + C_1\lambda^{n-1} + \cdots + C_n = 0 \tag{5.73}$$

of order n. If the real parts of all λ's are negative, then all possible solutions (5.71) decay with time so that the singular point is per definition asymptotically stable. If any one of the real parts of the λ's is positive, we can always find an initial trajectory which departs from $u_j(t)$ so that the singular point is unstable. We have so-called marginal stability if the λ's have nonpositive real parts but one or some of them are equal zero. The following should be mentioned for practical applications. Fortunately it is not necessary to determine the solutions of (5.73). A number of criteria have been developed, among them the *Hurwitz criterion* which allows us to check directly from the properties of A if $\mathrm{Re}\,\lambda < 0$.

Hurwitz Criterion

We quote this criterion without proof. All zeros of the polynomial $f(\lambda)$ (5.73) with real coefficients C_i lie on the left half of the complex λ-plane (i.e., $\operatorname{Re}\lambda < 0$) if and only if the following conditions are fulfilled

a) $$\frac{C_1}{C_0} > 0, \frac{C_2}{C_0} > 0, \cdots, \frac{C_n}{C_0} > 0$$

b) The principal subdeterminants H_j (Hurwitz determinants) of the quadratic scheme

$$\begin{matrix} C_1 & C_0 & 0 & 0 & \cdots\cdots & 0 & 0 \\ C_3 & C_2 & C_1 & C_0 & \cdots\cdots & 0 & 0 \\ C_5 & C_4 & C_3 & C_2 & \cdots\cdots & 0 & 0 \\ \cdot & \cdot & \cdot & \cdot & \cdot & \cdot & \cdot \\ 0 & 0 & 0 & 0 & \cdots\cdots & C_{n-1} & C_{n-2} \\ 0 & 0 & 0 & 0 & \cdots\cdots & 0 & C_n \end{matrix}$$

(e.g., $H_1 = C_1, H_2 = C_1C_2 - C_0C_3, \ldots, H_{n-1}, H_n = C_nH_{n-1}$) satisfy the inequalities $H_1 > 0; H_2 > 0; \ldots; H_n > 0$.

2) *Global Stability (Ljapunov Function)*

In the foregoing we have checked the stability by investigating the immediate vicinity of the point under discussion. On the other hand the potential curves depicted in Section 5.1 had allowed us to discuss the stability just by looking at the form of the potential, or in other words, we had a global criterion. So whenever we have a system which allows for a potential (compare Sect. 5.2) we can immediately discuss the stability of that system. However, there are quite a number of systems which do not possess a potential. In this case it is where Ljapunov's ideas come in. Ljapunov has defined a function which has the desirable properties of the potential to allow discussion of global stability, but which is not based on the requirement that the forces can be derived from a potential. We first define this Ljapunov function $V_L(\boldsymbol{q})$. (cf. Fig. 5.16).

1) $V_L(\boldsymbol{q})$ is continuous with its first partial derivatives in a certain region Ω about the origin. (5.74a)

2)

$$V_L(0) = 0. \tag{5.74b}$$

Without loss of generality we put the critical point into the origin.

3)

$$\text{For } \boldsymbol{q} \neq 0,\ V_L(\boldsymbol{q}) \text{ is positive in } \Omega. \tag{5.75}$$

4) We now take into account that $\boldsymbol{q}$ occurring in V_L is a time-dependent function

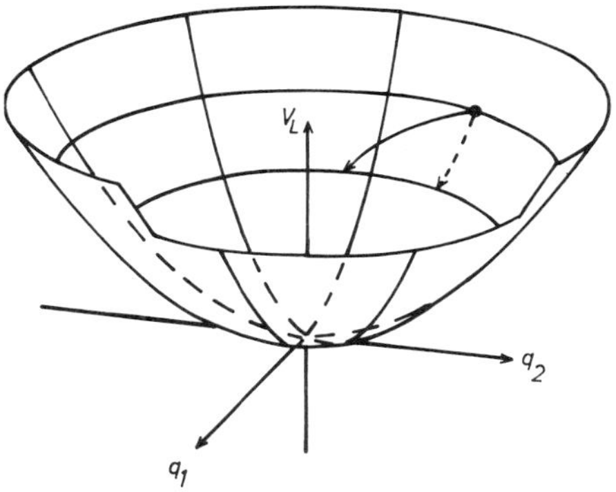

Fig. 5.16. Example of a Ljapunov function V_L in two dimensions. In contrast to the potential-case (cf. page 120) where the "particle" always follows the *steepest* descent (dashed line with arrow) it may follow here an arbitrary path (solid line with arrow), provided it does not go uphill (stability) or it goes everywhere downhill (asymptotic stability)

obeying the differential equations (5.62). Thus V_L becomes itself a time dependent function. Taking the derivative of V_L with respect to time we obtain

$$\dot{V}_L = \text{grad } V_L \cdot \dot{\boldsymbol{q}}, \tag{5.76}$$

where we can replace $\dot{\boldsymbol{q}}$ by means of (5.62). Now the requirement is

$$\dot{V}_L = \boldsymbol{F}(\boldsymbol{q}) \text{ grad } V_L \leqq 0 \quad \text{in } \Omega. \tag{5.77}$$

The great importance of V_L rests on the fact that in order to check (5.77) we need not solve the differential equations but just have to insert the Ljapunov function V_L into (5.77). We are now in a position to define Ljapunov's stability theorem.

A) *Stability Theorem*

If there exists in some neighborhood Ω of the origin a Ljapunov function $V_L(\boldsymbol{q})$, then the origin is stable. We furthermore have the

B) *Asymptotic Stability Theorem*

If $-\dot{V}_L$ is likewise positive definite in Ω, the stability is asymptotic. On the other hand one may also formulate theorems for instability. We mention two important cases. The first one is due to the Ljapunov and the next one represents a generalization due to Chetayev.

C) *Instability Theorem of Ljapunov*

Let $V(\boldsymbol{q})$ with $V(0) = 0$ have continuous first partial derivatives in Ω. Let $\dot{V}$ be positive definite and let V be able to assume positive values arbitrarily near the origin. Then the origin is unstable.

D) *Instability Theorem of Chetayev*

Let Ω be a neighborhood of the origin. Let there be given a function $V(\boldsymbol{q})$ and a region Ω_1 in Ω with the following properties:

1) $V(\boldsymbol{q})$ has continuous first partial derivatives in Ω_1

2) $V(q)$ and $\dot{V}(q)$ are positive in Ω_1
3) At the boundary points of Ω_1 inside Ω, $V(q) = 0$
4) The origin is a boundary point of Ω_1

Then the origin is unstable.

So far we have treated autonomous systems in which F in (5.62) does not depend on time t explicitly. We mention in conclusion that the Ljapunov theory can be extended in a simple manner to nonautonomous systems

$$\dot{q} = F(q, t). \tag{5.78}$$

Here it must be assumed that the solutions of (5.78) exist and are unique and that

$$F(0, t) = 0 \quad \text{for} \quad t \geq 0. \tag{5.79}$$

Though the reader will rather often meet the concepts of Ljapunov functions in the literature, there are only a few examples in which the Ljapunov function can be determined explicitly. Thus, though the theory is very beautiful its practical applicability has been rather limited so far.

Exercise on 5.3

1) Show that the information gain (3.38)

$$K(P, P') = \sum_m P(m) \ln (P(m)/P'(m)) \tag{E.1}$$

is a Ljapunov function for the master equation (4.111), where $P'(m)$ is the stationary solution of (4.111) and $P(m) = P(m, t)$ a time-dependent solution of it. Hint: Identify $P(m)$ with q_j of this chapter (i.e., $P \leftrightarrow q$, $m \leftrightarrow j$) and check that (E.1) fulfils the axioms of a Ljapunov function. Change $q = 0$ to $q = q^0 \equiv \{P'(m)\}$. To check (5.75) use the property (3.41). To check (5.77), use (4.111) and the inequality $\ln x \geq 1 - 1/x$. What does the result mean for $P(m, t)$? Show that $P'(m)$ is asymptotically stable.

2) If microreversibility holds, the transition probabilities $w(m, m')$ of (4.111) are symmetric:

$$w(m, m') = w(m', m).$$

Then the stationary solution of (4.111), $P'(m) = \text{const}$. Show that it follows from exercise 1) that the entropy S increases in such a system up to its maximal value (the famous Boltzmann's H-theorem).

5.4 Examples and Exercises on Bifurcation and Stability

Our mechanical example of a particle in a potential well is rather instructive because it allows us to explain quite a number of general features inherent in the original differential equation. On the other hand it does not shed any light on the importance of these considerations in other disciplines. To this end we present a few examples. Many more will follow in later chapters. The great importance of bifurcation rests in the fact that even a small change of a parameter, in our case the force constant k, leads to dramatic changes of the system.

Let us first consider a simple model of a laser. The laser is a device in which photons are produced by the process of stimulated emission. (For more details see Sect. 8.1). For our example we need to know only a few facts. The temporal change of photon number n, or, in other words, the photon production rate is determined by an equation of the form

$$\dot{n} = \text{gain} - \text{loss}. \tag{5.80}$$

The gain stems from the so-called stimulated emission. It is proportional to the number of photons present and to the number of excited atoms, N, (For the experts: We assume that the ground level, where the laser emission terminates, is kept empty) so that

$$\text{gain} = GNn. \tag{5.81}$$

G is a gain constant which can be derived from a microscopic theory but this is not our present concern. The loss term comes from the escape of photons through the endfaces of the laser. The only thing we need to assume is that the loss rate is proportional to the number of photons present. Therefore, we have

$$\text{loss} = 2\varkappa n. \tag{5.82}$$

$2\varkappa = 1/t_0$, where t_0 is the lifetime of a photon in the laser. Now an important point comes in which renders (5.80) nonlinear. The number of excited atoms N decreases by the emission of photons. Thus if we keep the number of excited atoms without laser action at a fixed number N_0 by an external pump, the actual number of excited atoms will be reduced due to the laser process. This reduction, ΔN, is proportional to the number of photons present, because all the time the photons force the atoms to return to their groundstates. Thus the number of excited atoms has the form

$$N = N_0 - \Delta N,\ \Delta N = \alpha n. \tag{5.83}$$

Inserting (5.81), (5.82) and (5.83) into (5.80) gives us the basic laser equation in our simplified model

$$\dot{n} = -kn - k_1 n^2, \tag{5.84}$$

where the constant k is given by

$$k = 2\varkappa - GN_0 \gtrless 0. \tag{5.85}$$

If there is only a small number of excited atoms, N_0, due to the pump, k is positive, whereas for sufficiently high N_0, k can become negative. The change of sign occurs at

$$GN_0 = 2\varkappa, \tag{5.86}$$

which is the laser threshold condition. Bifurcation theory now tells us that for $k > 0$ there is no laser light emission whereas for $k < 0$ the laser emits laser photons. The laser functions in a completely different way when operating below or above threshold. In later chapters we will discuss this point in much more detail and we refer the reader who is interested in a refined theory to these chapters.

Exactly the same (5.84) or (5.80) with the terms (5.81), (5.82) and (5.83) can be found in a completely different field, e.g., chemistry. Consider the autocatalytic reaction between two kinds of molecules, A, B, with concentrations n and N, respectively:

$A + B \rightarrow 2A$ production

$A \rightarrow C$ decay

The molecules A are created in a process in which the molecules themselves participate so that their production rate is proportional to n (compare (5.81)). Furthermore the production rate is proportional to N. If the supply of B-molecules is not infinitely fast, N will decrease again by an amount proportional to the number n of A-molecules present.

The same equations apply to certain problems of ecology and population dynamics. If n is the number of animals of a certain kind, N is a measure for the food supply available which is steadily renewed but only at a certain finite pace. Many more examples will be given in Chapters 9 and 10.

Exercise

Convince yourself that the differential equation (5.84) is solved by

$$n(t) = -\frac{k}{2k_1} - \frac{|k|}{2}\cdot\frac{c\cdot e^{-|k|t}-1}{c\cdot e^{-|k|t}+1}, \tag{5.87}$$

where

$$c = \frac{|k/k_1| - k/k_1 - 2n_0}{|k/k_1| - k/k_1 + 2n_0}$$

and $n_0 = n(0)$ is the initial value. Discuss the temporal behavior of this function and show that it approaches the stationary state $n = 0$ or $n = |k/k_1|$ irrespective of the initial value n_0, but depending on the signs of k, k_1. Discuss this dependence!

In a two-mode laser, two different kinds of photons 1 and 2 with numbers n_1

and n_2 are produced. In analogy to (5.80) [with (5.81,2,3)] the rate equations read

$$\dot{n}_1 = G_1 N n_1 - 2\varkappa_1 n_1, \tag{5.88}$$

$$\dot{n}_2 = G_2 N n_2 - 2\varkappa_2 n_2, \tag{5.89}$$

where the actual number of excited atoms is given by

$$N = N_0 - \alpha_1 n_1 - \alpha_2 n_2. \tag{5.90}$$

The stationary state

$$\dot{n}_1 = \dot{n}_2 = 0 \tag{5.91}$$

implies that for

$$\frac{G_1}{2\varkappa_1} \neq \frac{G_2}{2\varkappa_2} \tag{5.92}$$

at least n_1 or n_2 must vanish (proof?).

Exercise

What happens if

$$\frac{G_1}{2\varkappa_1} = \frac{G_2}{2\varkappa_2}. \tag{5.93}$$

Discuss the stability of (5.88) and (5.89). Does there exist a potential?
Discuss the critical point $n_1 = n_2 = 0$.
Are there further critical points?
Eqs. (5.88), (5.89) and (5.90) have a simple but important interpretation in ecology. Let n_1, n_2 be the numbers of two kinds of species which live on the same food supply N_0. Under the condition (5.92) only one species survives while the other dies out, because the species with the greater growth rate, G_1, eats the food much more quickly than the other species and finally eats all the food. It should be mentioned that, as in (5.83) the food supply is not given at an initial time but is kept at a certain rate. Coexistence of species becomes possible, however, if the food supply is at least partly different for different species. (Compare Sect. 10.1).

Exercise

Discuss the general equations

$$0 = \dot{n}_1 = a_{11} n_1 - a_{12} n_2, \tag{5.94}$$

$$0 = \dot{n}_2 = a_{21} n_1 - a_{22} n_2. \tag{5.95}$$

Another interesting example of population dynamics is the *Lotka-Volterra model.* It was originally devised to explain temporal oscillations in the occurrence of fish in the Adriatic sea. Here two kinds of fish are treated, namely, predator fishes and their preyfishes. The rate equations have again the form

$$\dot{n}_j = \text{gain}_j - \text{loss}_j; \quad j = 1, 2. \tag{5.96}$$

We identify the prey fishes with the index 1. If there are no predator fishes, the prey fishes will multiply according to the law

$$\text{gain}_1 = \alpha_1 n_1. \tag{5.97}$$

The prey fishes suffer, however, losses by being eaten by the predators. The loss rate is proportional to the number of preys and predators

$$\text{loss}_1 = \alpha n_1 n_2. \tag{5.98}$$

We now turn to the equation $j = 2$, for the predators. Evidently we obtain

$$\text{gain}_2 = \beta n_1 n_2. \tag{5.99}$$

Because they live on the prey, predator multiplication rate is proportional to their own number and to that of the prey fishes. Since predators may suffer losses by death, the loss term is proportional to the number of predator fishes present

$$\text{loss}_2 = 2\varkappa_2 n_2. \tag{5.100}$$

The equations of the *Lotka-Volterra model* therefore read

$$\begin{aligned} \dot{n}_1 &= \alpha_1 n_1 - \alpha n_1 n_2, \\ \dot{n}_2 &= \beta n_1 n_2 - 2\varkappa_2 n_2. \end{aligned} \tag{5.101}$$

Exercise

1) Make (5.101) dimensionless by casting them into the form

$$\begin{aligned} \frac{dn'_1}{dt'} &= n'_1 - n'_1 n'_2, \\ \frac{dn'_2}{dt'} &= a(-n'_2 + n'_1 n'_2). \end{aligned} \tag{5.102}$$

(Hint: Put

$$n_1 = 2\frac{\varkappa_2}{\beta} n'_1; \, n_2 = \frac{\alpha_1}{\alpha} n'_2; \, t = \frac{1}{\alpha_1} t') \tag{5.103}$$

2) Determine the stationary state $\dot{n}_1 = \dot{n}_2 = 0$. Prove the following conservation law

$$\sum_{j=1,2} \frac{1}{a_j}(n_j' - \ln n_j') = \text{const}, \qquad \begin{pmatrix} a_1 = 1 \\ a_2 = a \end{pmatrix} \tag{5.104}$$

or

$$\prod_{j=1,2} (n_j' \cdot e^{-n_j'})^{1/a_j} = \text{const}. \tag{5.105}$$

Hint: Introduce new variables $v_j = \ln n_j'$, $j = 1, 2$ and form

$$\sum_{j=1}^{2} \frac{1}{a_j} \frac{dv_j}{dt}(e^{v_j} - 1) \tag{5.106}$$

and use (5.102). From Fig. 5.17 it follows that the motion of n_1, n_2 is periodic. cf. Fig. 5.18. Why?

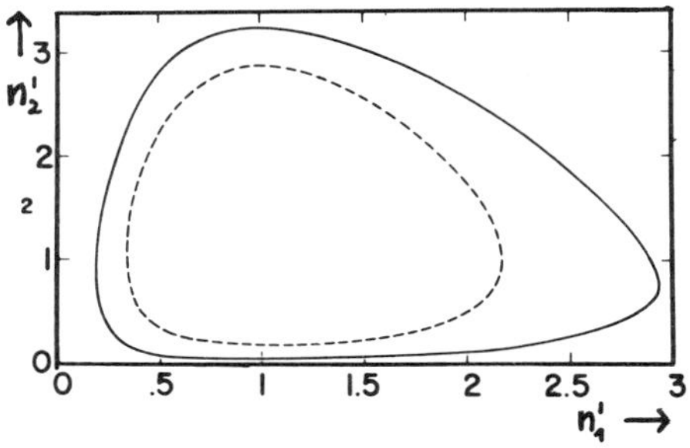

Fig. 5.17. Two typical trajectories in the n_1-n_2 phase plane of the Lotka-Volterra model (after Goel, N. S., S. C. Maitra, E. W. Montroll: Rev. Mod. Phys. *43*, 231 (1971)) for fixed parameters

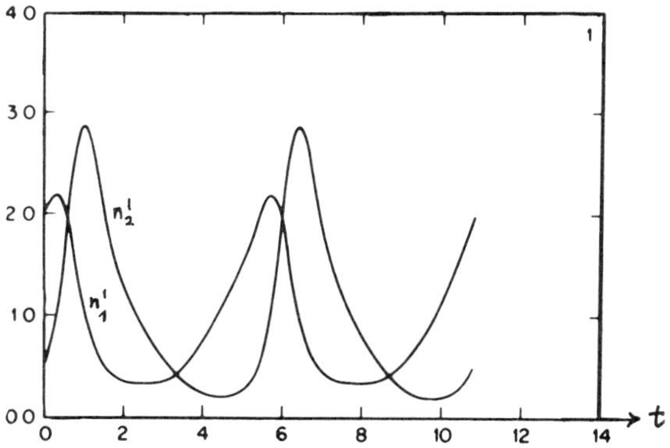

Fig. 5.18. Time variation of the two populations n_1, n_2 corresponding to a trajectory of Fig. 5.17

Hint: Use Section 5.1. Are there singular points? Are the trajectories stable or asymptotically stable?

Answer: The trajectories are stable, but not asymptotically stable. Why?

By means of a suitable Ljapunov function check the stability or instability of the following systems ($\lambda > 0$, $\mu > 0$):

1)

$$\dot{q}_1 = -\lambda q_1,$$
$$\dot{q}_2 = -\mu q_2. \tag{5.107}$$

Hint: Take as Ljapunov function

$$V_{\mathrm{L}} = q_1^2 + q_2^2. \tag{5.108}$$

2)

$$\dot{q}_1 = \lambda q_1,$$
$$\dot{q}_2 = -\mu q_2. \tag{5.109}$$

Hint: Take as Ljapunov function

$$V_{\mathrm{L}} = q_1^2 + q_2^2. \tag{5.110}$$

3) For complex variables q, the equations are

$$\dot{q} = (a + bi)q,$$
$$\dot{q}^* = (a - bi)q^*. \tag{5.111}$$

with

$$a,b \neq 0 \tag{5.112}$$

and

$$a < 0. \tag{5.113}$$

Hint: Take as Ljapunov function

$$V = qq^*. \tag{5.114}$$

Show in the above examples that the functions (5.108), (5.110) and (5.114) are Ljapunov functions. Compare the results obtained with those of Section 5.2 (and identify the above results with the case of stable node, unstable saddle point, and stable focus).
Show that the potential occurring in anharmonic oscillators has the properties of a Ljapunov function (within a certain region, Ω).

Exercise: Van der Pol Equation

This equation which has played a fundamental role in the discussion of the performance of radio tubes has the following form:

$$\ddot{q} + \varepsilon(q^2 - 1)\dot{q} + q = 0 \tag{5.115}$$

with

$$\varepsilon > 0. \tag{5.116}$$

Show by means of the equivalent equations

$$\begin{aligned} \dot{q} &= p - \varepsilon(q^3/3 - q), \\ \dot{p} &= -q, \end{aligned} \tag{5.117}$$

that the origin is the only critical point which is a source. For which ε's is it an unstable focus (node)?

Show that (5.117) allows for (at least) one limit cycle.

Hint: Use the discussion following the Poincaré-Bendixson theorem on p. 119. Draw a large enough circle around the origin, $q = 0$, $p = 0$, and show that all trajectories enter its interior. To this end, consider the rhs of (5.117) as components of a vector giving the local direction of the streamline passing through q, p. Now form the scalar product of that vector and the vector $\boldsymbol{q} = (q, p)$ pointing from the origin to that point. The sign of this scalar product tells you into which direction the streamlines point.

5.5* Classification of Static Instabilities, or an Elementary Approach to Thom's Theory of Catastrophes

In the foregoing we have encountered examples where the potential curve shows transitions from one minimum to two minima which led to the phenomenon of bifurcation. In this chapter we want to discuss the potential close to those points where linear stability is lost. To this end we start with the one-dimensional case and will eventually treat the general n-dimensional case. Our goal is to find a classification of critical points.

A) *One-Dimensional Case*

We consider the potential $V(q)$ and assume that it can be expanded into a Taylor series:

$$V(q) = c^{(0)} + c^{(1)}q + c^{(2)}q^2 + c^{(3)}q^3 + \cdots + c^{(m)}q^m + \cdots \tag{5.118}$$

The coefficients of the Taylor series of (5.118) are given as usual by

$$c^{(0)} = V(0), \tag{5.119}$$

$$c^{(1)} = \left.\frac{dV}{dq}\right|_{q=0}, \tag{5.120}$$

$$c^{(2)} = \left.\frac{1}{2}\frac{d^2V}{dq^2}\right|_{q=0}, \tag{5.121}$$

and quite generally by

$$c^{(l)} = \frac{1}{l!} \frac{d^l V}{dq^l}\bigg|_{q=0}, \tag{5.122}$$

provided that the expansion is taken at $q = 0$. Since the form of the potential curve does not change if we shift the curve by a constant amount, we may always put

$$c^{(0)} = 0. \tag{5.123}$$

We now assume that we are dealing with a point of equilibrium (which may be stable, unstable, or metastable)

$$\frac{dV}{dq} = 0. \tag{5.124}$$

From (5.124) it follows

$$c^{(1)} = 0. \tag{5.125}$$

Before going on let us make some simple but fundamental remarks on smallness. In what follows in this section we always assume that we are dealing with dimensionless quantities. We now compare the smallness of different powers of q. Choosing $q = 0.1$, q^2 gives 0.01, i.e., q^2 is only 10% of q. Choosing as a further example $q = 0.01$, $q^2 = 0.0001$, i.e., only 1% of q. The same is evidently true for consecutive powers, say q^n and q^{n+1}. When we go from one power to the next, choosing q sufficiently small allows us to neglect q^{n+1} compared to q^n. Therefore we can confine ourselves in the following to the leading terms of the expansion (5.118). The potential shows a local minimum provided

$$\frac{1}{2} \frac{d^2 V}{dq^2}\bigg|_{q=0}^{\cdot} \equiv c^{(2)} > 0 \tag{5.126}$$

(compare Fig. 5.3a).

For the following we introduce a slightly different notation

$$c^{(2)} = \mu. \tag{5.127}$$

As we will substantiate later on by many explicit examples, μ may change its sign when certain parameters of the system under consideration are changed. This turns the stable point $q = 0$ into an unstable point for $\mu < 0$ or into a point of neutral stability for $\mu = 0$. In the neighborhood of such a point the behavior of $V(q)$ is determined by the next nonvanishing power of q. We will call a point where $\mu = 0$ an instability point. We first assume

1) $c^{(3)} \neq 0$ so that $V(q) = c^{(3)} q^3 + \cdots$ (5.128)

We shall show later on in practical cases, $V(q)$ may be disturbed either by external causes, which in mechanical engineering may be loads, or internally by imperfections (compare the examples of Chapter 8). Let us assume that these perturbations are small. Which of them will change the character of (5.128) the most? Very close to $q = 0$ higher powers of q e.g., q^4, are much smaller than (5.128), so that such a term presents an unimportant change of (5.128). On the other hand, imperfections or other perturbations may lead to lower powers of q than cubic so that these can become dangerous in perturbing (5.128). Here we mean by "dangerous" that the state of the system is changed appreciably.

The most general case would be to include all lower powers leading to

$$V(q) = \alpha + \beta q + \gamma q^2 + c^{(3)} q^3. \tag{5.129}$$

Adding all perturbations which change the original singularity (5.128) in a non-trivial way are called according to Thom "unfoldings". In order to classify all possible unfoldings of (5.128) we must do away superfluous constants. First, by an appropriate choice of the scale of the q-axis we can choose the coefficient $c^{(3)}$ in (5.128) equal to 1. Furthermore we may shift the origin of the q axis by the transformation

$$q = q' + \delta \tag{5.130}$$

to do away the quadratic term in (5.129). Finally we may shift the zero point of the potential so that the constant term in (5.129) vanishes. We are thus left with the "normal" form $V(q)$

$$V(q) = q^3 + uq. \tag{5.131}$$

This form depends on a single free parameter, u. The potentials for $u = 0$ and $u < 0$ are exhibited in Fig. 5.19. For $u \to 0$ the maximum and minimum coincide to a single turning point.

2) $c^{(3)} = 0$, but $c^{(4)} \neq 0$.

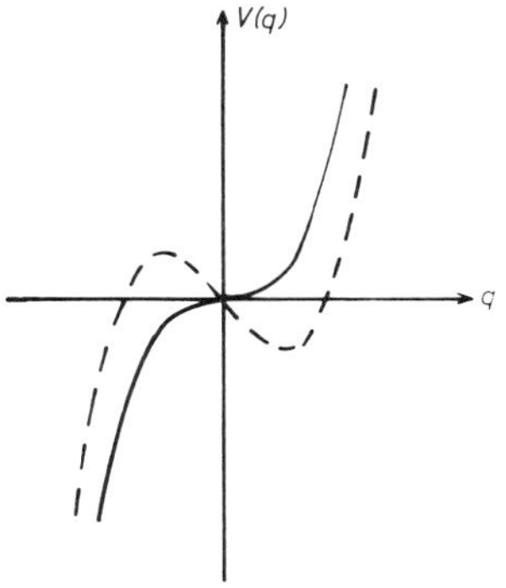

Fig. 5.19. The potential curve $V(q)=q^3+uq$ and its unfolding (5.131) for $u<0$

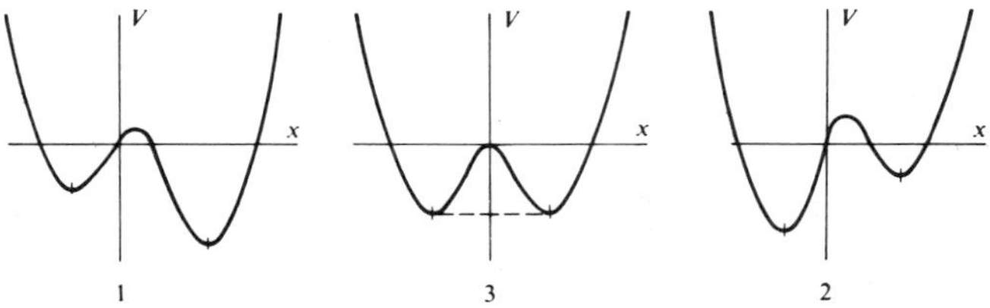

Fig. 5.20. The potential (5.133) for several values of u and v (after Thom)

The potential now begins with

$$V = q^4. \tag{5.132}$$

The unfolding of this potential is given by (cf. Fig. 5.20)

$$V(q) = \frac{q^4}{4} + \frac{uq^2}{2} + vq, \tag{5.133}$$

where we have already shifted the origin of the $(q - V)$-coordinate system appropriately. The factors 1/4 and 1/2 in the first or second term in (5.133) are chosen in such a manner that the derivative of V acquires a simple form

$$\frac{dV}{dq} = q^3 + uq + v. \tag{5.134}$$

If we put (5.134) = 0, we obtain an equation whose solution gives us the positions of the three extrema of the potential curve. Depending on the parameters u and v, we may now distinguish between different regions. If $u^3/27 + v^2/4 > 0$, there is only one minimum, whereas for $u^3/27 + v^2/4 < 0$ we obtain two minima which may differ in depth depending on the size of v. Thinking of physical systems, we may imagine that only that state is realized which has the lowest minimum. Therefore, if we change the parameters u and v, the state of the system may jump from one minimum to the other. This leads to different regions in the u-v plane depending on which minimum is realized. (For a critical comment on this jumping see the end of this section) (Fig. 5.21).

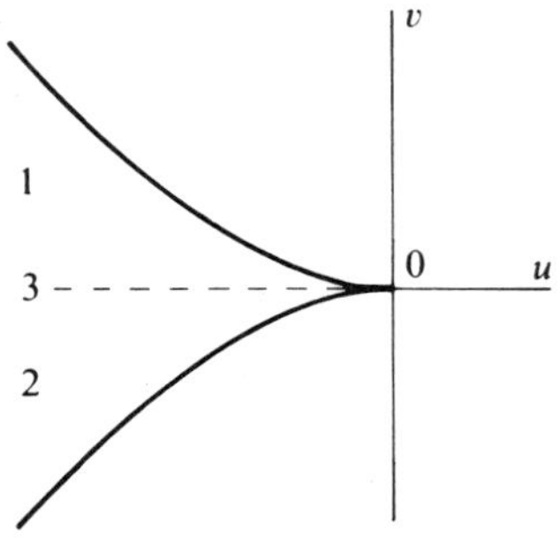

Fig. 5.21. In the u-v plane, the solid curve separates the region with one potential minimum (right region) from the region with two potential minima. The solid line represents in the sense of Thom the *catastrophic set* consisting of *bifurcation points*. The dotted line indicates where the two minima have the same value. This line represents the catastrophic set consisting of *conflict points* (after Thom)

3) If $c^{(4)} = 0$ but the next coefficient does not vanish we find as potential of the critical point

$$V = q^5. \tag{5.135}$$

Normalizing q appropriately, the unfolding of V reads

$$V = \frac{q^5}{5} + \frac{uq^3}{3} + \frac{vq^2}{2} + wq, \tag{5.136}$$

where the extrema are determined by

$$\frac{\partial V}{\partial q} = q^4 + uq^2 + vq + w = 0 \tag{5.137}$$

allowing for zero, two or four extrema implying zero, one or two minima. If we change u, or v, or w or some of them simultaneously it may happen that the number of minima changes (bifurcation points) or that their depths become equal (conflict points).

It turns out that such changes happen in general at certain surfaces in u-v-w space and that these surfaces have very strange shapes. (Compare Fig. 5.22). Then, as last example we mention the potential

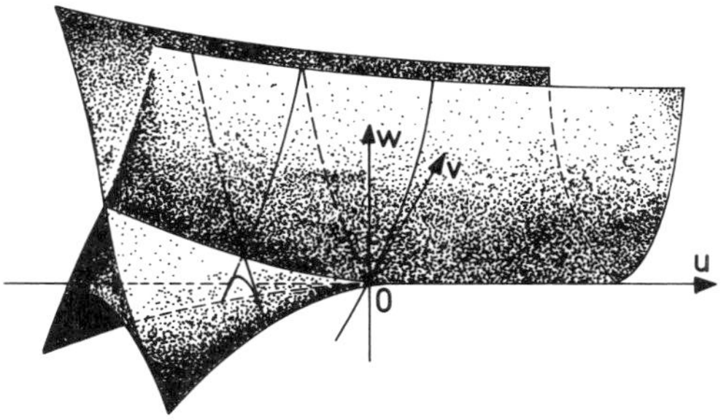

Fig. 5.22. The u, v, w space decays into regions determined by surfaces ("catastrophic sets") where the number of potential minima changes. The surfaces separating regions with one minimum from those with two minima have the form of a swallowtail (after Thom)

4) $V = q^6$ and its unfolding

$$V = \frac{q^6}{6} + \frac{tq^4}{4} + \frac{uq^3}{3} + \frac{vq^2}{2} + wq. \tag{5.138}$$

Let us now proceed to the

B) *Two-Dimensional Case*

Expanding the potential V into a Taylor series yields

$$V(q_1, q_2) = c^{(0)} + c_1^{(1)}q_1 + c_2^{(1)}q_2 + c_{11}^{(2)}q_1^2 + (c_{12}^{(2)} + c_{21}^{(2)})q_1q_2 + c_{22}^{(2)}q_2^2 + c_{111}^{(3)}q_1^3 + c_{112}^{(3)}q_1^2q_2 + \cdots \tag{5.139}$$

where we may assume

$$c_{12}^{(2)} = c_{21}^{(2)}. \tag{5.140}$$

Again by a shift of the V-coordinate we may ensure that

$$c^{(0)} = 0. \tag{5.141}$$

Furthermore, we assume that we are at the position of a local extremum, i.e., we have

$$\left.\frac{\partial V}{\partial q_1}\right|_{q_1, q_2 = 0} = 0 \tag{5.142}$$

and

$$\left.\frac{\partial V}{\partial q_2}\right|_{q_1, q_2 = 0} = 0. \tag{5.143}$$

Thus the leading term of (5.139) has the form

$$V_{\text{tr}} = b_{11}q_1^2 + 2b_{12}q_1q_2 + b_{22}q_2^2. \tag{5.144}$$

As we know from high-school mathematics, putting V_{tr} = constant, (5.144) defines a hyperbola, a parabola, or an ellipse. By a linear orthogonal transformation of the coordinates q_1 and q_2 we can make the axis of the coordinate system coincide with the principal axis of the ellipse, etc. Thus applying the transformation

$$\begin{aligned} q_1 &= A_{11}u_1 + A_{12}u_2, \\ q_2 &= A_{21}u_1 + A_{22}u_2 \end{aligned} \tag{5.145}$$

V_{tr} acquires the form

$$V_{\text{tr}} = \mu_1 u_2^2 + \mu_2 u_1^2. \tag{5.146}$$

If we apply the transformation (5.145) not only to the truncated form of V (5.144) but to the total form (5.139) (however with $c^{(0)} = c_1^{(1)} = c_2^{(1)} = 0$), we obtain a new potential in the form

$$V = \mu_1 u_1^2 + \mu_2 u_2^2 + \tilde{c}_{111}u_1^3 + \tilde{c}_{112}u_1^2u_2 + \cdots \tag{5.147}$$

This form allows us, again in a simple way, to discuss instabilities. Those occur if by a change of external parameters, μ_1 or μ_2 or both become $= 0$.

For further discussion we first consider $\mu_1 = 0$ and $\mu_2 > 0$, or, in other words, the system loses its stability in one coordinate. In what follows we shall denote the coordinate in the "unstable direction" by x, in the "stable direction" by y. Thus we discuss V in the form

$$V = V_1(x) + \mu_2 y^2 + yg(x) + y^2 h(x) + y^3 f(x), \tag{5.148}$$

where we have in particular

$$\begin{aligned} &V_1(x) \propto x^3 + \text{higher order},\\ &g(x) \propto x^2 + \text{higher order}, \end{aligned} \tag{5.149}$$

$$\begin{aligned} &h(x) \propto x + \text{higher order},\\ &f(x) \propto 1 + \text{higher order}. \end{aligned} \tag{5.150}$$

Since we want to discuss the immediate neighborhood of the instability point, $x = 0$, we may assume that x is small and we may even assume that it is so small that

$$h \ll \mu_2 \tag{5.151}$$

is fulfilled. Furthermore, we can restrict our analysis to a neighborhood of $y = 0$ so that higher order terms of y can be neglected. A typical situation is depicted in Figs. 5.23 and 5.24. Due to the term $g(x)$ in (5.148) the y minimum may be shifted

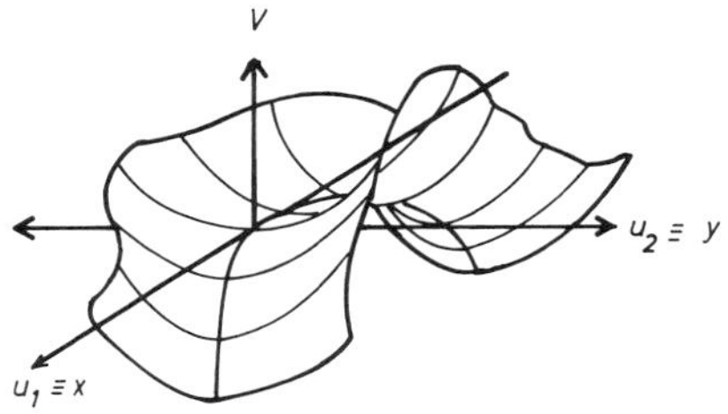

Fig. 5.23. The potential (5.147) for $\mu_1 < 0$ represents a distorted saddle (compare also Fig. 5.24)

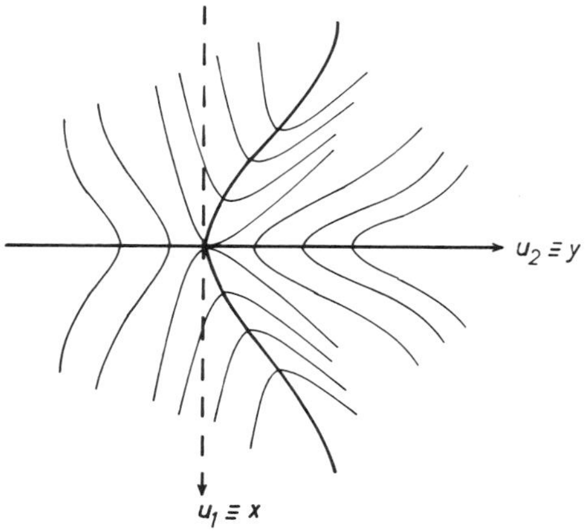

Fig. 5.24. Equipotential lines of the potential of Fig. 5.23

in y direction. For further discussion we exhibit the leading terms of (5.148) in the form

$$\hat{\mu}_2(y^2 + \hat{\mu}_2^{-1}gy + \tfrac{1}{4}\hat{\mu}_2^{-2}g^2) - \tfrac{1}{4}\hat{\mu}_2^{-2}g^2(x), \tag{5.152}$$

where we have added and subtracted the quadratic complement and where we have used the abbreviation

$$\hat{\mu}_2 = \mu_2 + h(x). \tag{5.153}$$

Introducing the new coordinate

$$\tilde{y} = y + \underbrace{\mu_2^{-1}g\cdot\tfrac{1}{2}}_{-y^{(0)}} \equiv y - y^{(0)} \tag{5.154}$$

helps us to cast (5.148) into the form (compare (5.151))

$$V = V_1(x) - \tfrac{1}{4}\hat{\mu}_2^{-2}g^2(x) + \hat{\mu}_2\tilde{y}^2 + \text{“higher order”}. \tag{5.155}$$

Provided the higher order terms are small enough we see that we have now found a complete decomposition of the potential V into a term which depends only on x and a second term which depends only on $\tilde{y}$. We now investigate in which way the so-called higher order terms are affected by the transformation (5.154). Consider for example the next term going with y^3. It reads

$$f(x)(\tilde{y} + y^{(0)})^3 \equiv f(x)\tilde{y}^3 + 3f(x)\tilde{y}^2y^{(0)} + 3f(x)\tilde{y}y^{(0)2} + f(x)y^{(0)3}. \tag{5.156}$$

Since $y^{(0)} \sim g(x)$, and $g(x)$ is a small quantity, the higher order terms in lowest approximation in y contribute

$$\propto f(x)g^3(x).$$

As (5.156) contains thus $g^3(x)$, and g is a small quantity, we find that the higher order terms give rise to a correction to the x-dependent part of the potential V of higher order. Of course, if the leading terms of the x-dependent part of the potential vanish, this term may be important. However, we can evidently devise an iteration procedure by which we repeat the procedure (5.148) to (5.155) with the higher order terms, each term leading to a correction term of decreasing importance. Such iteration procedure may be tedious, but we have convinced ourselves, at least in principle, that this procedure allows us to decompose V into an x and into a y (or $\tilde{y}$-) dependent part provided that we neglect terms of higher order within a well-defined procedure.

We now treat the problem quite generally without resorting to that iteration procedure. In the potential $V(x, y)$ we put

$$y = y^{(0)} + \tilde{y}. \tag{5.157}$$

We now require that $y^{(0)}$ is chosen so that

$$V(x, y^{(0)} + \tilde{y}) \tag{5.158}$$

has its minimum for $\tilde{y} = 0$, or, in other words, that

$$\left.\frac{\partial V}{\partial \tilde{y}}\right|_{\tilde{y}=0} = 0 \tag{5.159}$$

holds. This may be considered as an equation for $y^{(0)}$ which is of the form

$$W(x, y^{(0)}) = 0. \tag{5.160}$$

For any given x we may thus determine $y^{(0)}$ so that

$$y^{(0)} = y^{(0)}(x). \tag{5.161}$$

For $\tilde{y} \neq 0$ but small we may use the expansion

$$V(x, y^{(0)} + \tilde{y}) = \tilde{V}_1(x) + \tilde{y}^2\tilde{\mu}_2 + \tilde{y}^3\tilde{f}(x) + \cdots, \tag{5.162}$$

where the linear term is lacking on account of (5.159). By the above more explicit analysis (compare (5.155)) we have seen that in the neighborhood of $x = 0$ the potential retains its stability in y-direction. In other words, using the decomposition

$$\tilde{\mu}_2 = \mu_2 + \tilde{h}(x) \tag{5.163}$$

we may be sure that (5.163) remains positive. Thus the only instability we have to discuss is that inherent in $V_1(x)$ so that we are back with the one-dimensional case discussed above.

We now come to the *two-dimensional case*, $\mu_1 = 0$, $\mu_2 = 0$. The first, in general nonvanishing, terms of the potential are thus given by

$$V(x, y) = \tilde{c}_{111}^{(3)}x^3 + 3\tilde{c}_{112}^{(3)}x^2y + 3\tilde{c}_{122}^{(3)}xy^2 + \tilde{c}_{222}^{(3)}y^3 + \cdots \tag{5.164}$$

Provided that one or several of the coefficients $c^{(3)}$ are unequal zero, we may stop with the discussion of the form (5.164). Quite similar to the one-dimensional case we may assume that certain perturbations may become dangerous to the form (5.164). In general these are terms of a smaller power than 3. Thus the most general unfolding of (5.164) would be given by adding to (5.164) terms of the form

$$b_{11}x^2 + 2b_{12}xy + b_{22}y^2 + a_1x + a_2y, \tag{5.165}$$

(where we have already dropped the constant term).

We now describe qualitatively how we can cut down the formulas (5.164) and

(5.165) by a linear transformation of the form

$$x = A_{11}x_1 + A_{12}x_2 + B_1,$$
$$y = A_{21}x_1 + A_{22}x_2 + B_2. \tag{5.166}$$

We may cast (5.164) into simple normal forms quite similar to ellipses etc. This can be achieved by a proper choice of A's. By proper choice of B's we can further cast the quadratic or bilinear terms of (5.165) into a simpler form. If we take the most general form (5.164) and (5.165) including the constant term V_0, there are 10 constants. On the other hand the transformation (5.166) introduces 6 constants which together with $c^{(0)}$ yields 7 constants. Taking the total number of possible constants minus the seven, we still need three constants which can be attached to certain coefficients (5.165). With these considerations and after some analysis we obtain three basic forms which we denote according to Thom, in the following manner:

Hyperbolic umbilic

$$V = x_1^3 + x_2^3 + wx_1x_2 - ux_1 - vx_2, \tag{5.167}$$

Elliptic umbilic

$$V = x_1^3 - 3x_1x_2^2 + w(x_1^2 + x_2^2) - ux_1 - vx_2, \tag{5.168}$$

Parabolic umbilic

$$V = x_1^2x_2 + wx_2^2 + tx_2^3 - ux_1 - vx_2 + \tfrac{1}{4}(x_1^4 + x_2^4). \tag{5.169}$$

(umbilic = umbilicus = navel)

For the example of the parabolic umbilic a section through the potential curve is drawn in Fig. 5.25 for different values of the parameters u, v, t. We note that for $t < 0$ the parabolic umbilic goes over to the elliptic umbilic whereas for $t > 0$ we obtain the hyperbolic umbilic. Again for each set of u, v, w, t, the potential curve may exhibit different minima, one of which is the deepest (or several are simultaneously the deepest) representing a state of the system. Since one deepest minimum may be replaced by another one by changing the parameters u, v, w, t, the u,v,w,t-space is separated by different hyper-surfaces and thus decomposed into subspaces. A note should be added about the occurrence of quartic terms in (5.169) which seems to contradict our general considerations on page 134 where we stated that only powers smaller than that of the original singularity can be important. The reason is that the unfolding in (5.169) includes terms tx_2^3 which have the same power as the original terms, e.g., x_1^3 or $x_1^2x_2$. If we now seek the minimum of V,

$$\frac{\partial}{\partial x_2} V = 0 \tag{5.170}$$

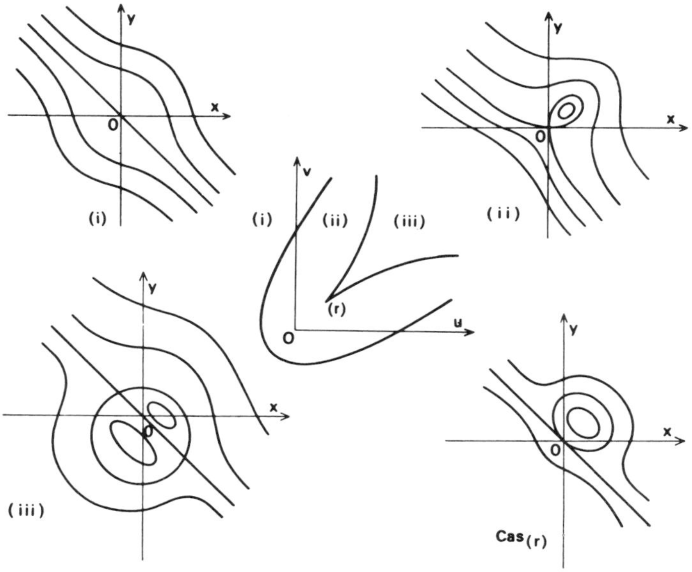

Fig. 5.25. The universal unfolding of the hyperbolic umbilic (at center) surrounded by the local potentials in regions 1, 2 and 3 and at the cusp C (after Thom)

and take $u = v = 0$ we have to solve the equation

$$x_1^2 + 2wx_2 + 3tx_2^2 = 0. \tag{5.171}$$

At least one solution of this quadratic equation (5.171), however, tends to infinity for $t \to 0$. This contradicts our original assumption that we are only considering the immediate vicinity of $x_1 = x_2 = 0$. The zeros of (5.171), however, are bounded to that minimum if we take quartic terms into account according to (5.169).

C) *The General n-Dimensional Case*

We assume from the very beginning that the potential function depending on the coordinate $q_1 \ldots q_n$ is expanded into a Taylor series

$$\begin{aligned} V(q_1, \ldots, q_n) = c^{(0)} + \textstyle\sum_j c_j^{(1)} q_j + \sum_{jj'} c_{jj'}^{(2)} q_j q_{j'} \\ + \textstyle\sum_{jj'j''} c_{jj'j''}^{(3)} q_j q_{j'} q_{j''} + \cdots, \end{aligned} \tag{5.172}$$

where the first coefficients are given by

$$c_j^{(1)} = \left.\frac{\partial V}{\partial q_j}\right|_0 \tag{5.173}$$

and

$$c_{jj'}^{(2)} = \frac{1}{2} \frac{\partial^2 V}{\partial q_j \partial q_{j'}}\bigg|_0 . \tag{5.174}$$

We assume that the minimum of V lies at $q_j = 0$, i.e.,

$$c_j^{(1)} = \frac{\partial V}{\partial q_j}\bigg|_{\boldsymbol{q}=0} = 0. \tag{5.175}$$

Because the (negative) derivative of V with respect to q_j gives us the forces $\boldsymbol{F}_j = -\partial V/\partial q_j$, the equilibrium is characterized by such a state where no force acts on the "particle." Again $c^{(0)}$ will be done away with and due to (5.175) the leading terms of (5.172) are now given by

$$\textstyle\sum_{jj'} c_{jj'}^{(2)} q_j q_{j'}, \tag{5.176}$$

where we may choose the c's always in a symmetric form

$$c_{jj'}^{(2)} = c_{j'j}^{(2)}. \tag{5.177}$$

This allows us to perform a principal-axis transformation

$$q_j = \sum A_{jk} u_k. \tag{5.178}$$

Linear algebra tells us that the resulting quadratic form

$$\textstyle\sum_j \mu_j u_j^2 \tag{5.179}$$

has only real values μ_j. Provided that all $\mu_j > 0$, the state $\boldsymbol{q} = 0$ is stable. We now assume that by a change of external parameters we reach a state where a certain set of the μ's vanishes. We number them in such a manner that they are the first, $j = 1, \ldots, k$, so that

$$\mu_1 = 0, \mu_2 = 0, \ldots, \mu_k = 0. \tag{5.180}$$

Thus we have now two groups of coordinates, namely, those associated with indices 1 to k which are coordinates in which the potential shows a critical behavior, while for $k + 1, \ldots, n$ the coordinates are those for which the potential remains uncritical. To simplify the notation we denote the coordinates so as to distinguish between these two sets by putting

$$\begin{aligned} u_1, \ldots, u_k &= x_1, \ldots, x_k, \\ u_{k+1}, \ldots, u_n &= y_1, \ldots, y_{n-k}. \end{aligned} \tag{5.181}$$

The potential we have to investigate is reduced to the form

$$V = \sum_{j=1}^{n-k} \mu_j y_j^2 + \sum_{j=1}^{n-k} y_j g_j(x_1, \ldots, x_k) + V_1(x_1, \ldots, x_k)$$
$$+ \sum_{j,j'}^{n-k} y_j y_{j'} h_{jj'}(x_1, \ldots, x_k) + \text{higher order in } y_1, \ldots \text{ with}$$
$$\text{coefficients still being functions of } x_1, \ldots, x_k \tag{5.182}$$

Our first goal is to get rid of the terms linear in y_j which can be achieved in the following way. We introduce new coordinates y_s by

$$y_s = y_s^{(0)} + \tilde{y}_s, \quad s = 1, \ldots, n. \tag{5.183}$$

The $y^{(0)}$'s are determined by the requirement that for $y_s = y_s^{(0)}$ the potential acquires a minimum

$$\left.\frac{\partial V}{\partial y_s}\right|_{y^{(0)}} = 0. \tag{5.184}$$

Expressed in the new coordinates $\tilde{y}_s$, V acquires the form

$$V = \tilde{V}(x_1, \ldots, x_k) + \sum_{jj'} \tilde{y}_j \tilde{y}_{j'} h_{jj'}(x_1, \ldots, x_k) + h.o. \text{ in } \tilde{y}$$
$$h_{jj'} = \delta_{jj'}\mu_j + \underset{\text{small}}{\hat{h}_{jj'}(x_1, \ldots, x_k)}. \tag{5.185}$$

where $h_{jj'}$ contains $\mu_j > 0$ stemming from the first term in (5.182) and further correction terms which depend on $x_1, \ldots, x_k$. This can be, of course, derived in still much more detail; but a glance at the two-dimensional case on pages 137–141 teaches us how the whole procedure works. (5.185) contains higher-order terms, i.e., higher than second order in the $\tilde{y}$'s. Confining ourselves to the stable region in the $\tilde{y}$-direction, we see that V is decomposed into a critical $\tilde{V}$ depending only on $x_1, \ldots, x_k$, and a second noncritical expression depending on y and x. Since all the critical behavior is contained in the first term depending on k coordinates $x_1, \ldots, x_k$, we have thus reduced the problem of instability to one which is of a dimension k, usually much smaller than n. This is the fundamental essence of our present discussion.

We conclude this chapter with a few general comments: If we treat, as everywhere in Chapter 5, completely deterministic processes, a system cannot jump from a minimum to another, deeper, minimum if a potential barrier lies between. We will come back to this question later on in Section 7.3. The usefulness of the above considerations lies mainly in the discussion of the impact of parameter changes on bifurcation. As will become evident later, the eigenvalues μ_j in (5.182) are identical with the (imaginary) frequencies occurring in linearized equations for the u's. Since $(c_{jj'}^{(2)})$ is a real, symmetric matrix, according to linear algebra the μ_j's are real. Therefore no oscillations occur and the critical modes are only *soft modes*.

6. Chance and Necessity

Reality Needs Both

6.1 Langevin Equations: An Example

Consider a football dribbled ahead over the grass by a football (soccer) player. Its velocity v changes due to two causes. The grass continuously slows the ball down by a friction force whereas the football player randomly increases the velocity of the ball by his kicks. The equation of motion of the football is precisely given by Newton's law: Mass · acceleration = force, i.e.,

$$m \cdot \dot{v} = F. \tag{6.1}$$

We determine the explicit form of F as follows:
We assume as usual in physics that the friction force is proportional to the velocity. We denote the friction constant by γ, so that the friction force reads $-\gamma v$. The minus sign takes into account that the friction force is opposite to the velocity of the particle. Now we consider the effects of the single kicks. Since a kick impulse lasts only a very short time, we represent the corresponding force by a δ-function of strength φ

$$\Phi_j = \varphi\delta(t - t_j), \tag{6.2}$$

where t_j is the moment a kick occurs. The effect of this kick on the change of velocity can be determined as follows: We insert (6.2) into (6.1):

$$m\dot{v} = \varphi\delta(t - t_j). \tag{6.3}$$

Integration over a short time interval around $t = t_j$ on both sides yields

$$\int_{t_j-0}^{t_j+0} m\dot{v}\, d\tau = \int_{t_j-0}^{t_j+0} \varphi\delta(t - t_j)\, d\tau. \tag{6.4}$$

Performing the integration, we obtain

$$mv(t_j + 0) - mv(t_j - 0) \equiv m\Delta v = \varphi, \tag{6.5}$$

which describes that at time t_j the velocity v is suddenly increased by the amount φ/m. The total force exerted by the kicker in the course of time is obtained by summing up (6.2) over the sequence j of pushes:

$$\Phi(t) = \varphi \sum_j \delta(t - t_j). \tag{6.6}$$

To come to realistic applications in physics and many other disciplines we have to change the whole consideration by just a minor point, which, by the way, happens rather often in football games. The impulses are not only exerted in one direction but randomly also in the reverse direction. Thus we replace Φ given by (6.6) by the function

$$\Psi(t) = \varphi \sum_j \delta(t - t_j)(\pm 1)_j \tag{6.7}$$

in which the sequence of plus and minus signs is a random sequence in the sense discussed earlier in this book with respect to tossing coins. Taking into account both the continuously acting friction force due to the grass *and* the random kicks of the football player, the total equation of motion of the football reads

$$m\dot{v} = -\gamma v + \Psi(t)$$

or, after dividing it by m

$$\dot{v} = -\alpha v + F(t), \tag{6.8}$$

where

$$\alpha = \gamma/m \tag{6.9a}$$

and

$$F(t) \equiv \frac{1}{m}\Psi = \frac{\varphi}{m}\sum_j \delta(t - t_j)(\pm 1)_j. \tag{6.9b}$$

How to perform the statistical average requires some thought. In one experiment, say a football game, the particle (ball) is moved at certain times t_j in a forward or backward direction so that during this experiment the particle follows a definite path. Compare Fig. 6.1 which shows the change of velocity during the time due to the impulses (abrupt changes) and due to friction force (continuous decreases inbetween the impulses). Now, in a second experiment, the times at which the particle is moved are different. The sequence of directions may be different so that another path arises (compare Fig. 6.2). Because the sequences of times and directions are random events, we cannot predict the single path but only averages. These averages over the different time sequences and directions of impulses will be performed below for several examples. We now imagine that we average F over the random sequence of plus and minus signs. Since they occur with equal probability, we immediately find

$$\langle F(t) \rangle = 0. \tag{6.10}$$

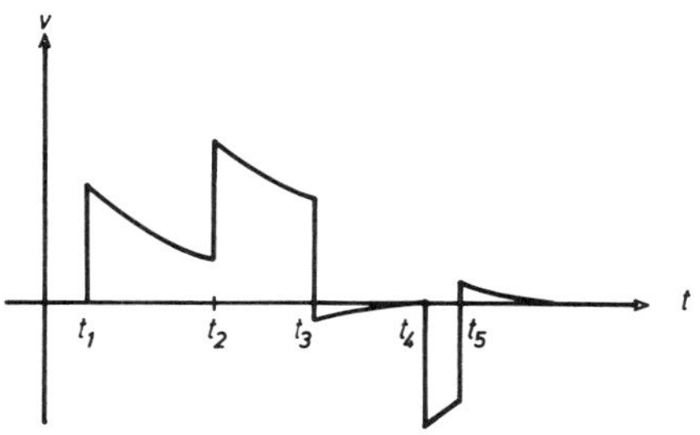

Fig. 6.1. The velocity v changes due to pushes (random force) and friction force

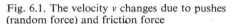

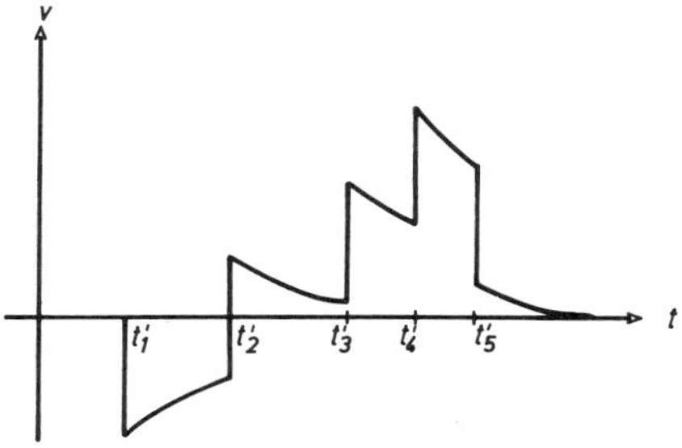

Fig. 6.2. Same as in Fig. 6.1 but with a different realization

We further form the product of F at a time t with F at another time t' and take the average over the times of the pushes and their directions. We leave the evaluation to the reader as an exercise (see end of this paragraph). Adopting a Poisson process (compare Sect. 2.12) we find the correlation function

$$\langle F(t)F(t')\rangle = \frac{\varphi^2}{m^2 t_0}\,\delta(t - t') = C\delta(t - t'). \tag{6.11}$$

Equation (6.8) together with (6.10) and (6.11) describes a physical process which is well known under the name of Brownian movement (Chapt. 4). Here a large particle immersed in a liquid is pushed around by the random action of particles of the liquid due to their thermal motion. The theory of Brownian movement plays a fundamental role not only in mechanics but in many other parts of physics and other disciplines, as we shall demonstrate later in this book (cf. Chapt. 8). The differential equation (6.8) can be solved immediately by the method of variation of the constant. The solution is given by

$$v(t) = \int_0^t e^{-\alpha(t-\tau)}F(\tau)\,d\tau + v(0)e^{-t\alpha} \tag{6.12}$$

In the following we neglect "switching on" effects. Therefore we neglect the last term in (6.12), which decays rapidly.

Let us now calculate the mean kinetic energy defined by

$$\frac{m}{2}\langle v^2\rangle. \tag{6.13}$$

Again the brackets denote averaging over all the pushes. Inserting (6.12) into (6.13) we obtain

$$\frac{m}{2}\left\langle \int_0^t e^{-\alpha(t-\tau)}F(\tau)\,d\tau \int_0^t e^{-\alpha(t-\tau')}F(\tau')\,d\tau' \right\rangle. \tag{6.14}$$

Since the averaging process over the pushes has nothing to do with the integration, we may exchange the sequence of averaging and integration and perform the average first. Exploiting (6.11), we find for (6.14) immediately

$$\frac{m}{2}\langle v^2\rangle = \frac{m}{2}\int_0^t\int_0^t d\tau d\tau'\, e^{-2\alpha t+\alpha(\tau+\tau')}\delta(\tau-\tau')C. \tag{6.15}$$

Due to the δ-function the double integration reduces to a single integral which can be immediately evaluated and yields

$$\frac{m}{2}\langle v^2\rangle = \frac{m}{2}C\frac{1}{2\alpha}(1 - e^{-2\alpha t}). \tag{6.16}$$

Since we want to consider the stationary state we consider large times, t, so that we can drop the exponential function in (6.16) leaving us with the result

$$\frac{m}{2}\langle v^2\rangle = \frac{m^2 C}{4\gamma} \quad \left(=\frac{mC}{4\alpha}\right). \tag{6.17}$$

Now, a fundamental point due to Einstein comes in. We may assume that the particle immersed in the liquid is in thermal equilibrium with its surroundings. Since the one-dimensional motion of the particle has one degree of freedom, according to the equipartition theorem of thermodynamics it must have the mean energy (1/2) $k_B T$, where k_B is Boltzmann's constant and T the absolute temperature. A comparison of the resulting equation

$$\frac{m}{2}\langle v^2\rangle = \tfrac{1}{2}k_B T \tag{6.18}$$

with (6.17) leads to the relation

$$C = \frac{2\gamma}{m^2}k_B T. \tag{6.19}$$

Recall that the constant C appears, in the correlation function of the fluctuating forces (6.11). Because C contains the damping constant γ, the correlation function is intrinsically connected with the damping or, in other words, with the dissipation of the system. (6.19) represents one of the simplest examples of a dissipation-fluctuation theorem: The size of the fluctuation ($\sim C$) is determined by the size of dissipation ($\sim \gamma$). The essential proportionality factor is the absolute temperature. The following section will derive (6.8), not merely by plausibility arguments but from first principles.

As we have discussed above we cannot predict a single path but only averages. One of the most important averages is the two-time correlation function

$$\langle v(t)v(t')\rangle \tag{6.20}$$

which is a measure how fast the velocity loses its memory or, in other words, if we have fixed the velocity at a certain time t', how long does it take the velocity to differ appreciably from its original value? To evaluate (6.20) we insert (6.12)

$$v(t) = \int_0^t e^{-\alpha(t-\tau)} F(\tau)\, d\tau \tag{6.21}$$

into (6.20) which yields

$$\int_0^t \int_0^{t'} e^{-\alpha(t-\tau)} e^{-\alpha(t'-\tau')} \langle F(\tau) F(\tau') \rangle \, d\tau \, d\tau'. \tag{6.22}$$

This reduces on account of (6.11) to

$$C \int_0^t \int_0^{t'} e^{-\alpha(t-\tau)-\alpha(t'-\tau')} \delta(\tau - \tau') \, d\tau \, d\tau'. \tag{6.23}$$

The integrals can be immediately evaluated. The result reads, for $t > t'$,

$$\frac{C}{2\alpha} \{ e^{-\alpha(t-t')} - e^{-\alpha(t+t')} \}. \tag{6.24}$$

If we consider a stationary process, we may assume t and t' large but $t - t'$ small. This leaves us with the final result

$$\langle v(t) v(t') \rangle = \frac{C}{2\alpha} e^{-\alpha|t-t'|}. \tag{6.25}$$

Thus the time T after which the velocity loses its memory is $T = 1/\alpha$. The case $\alpha = 0$ leads evidently to a divergence in (6.25). The fluctuations become very large or, in other words, we deal with critical fluctuations. Extending (6.20) we may define higher-order correlation functions containing several times by

$$\langle v(t_n) v(t_{n-1}) \cdots v(t_1) \rangle. \tag{6.26}$$

When we try to evaluate these correlation functions in the same manner as before, we have to insert (6.21) into (6.26). Evidently we must know the correlation functions

$$\langle F(\tau_n) F(\tau_{n-1}) \cdots F(\tau_1) \rangle. \tag{6.27}$$

The evaluation of (6.27) requires additional assumptions about the F's. In many practical cases the F's may be assumed to be Gaussian distributed (Sect. 4.4). In that case one finds

$$\begin{aligned} &\langle F(\tau_n) F(\tau_{n-1}) \cdots F(\tau_1) \rangle = 0 \text{ for } n \text{ odd}, \\ &\langle F(\tau_n) F(\tau_{n-1}) \cdots F(\tau_1) \rangle = \textstyle\sum_P \langle F(\tau_{\lambda_1}) F(\tau_{\lambda_2}) \rangle \cdots \langle F(\tau_{\lambda_{n-1}}) F(\tau_{\lambda_n}) \rangle \text{ for } n \text{ even}. \end{aligned} \tag{6.28}$$

where $\sum_P$ runs over all permutations $(\lambda_1, \ldots, \lambda_n)$ of $(1, \ldots, n)$.

Exercise on 6.1

1) Evaluate the lhs of (6.11) by assuming for the t_j's a Poisson process.

Hint: $\langle \delta(t - t_j)\delta(t' - t_j) \rangle = \frac{1}{t_0}\delta(t - t')$.

2) Eq. (6.28) provides an excellent example to show the significance of cumulants. Prove the following relations between moment functions $m_n(t_1, \ldots, t_n)$ and cumulants $k_n(t_1, \ldots, t_n)$ by use of (4.101) and (4.102).

$$\begin{aligned} m_1(t_1) &= k_1(t_1), \\ m_2(t_1, t_2) &= k_2(t_1, t_2) + k_1(t_1)k_1(t_2), \\ m_3(t_1, t_2, t_3) &= k_3(t_1, t_2, t_3) + 3\{k_1(t_1)k_2(t_2, t_3)\}_s + k_1(t_1)k_1(t_2)k_1(t_3), \\ m_4(t_1, t_2, t_3, t_4) &= k_4(t_1, t_2, t_3, t_4) + 3\{k_2(t_1, t_2)k_2(t_3, t_4)\}_s \\ &\quad + 4\{k_1(t_1)k_3(t_2, t_3, t_4)\}_s + 6\{k_1(t_1)k_1(t_2)k_2(t_3, t_4)\}_s \\ &\quad + k_1(t_1)k_1(t_2)k_1(t_3)k_1(t_4). \end{aligned} \tag{E.1}$$

$\{\cdots\}_s$ are symmetrized products.
These relations can be taken to rewrite (6.28) using cumulants. The resulting form is more simple and compact.

3) Derive (6.28) with help of (4.101) and (4.104), $k_1 = 0$.

6.2* Reservoirs and Random Forces

In the preceding chapter we have introduced the random force F and its properties as well as the friction force by plausibility arguments. We now want to show how both quantities can be derived consistently by a detailed physical model. We make the derivation in a manner which shows how the whole procedure can be extended to more general cases, not necessarily restricted to applications in physics. Instead of a free particle we consider a harmonic oscillator, i.e., a point mass fixed to a spring. The elongation of the point mass from its equilibrium position is called q. Denoting Hooke's constant by k, the equation of the harmonic oscillator reads ($v = \dot{q}$)

$$m\ddot{q} = -kq. \tag{6.29}$$

For what follows we introduce the abbreviation

$$\frac{k}{m} = \omega_0^2, \tag{6.30}$$

where ω_0 is the frequency of the oscillator. The equation of motion (6.29) then reads

$$\ddot{q} = -\omega_0^2 q. \tag{6.31}$$

Equation (6.31) can be replaced by a set of two other equations if we put, at first,

$$\dot{q} = p \tag{6.32}$$

and then replace $\dot{q}$ by p in (6.31) which gives

$$\dot{p} = -\omega_0^2 q \tag{6.33}$$

(compare also Sect. 5.2). It is now our goal to transform the pair of equations (6.32) and (6.33) into a pair of equations which are complex conjugate to each other. To this end we introduce the new variable $b(t)$ and its conjugate complex $b^*(t)$, which shall be connected with p and q by the relations

$$\frac{1}{\sqrt{2}}(\sqrt{\omega_0}q + ip/\sqrt{\omega_0}) = b \tag{6.34}$$

and

$$\frac{1}{\sqrt{2}}(\sqrt{\omega_0}q - ip/\sqrt{\omega_0}) = b^*. \tag{6.35}$$

Multiplying (6.32) by $\sqrt{\omega_0}$ and (6.33) by $i/\sqrt{\omega_0}$ and adding the resulting equations we find after an elementary step

$$b^{\cdot} = -i\omega_0 b, \quad \text{where } b^{\cdot} \equiv db/dt. \tag{6.36}$$

Similarly the subtraction of (6.33) from (6.32) yields an equation which is just the conjugate complex of (6.36).

After these preliminary steps we return to our original task, namely, to set up a physically realistic model which eventually leads us to dissipation and fluctuation. The reason why we hesitate to introduce the damping force $-\gamma v$ from the very beginning is the following. All fundamental (microscopic) equations for the motion of particles are time reversal invariant, i.e. the motion is completely reversible. Originally there is no place in these equations for a friction force which violates time reversal invariance. Therefore, we want to start from equations which are the usual mechanical equations having time reversal invariance. As we have mentioned above in connection with Brownian movement the large particle interacts with (many) other particles in the liquid. These particles act as a "reservoir" or "heat-bath"; they maintain the mean kinetic energy of the large particle at $1/2\ k_B T$ (per one degree of freedom). In our model we want to mimic the effect of the "small" particles by a set of very many harmonic oscillators, acting on the "large" particle, which we treat as an oscillator by (6.36) and its conjugate complex. We assume that the reservoir oscillators have different frequencies ω out of a range $\Delta\omega$ (also called bandwidth). In analogy to the description (6.34) and (6.35), we use complex amplitudes B and B^* for the reservoir oscillators, which we distinguish by an index ω. In our model we describe the joint action of the B's on the "large" oscillator as a sum over the individual B's and assume that each contributes linearly. (This assumption amounts to a linear coupling between oscillators). For these reasons we obtain as our starting equation

$$b^{\cdot} = -i\omega_0 b + i\sum_\omega g_\omega B_\omega. \tag{6.37}$$

The coefficients g_ω describe the strength of the coupling between the other oscillators and the one under consideration. Now the "big" oscillator reacts on all the other oscillators which is described by

$$\dot{B}_\omega = -i\omega B_\omega + ibg_\omega. \tag{6.38}$$

(Readers interested in how to obtain (6.37) and (6.38) in the usual framework of mechanics are referred to the exercises). The solution of (6.38) consists of two parts, namely, the solution of the homogeneous equation (where $ibg_\omega = 0$) and a particular solution of the inhomogeneous equation. One readily verifies that the solution reads

$$B_\omega(t) = \mathrm{e}^{-i\omega t}B_\omega(0) + i\int_0^t \mathrm{e}^{-i\omega(t-\tau)}b(\tau)g_\omega\, d\tau, \tag{6.39}$$

where $B_\omega(0)$ is the initial value, at $t = 0$, of the oscillator amplitude B_ω. Inserting (6.39) into (6.37) we find an equation for b

$$b^{\cdot}(t) = -i\omega_0 b(t) - \int_0^t \sum_\omega g_\omega^2\, \mathrm{e}^{-i\omega(t-\tau)}b(\tau)\, d\tau + i\sum_\omega g_\omega \mathrm{e}^{-i\omega t}B_\omega(0). \tag{6.40}$$

The only reminiscence from the B's stems from the last term in (6.40). For further discussion, we eliminate the term $-i\omega_0 b(t)$ by introducing the new variable $\tilde{b}$,

$$b = \tilde{b}\mathrm{e}^{-i\omega_0 t}. \tag{6.41}$$

Using the abbreviation

$$\tilde{\omega} = \omega - \omega_0 \tag{6.42}$$

we may write (6.40) in the form

$$\tilde{b}^{\cdot} = -\int_0^t \sum_\omega g_\omega^2\, \mathrm{e}^{-i\tilde{\omega}(t-\tau)}\tilde{b}(\tau)\mathrm{d}\tau + i\sum_\omega g_\omega\, \mathrm{e}^{-i\tilde{\omega}t}B_\omega(0). \tag{6.43}$$

To illustrate the significance of this equation let us identify, for the moment, b with the velocity v which occurred in (6.8). The integral contains b linearly which suggests that a connection with a damping term $-\gamma b$ might exist. Similarly the last term in (6.43) is a given function of time which we want to identify with a random force F. How do we get from (6.43) to the form of (6.8)? Apparently the friction force in (6.43) not only depends on time, t, but also on previous times τ. Under which circumstances does this memory vanish? To this end we consider a transition from discrete values ω to continuously varying values, i.e., we replace the sum over ω by an integral

$$\sum_\omega g_\omega^2 e^{-i\tilde{\omega}(t-\tau)} \approx \int_{-\Delta\omega/2}^{+\Delta\omega/2} g_{\omega_0+\tilde{\omega}}^2 e^{-i\tilde{\omega}(t-\tau)}\, d\tilde{\omega}. \tag{6.44}$$

For time differences $t - \tau$ not too short, the exponential function oscillates rapidly for $\tilde{\omega} \neq 0$ so that only contributions of $\tilde{\omega} \approx 0$ to the integral are important. Now assume that the coupling coefficients g vary in the neighborhood of ω_0 (i.e., $\tilde{\omega} = 0$) only slightly. Because only small values of $\tilde{\omega}$ are important, we may extend the boundaries of the integral to infinity. This allows us to evaluate the integral (6.44)

$$(6.44) = 2\pi g^2 \delta(t - \tau), \quad (g = g_{\omega_0}). \tag{6.45}$$

We now insert (6.45) into the integral over the time occurring in (6.43). On account of the δ-function we may put $\tilde{b}(\tau) = \tilde{b}(t)$ and put it in front of the integral. Furthermore, we observe that the δ-function contributes to the integral only with the factor 1/2 because the integration runs only to $\tau = t$ and not further. Since the δ-function is intrinsically a symmetric function, only 1/2 of the δ-function is covered. We thus obtain

$$\int_0^t 2\pi g^2 \delta(t - \tau)\, d\tau = \pi g^2 = \varkappa \tag{6.46}$$

where $\varkappa$ has been introduced as an abbreviation. The last term in (6.43) will now be abbreviated by $\tilde{F}$

$$\tilde{F} = e^{i\omega_0 t} \underbrace{i \sum_\omega g_\omega e^{-i\omega t} B_\omega(0)}_{F}. \tag{6.47}$$

By this intermediate steps we may rewrite our original (6.43) in the form

$$\dot{\tilde{b}} = -\varkappa\tilde{b} + \tilde{F}(t). \tag{6.48}$$

When returning from $\tilde{b}$ to b (cf. (6.41)) we find as fundamental equation

$$\dot{b} = -i\omega b - \varkappa b + F(t). \tag{6.49}$$

The result looks very fine because (6.49) has exactly the form we have been searching. There are, however, a few points which need a careful discussion. In the model of the foregoing Section 6.1, we assumed that the force F is random. How does the randomness come into our present model? A look at (6.47) reveals that the only quantities which can introduce any randomness are the initial values of B_ω. Thus we adopt the following attitude (which closely follows information theory). The precise initial values of the reservoir oscillators are not known except in a statistical sense; that means we know, for instance, their distribution functions. Therefore we will use only some statistical properties of the reservoirs at the initial time. First we may assume that the average over the amplitudes B_ω vanishes because otherwise the nonvanishing part could always be subtracted as a deterministic force. We furthermore assume that the B_ω's are Gaussian distributed, which can be motivated in different manners. If we consider physics, we may assume that the reservoir oscillators are all kept at thermal equilibrium. Then the probability distribution is given by the Boltzmann distribution (3.71) $P_\omega = \mathcal{N}_\omega e^{-E_\omega/k_B T}$ where E_ω is the energy of oscillator ω and $\mathcal{N}_\omega \equiv Z_\omega^{-1}$ is the normalization factor. We assume, as usual for harmonic oscillators, that the energy is proportional to $p^2 + q^2$ (with appropriate factors), or in our formalism, proportional to $B_\omega B_\omega^*$, i.e., $E_\omega = c_\omega B_\omega^* B_\omega$. Then we immediately find this announced Gaussian distribution for $B, B^*, f(B, B^*) = \mathcal{N}_\omega \exp(-c_\omega B_\omega^* B_\omega / k_B T)$. Unfortunately there is no unique relation between frequency and energy, which would allow us to determine c_ω. (This gap can be filled only by quantum theory).

Let us now investigate if the force (6.47) has the properties (6.10), (6.11) we expect according to the preceding section. Because we constructed the forces that time in a completely different manner out of individual impulses, it is by no means obvious that the forces (6.47) fulfil relations of the form (6.11). But let us check it. We form

$$\langle \tilde{F}^*(t)\tilde{F}(t')\rangle. \tag{6.50}$$

(Note that the forces are now complex quantities). Inserting (6.47) into (6.50) and taking the average yields

$$\sum_{\tilde{\omega}} \sum_{\tilde{\omega}'} g_{\tilde{\omega}+\omega_0} g_{\tilde{\omega}'+\omega_0} e^{i\tilde{\omega}t - i\tilde{\omega}'t'} \langle B_{\tilde{\omega}+\omega_0}^*(0) B_{\tilde{\omega}'+\omega_0}(0)\rangle. \tag{6.51}$$

We assume that initially the B's have been uncorrelated, i.e.,

$$\langle B_\omega^*(0) B_{\omega'}(0)\rangle = N_\omega \delta_{\omega\omega'}, \tag{6.52}$$

where we have abbreviated $\langle B_\omega^*(0) B_\omega(0)\rangle$ by N_ω. Thus (6.51) reduces to

$$\sum_{\tilde{\omega}} g_{\tilde{\omega}+\omega_0}^2 e^{i\tilde{\omega}(t-t')} \langle B_{\tilde{\omega}+\omega_0}^*(0) B_{\tilde{\omega}+\omega_0}(0)\rangle. \tag{6.53}$$

The sum occurring in (6.53) is strongly reminiscent of that of the lhs of (6.44) with the only difference that now an additional factor, namely, the average $\langle \cdots \rangle$ occurs. Evaluating (6.53) by exactly the same considerations which have led us from (6.44) to (6.45), we now obtain

$$\int_{-\infty}^{+\infty} g_{\omega_0}^2 \, d\tilde{\omega} \, e^{i\tilde{\omega}(t-t')} N_{\omega_0} = g_{\omega_0}^2 N_{\omega_0} 2\pi\delta(t - t'). \tag{6.54}$$

Using this final result (6.54) instead of (6.51) we obtain the desired correlation function as

$$\langle \tilde{F}^*(t)\tilde{F}(t')\rangle = 2\varkappa N_{\omega_0} \delta(t - t'). \tag{6.55}$$

This relation may be supplemented by

$$\langle F^*(t)F^*(t')\rangle = \langle F(t)F(t')\rangle = 0, \tag{6.56}$$

if we make the corresponding assumptions about the original average values of $B_\omega^* B_\omega^*$ or $B_\omega B_\omega$. How to determine the constant N_{ω_0}? Adopting the thermal distribution function $\propto \exp(-c_\omega B_\omega^* B_\omega / k_B T)$ it is clear that N_{ω_0} must be proportional to $k_B T$. However, the proportionality factor remains open or could be fixed indirectly by Einstein's requirement (Sect. 6.1). (In quantum theory there is no difficulty at all. We can just identify N_{ω_0} with the number of thermal quanta of the oscillator ω_0).

In our above treatment a number of problems have been swept under the carpet. First of all, we have converted a set of equations which have complete time-reversal invariance into an equation violating this principle. The reason lies in passing from the sum in (6.44) to the integral and in approximating it by (6.45). What happens in practice is the following: First the heatbath amplitudes B_ω are out of phase, thus leading to a quick decay of (6.44) considered as a function of the time difference $t - \tau$. However, there remains some parts of the integrand of (6.40) which are very small but nevertheless decisive in maintaining reversibility jointly with the impact of the fluctuating forces (6.47).

Another problem which has been swept under the carpet is the question of why in (6.54) N occurs with the index ω_0. This was, of course, due to the fact that we have evaluated (6.53) in the neighborhood of $\tilde{\omega} = 0$ so that only terms with index ω_0 survive. We could have done exactly the same procedure not by starting from (6.47) but by starting from F alone. This then had led us to a result similar to (6.53), however, with $\omega_0 = 0$. The idea behind our approach is this: if we solve the original full equation (6.40) taking into account memory effects, then the eigenfrequency ω_0 occurs in an interation procedure picking up every time only those contributions ω which are close to resonance with ω_0. It must be stressed, however, that this procedure requires a good deal of additional thought and has not yet been performed in the literature, at least to our knowledge.

In conclusion we may state that we were, to some extent, able to derive the Langevin equation from first principles, however, with some additional assumptions which go beyond a purely mechanical treatment. Adopting now this formalism, we may generalize (6.49) to that of an oscillator coupled to reservoirs at different temperatures and giving rise to *different* damping constants $\varkappa_j$. A completely analogous treatment as above yields

$$\dot{b} = -i\omega_0 b - (\varkappa_1 + \varkappa_2 + \cdots + \varkappa_n)b + F_1 + F_2 + \cdots + F_n. \tag{6.57}$$

The correlation functions are given by

$$\langle F_j^*(t)F_k(t')\rangle = \delta_{jk} 2\varkappa_j N_j \delta(t - t'). \tag{6.58}$$

Adopting from quantum theory that at elevated temperatures

$$N_j \approx \frac{k_B T}{\hbar\omega_0} \tag{6.59}$$

we may show that

$$\hbar\omega_0\langle b^*b\rangle = \frac{k_B(\varkappa_1 T_1 + \cdots + \varkappa_n T_n)}{\varkappa_1 + \varkappa_2 + \cdots + \varkappa_n}, \tag{6.60}$$

i.e., instead of one temperature T we have now an average temperature

$$T = \frac{\varkappa_1 T_1 + \cdots + \varkappa_n T_n}{\varkappa_1 + \cdots + \varkappa_n}. \tag{6.61}$$

It is most important that the average temperature depends on the strength of the dissipation ($\sim \varkappa_j$) caused by the individual heatbath. Later on we will learn about problems in which coupling of a system to reservoirs at different temperatures indeed occurs. According to (6.59) the fluctuations should vanish when the temperature approaches absolute zero. However, from quantum theory it is known that the quantum fluctuations are important, so that a satisfactory derivation of the fluctuating forces must include quantum theory.

Exercises on 6.2

1) Derive (6.37), (6.38) from the Hamiltonian

$$H = \omega_0 b^*b + \textstyle\sum_\omega \omega B_\omega^* B_\omega - \sum_\omega g_\omega(bB_\omega^* + b^*B_\omega),$$
(g_ω real)

by means of the Hamiltonian equations

$$\dot{b} = -i\partial H/\partial b^*, \quad \dot{B} = -i\partial H/\partial B^*$$
(and their complex conjugates).

2) Perform the integration in (6.23) for the two different cases: $t > t'$ and $t < t'$. For large values t and t', one may confirm (6.25).

6.3 The Fokker-Planck Equation

We first consider

A) *Completely Deterministic Motion*

and treat the equation

$$\dot{q}(t) = K(q(t)), \tag{6.62}$$

which may be interpreted as usual in our book as the equation of overdamped motion of a particle under the force K. Since in this chapter we want to derive equations capable of describing both deterministic and random processes we try to treat the motion of the particle in a formalism which uses probability theory. In the course of time the particle proceeds along a path in the q-t plane. If we pick out a fixed time t, we can ask for the probability of finding the particle at a certain coordinate q. This probability is evidently zero if $q \neq q(t)$ where $q(t)$ is the solution of (6.62). What kind of probability function yields 1 if $q = q(t)$ and $= 0$ otherwise? This is achieved by introducing a probability density equal to the δ-function (compare Fig. 6.3).

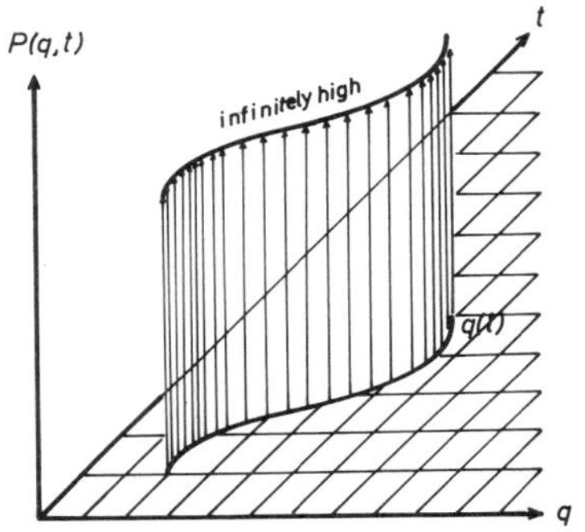

Fig. 6.3. Example of an infinitely peaked "probability" distribution

$$P(q, t) = \delta(q - q(t)). \tag{6.63}$$

Indeed we have seen earlier in Section 2.5, Exercise 1 that an integral over the function $\delta(q - q_0)$ vanishes, if the integration interval does not contain q_0, and that it yields 1 if that interval contains a surrounding of q_0:

$$\begin{aligned} \int_{q_0-\varepsilon}^{q_0+\varepsilon} \delta(q - q_0)\, dq &= 1 \\ &= 0 \text{ otherwise.} \end{aligned} \tag{6.64}$$

It is now our goal to derive an equation for this probability distribution P. To this end we differentiate P with respect to time. Since on the rhs the time dependence is inherent in $q(t)$, we differentiate first the δ-function with respect to $q(t)$, and then, using the chain rule, we multiply by $\dot{q}$

$$\dot{P}(q, t) = \frac{d}{dq(t)}\delta(q - q(t))\dot{q}(t). \tag{6.65}$$

The derivative of the δ-function with respect to $q(t)$ can be rewritten as

$$-\frac{d}{dq}\delta(q - q(t))\dot{q}(t). \tag{6.66}$$

Now we make use of the equation of motion (6.62) replacing $\dot{q}(t)$ by K. This yields the final formula

$$\dot{P}(q, t) = -\frac{d}{dq}(K(q)P). \tag{6.67}$$

It must be observed that the differentiation of q now implies the differentiation of the *product* of P *and* K. We refer the reader for a proof of this statement to the end of this chapter. Our result (6.67) can be readily generalized to a system of differential equations

$$\dot{q}_i = K_i(\boldsymbol{q}), \quad i = 1, \ldots, n, \tag{6.68}$$

where $\boldsymbol{q} = (q_1, \ldots, q_n)$. Instead of (6.68) we use the vector equation

$$\dot{\boldsymbol{q}} = \boldsymbol{K}(\boldsymbol{q}). \tag{6.69}$$

At time t the state of the total system is described by a point $q_1 = q_1(t)$, $q_2 = q_2(t) \ldots q_n = q_n(t)$ in the space of the variables $q_1 \ldots q_n$. Thus the obvious generalization of (6.63) to many variables is given by

$$P = \delta(q_1 - q_1(t))\delta(q_2 - q_2(t)) \ldots \delta(q_n - q_n(t)) \equiv \delta(\boldsymbol{q} - \boldsymbol{q}(t)), \tag{6.70}$$

where the last identity just serves a definition of the δ-function of a vector. We again take the time derivative of P. Because the time t occurs in each factor, we obtain a sum of derivatives

$$\dot{P}(\boldsymbol{q}, t) = -\frac{d}{dq_1}P\cdot\dot{q}_1(t) - \frac{d}{dq_2}P\cdot\dot{q}_2\cdots - \frac{d}{dq_n}P\dot{q}_n \tag{6.71}$$

which can be transformed, in analogy to (6.67), into

$$\dot{P}(\boldsymbol{q}, t) = -\frac{d}{dq_1}PK_1(\boldsymbol{q}) \cdots - \frac{d}{dq_n}PK_n(\boldsymbol{q}). \tag{6.72}$$

Writing the derivatives as an n-dimensional ∇ operator, (6.72) can be given the elegant form

$$\dot{P}(\boldsymbol{q}, t) = -\nabla_{\boldsymbol{q}}(P\boldsymbol{K}), \tag{6.73}$$

which permits a very simple interpretation by invoking fluid dynamics. If we identify P with a density in $\boldsymbol{q}$-space, the left hand side describes the temporal change of this density P, whereas the vector $\boldsymbol{K}P$ can be interpreted as the (probability) flux. $\boldsymbol{K}$ is the velocity in $\boldsymbol{q}$-space. (6.73) has thus the form of a continuity equation.

B) *Derivation of the Fokker-Planck Equation, One-Dimensional Motion*

We now combine what we have learned in Section 4.3 about Brownian movement and in the foregoing about a formal description of particle motion by means of a probability distribution. Let us again consider the example of a football which follows several different paths during several games. The probability distribution for a given path, 1, is

$$P_1(q, t) = \delta(q - q_1(t)) \tag{6.74}$$

for a given other path, 2,

$$P_2(q, t) = \delta(q - q_2(t)) \tag{6.75}$$

and so on. Now we take the average over all these paths, introducing the function

$$f(q, t) = \langle P(q, t) \rangle. \tag{6.76}$$

If the probability of the occurrence of a path i is p_i, this probability distribution can be written in the form

$$f(q, t) = \sum_i p_i \delta(q - q_i(t)) \tag{6.77}$$

or, by use of (6.74), (6.75), (6.76)

$$f(q, t) = \langle \delta(q - q(t)) \rangle. \tag{6.78}$$

$f dq$ gives us the probability of finding the particle at position q in the interval dq at time t. Of course, it would be a very tedious task to evaluate (6.77) which would require that we introduce a probability distribution of the pushes during the total course of time. This can be avoided, however, by deriving directly a differential equation for f. To this end we investigate the change of f in a time interval Δt

$$\Delta f(q, t) \equiv f(q, t + \Delta t) - f(q, t) \tag{6.79}$$

which by use of (6.78) takes the form

$$\Delta f(q, t) = \langle \delta(q - q(t + \Delta t)) \rangle - \langle \delta(q - q(t)) \rangle. \tag{6.80}$$

We put

$$q(t + \Delta t) = q(t) + \Delta q(t) \tag{6.81}$$

and expand the δ-function with respect to powers of Δq. We now have in mind that the motion of q is not determined by the deterministic equation (6.62) but rather by the Langevin equation of Section 6.1. As we shall see a little later, this new situation requires that we expand up to powers quadratic in Δq. This expansion thus yields

$$\Delta f(q, t) = \left\langle \left(-\frac{d}{dq} \delta(q - q(t)) \right) \Delta q(t) \right\rangle + \frac{1}{2} \left\langle \frac{d^2}{dq^2} \delta(q - q(t)) (\Delta q(t))^2 \right\rangle. \tag{6.82}$$

By means of the Langevin equation

$$\dot{q}(t) = -\gamma q(t) + F(t) \tag{6.83}$$

we find Δq by integration over the time interval Δt. In this integration we assume that q has changed very little but that many pushes have already occurred. We thus obtain by integrating (6.83)

$$\int_t^{t+\Delta t} \dot{q}(t')\, dt' = q(t + \Delta t) - q(t) \equiv \Delta q$$
$$= -\int_t^{t+\Delta t} \gamma q(t')\, dt' + \int_t^{t+\Delta t} F(t')\, dt' = -\gamma q(t)\Delta t + \Delta F(t). \tag{6.84}$$

We evaluate the first term on the right hand side of (6.82)

$$\left\langle \frac{d}{dq} \delta(q - q(t))\Delta q(t) \right\rangle. \tag{6.85}$$

Inserting the rhs of (6.84) into (6.85) yields

$$\frac{d}{dq}\{\langle \delta(q - q(t))(-\gamma q(t)\Delta t)\rangle + \langle \delta(q - q(t))\rangle\langle \Delta F\rangle\}. \tag{6.86}$$

The splitting of the average containing ΔF in the product of two averages requires a comment: ΔF contains all pushes which have occurred after the time t whereas $q(t)$ is determined by all pushes prior to this time. Due to the independence of the pushes, we may split the total average into the product of the averages as written down in (6.86). Since the average of F vanishes and so does that of ΔF, (6.86) reduces to

$$-\gamma\Delta t \frac{d}{dq}\{\langle \delta(q - q(t))q\rangle\}. \tag{6.87}$$

Note that we have replaced $q(t)$ by q, exactly as in (6.67).

We now come to the evaluation of the term

$$\left\langle \frac{d^2}{dq^2} \delta(q - q(t))(\Delta q(t))^2 \right\rangle \tag{6.88}$$

which using the same arguments as just now can be split into

$$\frac{d^2}{dq^2} \langle \delta(q - q(t))\rangle\langle \Delta q(t)^2\rangle. \tag{6.89}$$

When inserting Δq in the second part using (6.84) we find terms containing Δt^2, terms containing $\Delta t \Delta F$, and $(\Delta F)^2$. We will show that $\langle (\Delta F)^2\rangle$ goes with Δt. Since the average of ΔF vanishes, $\langle (\Delta F)^2\rangle$ is the only contribution to (6.89) which is linear in Δt. We evaluate

$$\langle \Delta F^2 \rangle \equiv$$

$$\langle \Delta F(t) \Delta F(t) \rangle = \int_t^{t+\Delta t} \int_t^{t+\Delta t} dt'\, dt''\, \langle F(t') F(t'') \rangle. \tag{6.90}$$

We assume that the correlation function between the F's is δ-correlated

$$\langle F(t) F(t') \rangle = Q \delta(t - t') \tag{6.91}$$

which permits the immediate evaluation of (6.90) yielding

$$Q \Delta t. \tag{6.92}$$

Thus we have finally found (6.88) in the form

$$\frac{d^2}{dq^2} \langle \delta(q - q(t)) \rangle Q \Delta t. \tag{6.93}$$

We now divide the original equation (6.82) by Δt and find with the results of (6.87) and (6.93)

$$\frac{df}{dt} = \frac{d}{dq}(\gamma q f) + \tfrac{1}{2} Q \frac{d^2}{dq^2} f \tag{6.94}$$

after the limit $\Delta t \to 0$ has been taken. This equation is the so-called *Fokker-Planck equation* which describes the change of the probability distribution of a particle during the course of time (cf. Fig. 6.4). $K = -\gamma q$ is called *drift-coefficient*, while Q is known as *diffusion coefficient*. Exactly the same method can be applied to the general case of many variables and to arbitrary forces $K_i(\boldsymbol{q})$, i.e., not necessarily simple friction forces. If we assume the corresponding Langevin equation in the

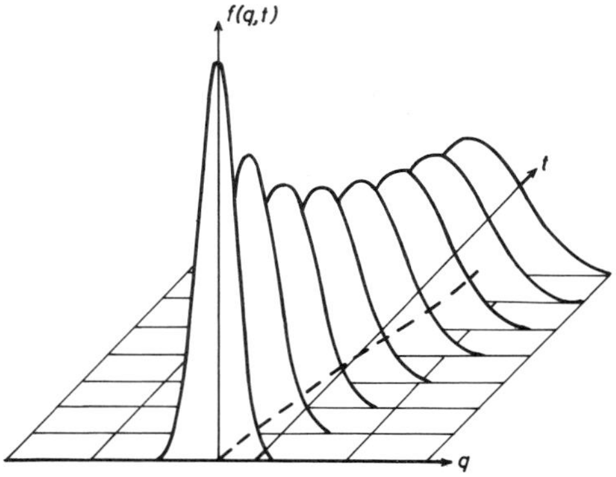

Fig. 6.4. Example of $f(q, t)$ as function of time t and variable q. Dashed line: most probable path

form

$$\dot{q}_i = K_i(\boldsymbol{q}) + F_i(t) \tag{6.95}$$

and δ-correlated fluctuating forces F

$$\langle F_i(t)F_j(t')\rangle = Q_{ij}\delta(t - t') \tag{6.96}$$

we may derive for the distribution function f in $\boldsymbol{q}$-space

$$f(q_1, \ldots, q_n; t) \equiv f(\boldsymbol{q}; t) \tag{6.97}$$

the following Fokker-Planck equation

$$f^{\cdot} = -\nabla_{\boldsymbol{q}}\{\boldsymbol{K}f\} + \tfrac{1}{2}\sum_{ij} Q_{ij}\frac{\partial^2}{\partial q_i \partial q_j}f, \quad \text{where } f^{\cdot} \equiv \frac{\partial f}{\partial t}. \tag{6.98}$$

In our above treatment we included derivatives of the δ-function up to second order. A detailed treatment shows that, in general, higher derivatives also yield contributions $\propto \Delta t$. An important exception is made if the fluctuating forces are Gaussian (cf. Sect. 4.4). In this case, the Fokker-Planck equations (6.94) and (6.98) are exact.

We want to prove that in (6.67) the differentiation with respect to q must also involve $K(q)$. To prove this we form the expression

$$\int_{q=q(t)-\varepsilon}^{q=q(t)+\varepsilon} h(q)\frac{d}{dq}\delta(q - q(t))K(q(t))\,dq \tag{6.99}$$

which is obtained from the rhs of (6.66) by replacing $\dot{q}$ by means of (6.62) and by multiplication with an arbitrary function $h(q)$. Note that this procedure must always be applied if a δ-function appears in a differential equation in a form like (6.65). Partial integration of (6.99) leads to

$$-\int_{q=q(t)-\varepsilon}^{q(t)+\varepsilon} h'(q)\delta(q - q(t))K(q(t))\,dq, \tag{6.100}$$

where the δ-function can now be evaluated. On the other hand we end up with the same result (6.100) if we start right away from

$$\int h(q)\frac{d}{dq}\{\delta(q - q(t))K(q)\}\,dq, \tag{6.101}$$

where the coordinate $q(t)$ in K is now replaced by q.

Exercises on 6.3

1) In classical mechanics the coordinates $q_j(t)$ and momenta $p_j(t)$ of particles obey

the Hamiltonian equations of motion

$$\dot{q}_j(t) = \partial H/\partial p_j,\ \dot{p}_j(t) = -\partial H/\partial q_j, j = 1, \ldots, n$$

where the Hamiltonian function H depends on all q_j's and p_j's: $H = H(\boldsymbol{q}, \boldsymbol{p})$. We define a distribution function $f(\boldsymbol{q}, \boldsymbol{p}; t)$ by $f = \delta(\boldsymbol{q} - \boldsymbol{q}(t))\delta(\boldsymbol{p} - \boldsymbol{p}(t))$. Show that f obeys the so-called Liouville equation

$$f^{\cdot} = \sum_j\left(\frac{\partial H}{\partial q_j}\frac{\partial}{\partial p_j} - \frac{\partial H}{\partial p_j}\frac{\partial}{\partial q_j}\right)f. \tag{E.1}$$

Hint: Repeat the steps (6.63)–(6.67) for q_j and p_j.

2) Functions f satisfying the Liouville equation E.1 with $f^{\cdot} = 0$ are called constants of motion.
Show a) $g = H(\boldsymbol{q}, \boldsymbol{p})$ is a constant of motion;
b) if $h_1(\boldsymbol{q}, \boldsymbol{p})$ and $h_2(\boldsymbol{q}, \boldsymbol{p})$ are constants of motion then also $h_1 + h_2$ and $h_1 \cdot h_2$ are constants of motion;
c) if $h_1, \ldots, h_l$ are such constants, then any function $G(h_1, \ldots, h_l)$ is also a constant of motion.
Hint: a) insert g into (E.1)
b) use the product rule of differentiation
c) use the chain rule.

3) Show by generalizing 2) that $f(g_1, \ldots, g_l)$ is a solution of (E.1), if the g_k's are solutions, $g_k = g_k(\boldsymbol{q}, \boldsymbol{p}; t)$, of (E.1).

4) Verify that the information entropy (3.42) satisfies (E.1) provided the following identifications are made:

$$\left.\begin{matrix} q_j \\ p_j \end{matrix}\right\} \rightarrow \text{index } i \text{ (value of "random variable")}$$

$$f(\boldsymbol{q}, \boldsymbol{p}) \rightarrow p_i.$$

Replace $\sum$ in (3.42) now by an integral $\int \cdots d^n p\, d^n q$. Why does also the coarse-grained information entropy fulfil (E.1)?

5) Verify that (3.42) with (3.48) is solution of (E.1) provided the f_k's are constants of motion (cf. exercise 2).

6.4 Some Properties and Stationary Solutions of the Fokker-Planck Equation

In this section we show how to find time-independent solutions of several types of Fokker-Planck equations which are often met in practical applications. We confine the following considerations to q-independent diffusion coefficients, Q_{jk}.

A) *The Fokker-Planck Equation as Continuity Equation*

1) One-dimensional example:

We write the one-dimensional Fokker-Planck equation (6.94) in the form

$$f^{\cdot} + \frac{d}{dq}\left(Kf - \tfrac{1}{2}Q\frac{df}{dq}\right) = 0, \quad K = K(q), \quad f = f(q, t). \tag{6.102}$$

By means of the abbreviation

$$j = \left(Kf - \tfrac{1}{2}Q\frac{df}{dq}\right), \tag{6.103}$$

(6.102) can be represented as

$$f^{\cdot} + \frac{d}{dq}j = 0. \tag{6.104}$$

This is the one-dimensional case of a continuity equation (cf. (6.73) and the exercise): The temporal change of the probability density $f(q)$ is equal to the negative divergence of the probability current j.

2) n-Dimensional case

The Fokker-Planck equation (6.98) may be cast in the form

$$f^{\cdot} + \sum_{k=1}^{n} \frac{\partial}{\partial q_k}\left((K_k f - \tfrac{1}{2}\sum_{l=1}^{n} Q_{kl}\frac{\partial f}{\partial q_l}\right) = 0. \tag{6.105}$$

We now define the probability current by

$$\boldsymbol{j} = (j_1, j_2, \ldots, j_k, \ldots, j_n),$$

where

$$j_k = K_k f - \tfrac{1}{2}\sum_{l=1}^{n} Q_{kl}\frac{\partial f}{\partial q_l}. \tag{6.106}$$

In analogy to (6.104) we then obtain

$$f^{\cdot} + \nabla_{\boldsymbol{q}} \cdot \boldsymbol{j} = 0, \tag{6.107}$$

where $\nabla_{\boldsymbol{q}} = (d/dq_1, \ldots, d/dq_n)$.

B) *Stationary Solutions of the Fokker-Planck Equation*

The stationary solution is defined by $f^{\cdot} = 0$, i.e., f is time-independent.

1) One-dimension

We obtain from (6.104) by simple integration

$$j = \text{const.} \tag{6.108}$$

In the following we impose the "natural boundary condition" on f which means that f vanishes for $q \to \pm\infty$. This implies (compare (6.103)) that $j \to 0$ for $q \to \pm\infty$, i.e., the constant in (6.108) must vanish. Using (6.103), we then have

$$\tfrac{1}{2}Q\frac{df}{dq} = Kf. \tag{6.109}$$

It is a simple matter to verify that (6.109) is solved by

$$f(q) = \mathcal{N} \exp\left(-2V(q)/Q\right), \tag{6.110}$$

where

$$V(q) = -\int_{q_0}^{q} K(q)\, dq \tag{6.111}$$

has the meaning of a potential, and the normalization constant $\mathcal{N}$ is determined by

$$\int_{-\infty}^{+\infty} f(q)\, dq = 1. \tag{6.112}$$

2) n dimensions

Here, (6.107) with $\dot{f} = 0$ reads

$$\nabla_q \boldsymbol{j} = 0. \tag{6.113}$$

Unfortunately, (6.113) does not always imply $\boldsymbol{j} = 0$, even for natural boundary conditions. However, a solution analogous to (6.110) obtains, if the drift coefficients $K_k(\boldsymbol{q})$ fulfil the so-called potential condition:

$$K_k = -\frac{\partial}{\partial q_k} V(\boldsymbol{q}). \tag{6.114}$$

If, furthermore, the diffusion coefficients obey the condition

$$Q_{kl} = \delta_{kl} Q, \tag{6.115}$$

then

$$f(\boldsymbol{q}) = \mathcal{N} \exp\{-2V(\boldsymbol{q})/Q\}. \tag{6.116}$$

It is assumed that $V(\boldsymbol{q})$ ensures that $f(\boldsymbol{q})$ vanishes for $|\boldsymbol{q}| \to \infty$.

C) *Examples*

To illustrate (6.110) we treat a few special cases:

a)

$$K(q) = -\alpha q. \tag{6.117}$$

We immediately find

$$V(q) = \frac{\alpha}{2} q^2$$

which is plotted in Fig 6.5. The corresponding probability density $f(q)$ is plotted in the same figure. To interpret $f(q)$ let us recall the Langevin equation corresponding to (6.102), (6.117)

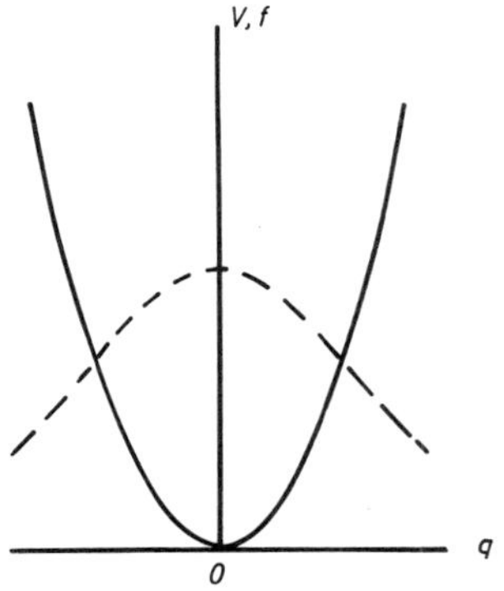

Fig. 6.5. Potential $V(q)$ (solid line) and probability density $f(q)$ (dashed line) for (6.117)

$$\dot{q} = -\alpha q + F(t).$$

What happens to our particle with coordinate q is as follows: The random force $F(t)$ pushes the particle up the potential slope (which stems from the systematic force, $K(q)$). After each push, the particle falls down the slope. Therefore, the most probable position is $q = 0$, but also other positions q are possible due to the random pushes. Since many pushes are necessary to drive the particle far from $q = 0$, the probability of finding it in those regions decreases rapidly. When we let α become smaller, the restoring force K becomes weaker. As a consequence, the potential curve becomes flatter and the probability density $f(q)$ is more spread out.

Once $f(q)$ is known, moments $\langle q^n \rangle = \int q^n f(q)\, dq$ may be calculated. In our present case, $\langle q \rangle = 0$, i.e., the center of $f(q)$ sits at the origin, and $\langle q^2 \rangle = (1/2)(Q/\alpha)$ is a measure for the width of $f(q)$ (compare Fig. 6.5).

b)

$$K(q) = -\alpha q - \beta q^3, \tag{6.118}$$

$$V(q) = \frac{\alpha}{2} q^2 + \frac{\beta}{4} q^4,$$

$$\dot{q} = -\alpha q - \beta q^3 + F(t).$$

The case $\alpha > 0$ is qualitatively the same as a), compare Fig. 6.6a. However, for $\alpha < 0$ a new situation arises (cf. Fig. 6.6b). While without fluctuations, the

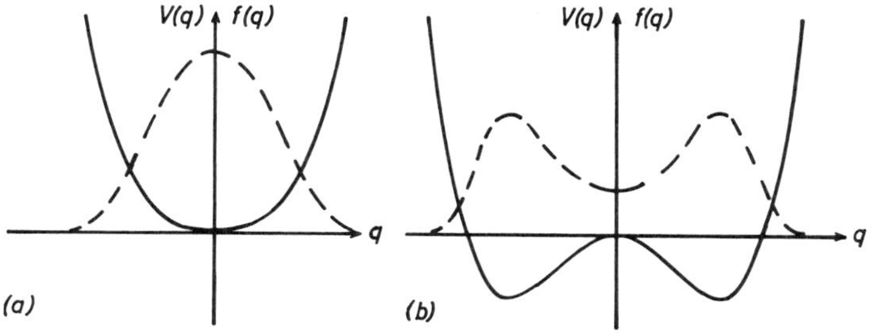

Fig. 6.6a and b. Potential $V(q)$ (solid line) and probability density (dashed line) for (6.118). (a) $\alpha > 0$, (b) $\alpha < 0$

particle coordinate occupies either the left *or* right valley (broken symmetry, compare Sect. 5.1), in the present case $f(q)$ is symmetric. The "particle" may be found with equal probability in both valleys. An important point should be mentioned, however. If the valleys are deep, and we put the particle initially at the bottom of one of them, it may stay there for a very long time. The determination of the time it takes to pass over to the other valley is called the "first passage time problem".

c)

$$K(q) = -\alpha q - \gamma q^2 - \beta q^3, \tag{6.119}$$

$$V(q) = \frac{\alpha}{2} q^2 + \frac{\gamma}{3} q^3 + \frac{\beta}{4} q^4,$$

$$\dot{q} = -\alpha q - \gamma q^2 - \beta q^3 + F(t).$$

We assume $\gamma > 0$, $\beta > 0$ as fixed, but let α vary from positive to negative values. Figs. 6.7a–d exhibit the corresponding potential curves a)–d) and probability densities. Note the pronounced jump in the probability density at $q = 0$ and $q = q_1$ when passing from Fig. c to Fig. d.

d) This and the following example illustrate the "potential case" in two dimensions.

$$\left.\begin{aligned} K_1(\boldsymbol{q}) &= -\alpha q_1 \\ K_2(\boldsymbol{q}) &= -\alpha q_2 \end{aligned}\right\} \text{force,} \tag{6.120}$$

$$V(\boldsymbol{q}) = \frac{\alpha}{2}(q_1^2 + q_2^2) \text{ potential,}$$

$$\dot{q}_i = -\alpha q_i + F_i(t), \; i = 1, 2; \text{ Langevin equation,}$$

where $\langle F_i(t) F_j(t') \rangle = Q_{ij} \delta(t - t') = \delta_{ij} Q \delta(t - t')$.
The potential surface V and the probability density $f(q)$ are qualitatively those of Figs. 6.8 and 6.9.

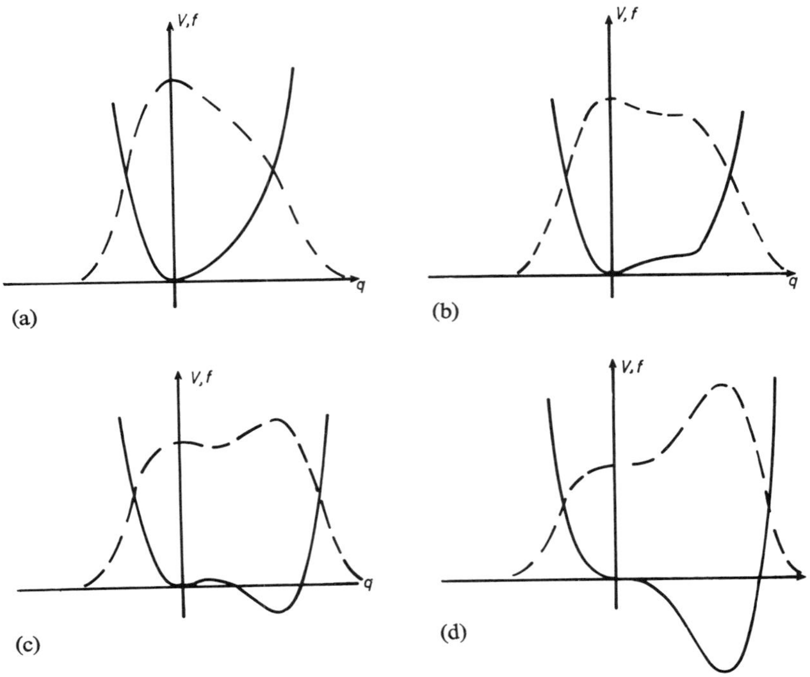

Fig. 6.7. (6.119) V(solid) and f(dashed) for varying α

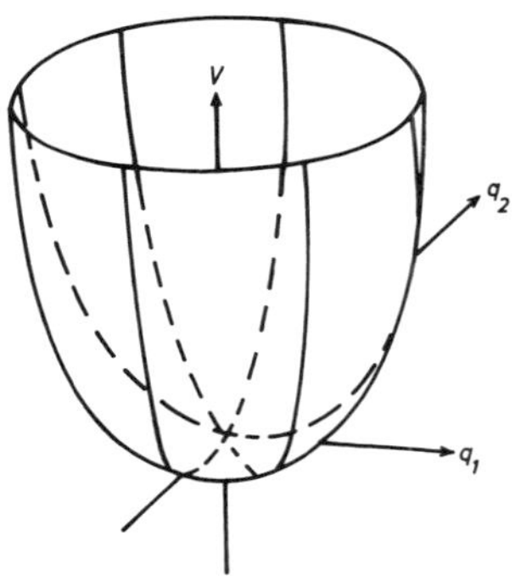

Fig. 6.8. The potential belonging to (6.121) for $\alpha>0$

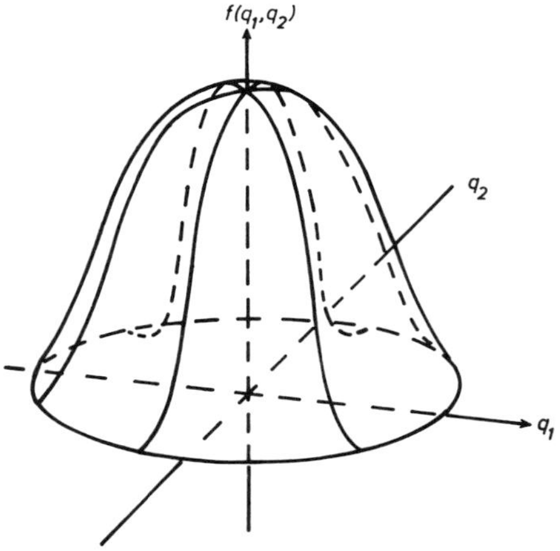

Fig. 6.9. The distribution function belonging to the potential (6.121), $\alpha>0$

e) We present a two-dimensional generalization of case b) above:

$$\left.\begin{aligned} K_1(\boldsymbol{q}) &= -\alpha q_1 - \beta(q_1^2 + q_2^2)q_1 \\ K_2(\boldsymbol{q}) &= -\alpha q_2 - \beta(q_1^2 + q_2^2)q_2 \end{aligned}\right\}\text{force,} \tag{6.121}$$

or, in short,

$$\boldsymbol{K}(\boldsymbol{q}) = -\alpha\boldsymbol{q} - \beta q^2 \cdot \boldsymbol{q} \quad \text{force,}$$

$$V(\boldsymbol{q}) = \frac{\alpha}{2}(q_1^2 + q_2^2) + \frac{\beta}{4}(q_1^2 + q_2^2)^2 \quad \text{potential.}$$

We assume $\beta > 0$. For $\alpha > 0$, potential surface and probability density $f(q)$ are shown in Figs. 6.8 and 6.9, respectively. For $\alpha < 0$, V and f are shown in Figs. 6.10 and 6.11, respectively. What is new compared to case b is the *continuously* broken symmetry. Without fluctuating forces, the particle could sit anywhere at the bottom of the valley in an (marginal) equilibrium position. Fluctuations drive the particle round the valley, completely analogous to Brownian movement in one dimension. In the stationary state, the particle may be found along the bottom with equal probability, i.e., the symmetry is restored.

f) In the general case of a known potential $V(\boldsymbol{q})$, a discussion in terms of Section 5.5 may be given. We leave it to the reader as an exercise to perform this "translation".

Exercise on 6.4

Convince yourself that (6.104) is a continuity equation.

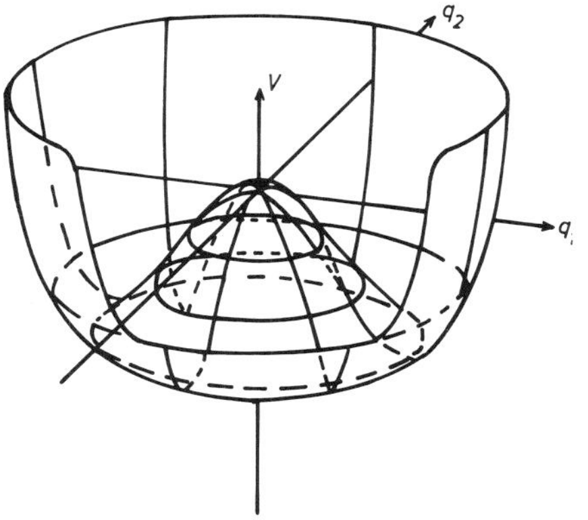

Fig. 6.10. The potential belonging to (6.121) for $\alpha<0$

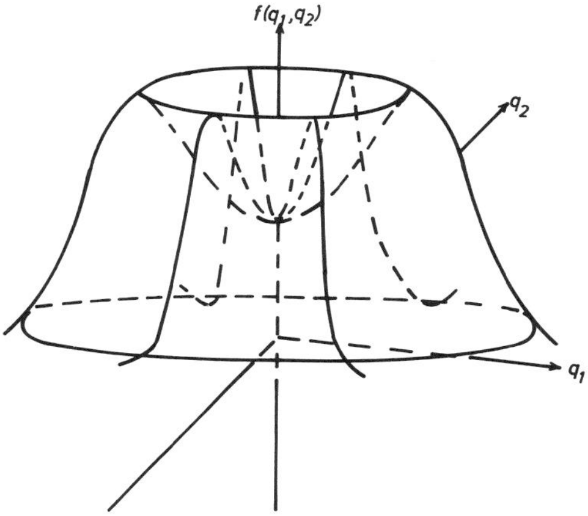

Fig. 6.11. The distribution function belonging to the potential (6.121) for $\alpha<0$

Hint: Integrate (6.103) from $q = q_1$ till $q = q_2$ and discuss the meaning of

$$\frac{d}{dt}\int_{q_1}^{q_2} f(q)\,dq \quad \text{etc.}$$

6.5 Time-Dependent Solutions of the Fokker-Planck Equation

1) *An Important Special Case: One-Dimensional Example*

The drift coefficient is linear in q:

$$K = -\alpha q$$

(by a simple shift of the origin of the q-coordinate we may cover also the case $K = c - \alpha q$). We present a more or less heuristic derivation of the corresponding solution. Because the stationary solution (6.110) has the form of a Gaussian distribution, provided that the drift coefficients are linear, we try a hypothesis in the form of a Gaussian distribution

$$f(q, t) = \mathcal{N}(t)\exp\left\{-\frac{q^2}{a} + \frac{2b}{a}q\right\}, \tag{6.122}$$

where we admit that the width of the Gaussian distribution, a, the displacement, b, and the normalization, $\mathcal{N}(t)$, are time-dependent functions. We insert (6.122) into the time-dependent Fokker-Planck equation (6.102). After performing the differentiation with respect to time t and coordinate q, we divide the resulting equation on both its sides by

$$\mathcal{N}(t)\exp\{-q^2/a + 2bq/a\}.$$

We are then left with an equation containing powers of q up to second order. Comparing in this equation the coefficients of the same powers of q, yields the following three equations (after some rearrangements)

$$\dot{a} = -2\alpha a + 2Q, \tag{6.123}$$

$$\dot{b} = -\alpha b, \tag{6.124}$$

$$\frac{\dot{\mathcal{N}}}{\mathcal{N}} = \alpha + Q\frac{2b^2}{a^2} - \frac{Q}{a}. \tag{6.125}$$

Eqs. (6.123) and (6.124) are linear differential equations for α and β which can be solved explicitly

$$a(t) = \frac{Q}{\alpha}(1 - \exp(-2\alpha t)) + a_0\cdot\exp(-2\alpha t), \tag{6.126}$$

$$b(t) = b_0\exp(-\alpha t). \tag{6.127}$$

Eq. (6.125) looks rather grim. It is a simple matter, however, to verify that it is solved by the hypothesis

$$\mathcal{N} = (\pi a)^{-1/2} \exp(-b^2/a), \tag{6.128}$$

which merely comes from the fact that (6.128) normalizes the distribution function (6.122) for all times. Inserting (6.128) into (6.122) we obtain

$$f(q, t) = (\pi a(t))^{-1/2} \exp\{-(q - b(t))^2/a(t)\}. \tag{6.129}$$

Fig. 6.4 shows an example of (6.129). If the solution (6.129) is subject to the initial condition $a \to 0$ (i.e., $a_0 = 0$) for the initial time $t \to t_0 \equiv 0$, (6.129) reduces to a δ-function, $\delta(q - b_0)$, at $t = 0$, or, in other words, it is a Green's function of the Fokker-Planck equation. The same type of solutions of a Fokker-Planck equation with linear drift and constant diffusion coefficients can be found also for *many* variables $\boldsymbol{q}$.

Examples for the Application of Time-Dependent Solutions

With help of the time-dependent solutions, we may calculate *time-dependent moments*, e.g.,

$$\langle q \rangle = \int q f(q, t)\, dq. \tag{6.130}$$

Inserting (6.129) into (6.130) we may evaluate the integral immediately (by passing over to a new coordinate $q = q' + b(t)$) and obtain

$$\langle q \rangle = b_0 \exp(-\alpha t). \tag{6.131}$$

Because $f(q, t)$ is determined uniquely only if the initial distribution $f(q, 0)$ is given, when evaluating (6.130) we must observe this condition. In many practical cases, $f(q, 0)$ is chosen as δ-function, $f(q, 0) = \delta(q - q_0)$, i.e., we know for sure that the particle was at $q = q_0$ for $t = 0$. To indicate this, (6.130) is then written as

$$\langle q \rangle_{q_0}. \tag{6.132}$$

In our above example, $b_0 \equiv q_0$.

In many practical cases two-time *correlation functions*

$$\langle q(t) q(t') \rangle$$

are important. They are defined by (compare Sect. 4.4)

$$\langle q(t) q(t') \rangle = \int q\, dq \int q'\, dq' f(q,t; q',t'), \tag{6.133}$$

where

$$f(q,t; q',t') \tag{6.134}$$

is a joint probability density. Since the Fokker-Planck equation describes only Markovian processes, we may decompose (6.134) into a probability density at time t', $f(q', t')$ and a conditional probability $f(q, t \mid q', t')$ according to Section 4.3

$$f(q,t; q',t') = f(q, t \mid q', t')f(q', t'). \tag{6.135}$$

In practical cases $f(q', t')$ is taken as the stationary solution $f(q)$, of the Fokker-Planck equation, if not otherwise stated. $f(q, t \mid q', t')$ is just that time-dependent solution of the Fokker-Planck equation which reduces to the δ-function $\delta(q - q')$ at time $t = t'$. Thus $f(q, t \mid q', t')$ is a Green's function. In our present example (6.117), we have $f(q) = (\alpha/\pi Q)^{1/2} \exp(-\alpha q^2/Q)$ and $f(q, t \mid q', t')$ given by (6.129) with $a_0 = 0$ and $b_0 = q'$. With these functions we may simply evaluate (6.133) where we put without loss of generality $t' = 0$.

$$(6.133) = \iint q(\pi a(t))^{-1/2} \exp\left\{-\frac{1}{a(t)}(q - q'e^{-\alpha t})^2\right\} dq$$

$$\cdot q'(\pi Q/\alpha)^{-1/2} \exp\left\{-\frac{\alpha}{Q} q'^2\right\} dq'. \tag{6.136}$$

Replacing q by $q + q' \exp(-\alpha t)$ we may simply perform the integrations (which are essentially over Gaussian densities),

$$\begin{aligned} \langle q(t)q(t')\rangle &= e^{-\alpha t}\langle q'^2\rangle \\ &= \frac{1}{2}\frac{Q}{\alpha} e^{-\alpha t}, \end{aligned} \tag{6.137}$$

which is in accordance with (6.25).

Now we turn to the general equation (6.105).

2)* *Reduction of the Time-Dependent Fokker-Planck Equation to a Time-Independent Equation*

We put

$$f(\boldsymbol{q}, t) = \exp(-\lambda t)\Psi(\boldsymbol{q}) \tag{6.138}$$

and insert it into (6.105). Performing the differentiation with respect to time and then multiplying both sides of (6.105) by $\exp(\lambda t)$ yields

$$-\lambda\Psi(\boldsymbol{q}) = -\sum_{k=1}^{n} \frac{\partial}{\partial q_k}\left(K_k\Psi - \tfrac{1}{2}\sum_{l=1}^{n} Q_{kl}\frac{\partial}{\partial q_l}\Psi\right). \tag{6.139}$$

We do not discuss methods of solution of (6.139). We simply mention a few important properties: (6.139) allows for an infinite set of solutions, $\Psi_m(\boldsymbol{q})$ and eigenvalues λ_m, $m = 0, 1, 2, \ldots$ provided that suitable boundary conditions are given, e.g., natural ones. The most general solution of (6.105) is obtained as linear combination of (6.138).

$$f(\boldsymbol{q}, t) = \sum_{m=0}^{\infty} c_m \exp(-\lambda_m t)\Psi_m(\boldsymbol{q}). \tag{6.140}$$

When a stationary solution of (6.105) exists, $\lambda_0 = 0$. The coefficients c_m may be fixed by prescribing the initial distribution, e.g., at time $t = 0$:

$$f(\boldsymbol{q}, 0) = f_0(\boldsymbol{q}). \tag{6.141}$$

Even in the one-dimensional case and for rather simple K's and Q's (6.139) can be solved only with help of computers.

3)* *A Formal Solution*

We start from the Fokker-Planck equation (6.105) which we write in the form

$$\dot{f} = Lf. \tag{6.142}$$

In it, L is the "operator"

$$L = -\sum_k \frac{\partial}{\partial q_k} K_k + \tfrac{1}{2} \sum_{kl} Q_{kl} \frac{\partial^2}{\partial q_k \partial q_l}. \tag{6.143}$$

If L were just a number, solving (6.142) would be a trivial matter and $f(t)$ would read

$$f(t) = e^{Lt} f(0). \tag{6.144}$$

By inserting (6.144) into (6.142) and differentiating it with respect to time t, one verifies immediately that (6.144) fulfils (6.142) even in the present case where L is an operator. To evaluate (6.144), we define $\exp(Lt)$ just by the usual power series expansion of the exponential function:

$$e^{Lt} = \sum_{\nu=0}^{\infty} \frac{1}{\nu!} L^\nu t^\nu. \tag{6.145}$$

L^ν means: apply the operator L ν times on a function standing on the right of it:

$$L^\nu f(\boldsymbol{q}, t) = \underbrace{L \cdot L \cdot L \cdots L}_{\nu} f(\boldsymbol{q}, t). \tag{6.146}$$

4)* *An Iteration Procedure*

In practical applications, one may try to solve (6.142) by iteration:
Be $f(\boldsymbol{q}, t)$ given at time $t = t_0$, we wish to construct f at a slightly later time, $t + \tau$. To this end we recall the definition

$$\dot{f} = \lim_{\tau \to 0} \frac{1}{\tau} (f(t + \tau) - f(t)).$$

Leaving τ finite (but very small), we recast (6.142) into the form:

$$f(\mathbf{q}, t_0 + \tau) = f(\mathbf{q}, t_0) + \tau L f(\mathbf{q}, t_0) \equiv (1 + \tau L) f(\mathbf{q}, t_0). \tag{6.147}$$

Repeating this procedure at times $t_2 = t_0 + 2\tau, \ldots, t_n = t_0 + N\tau$, we find $(t_N = t)$

$$f(\mathbf{q}, t) = (1 + \tau L)^N f(\mathbf{q}, t_0). \tag{6.148}$$

(6.148) becomes an exact solution of (6.142) in the limit $\tau \to 0$, $N \to \infty$ but $N\tau = t - t_0$. (6.148) is an alternative form to (6.144).

Exercises on 6.5

Verify that (6.129) with (6.126), (6.127), $a_0 = 0$, reduces to a δ-function, $\delta(q - q_0)$ for $t \to 0$.

6.6* Solution of the Fokker-Planck Equation by Path Integrals

1) *One-Dimensional Case*

In Section 4.3 we solved a very special Fokker-Planck equation by a path integral, namely in the case of a vanishing drift coefficient, $K \equiv 0$. Here we present the idea of how to generalize this result to $K \neq 0$. In the present case, L of (6.142) is given by

$$L = \frac{d}{dq} K(q) + \frac{Q}{2} \frac{d^2}{dq^2}. \tag{6.149}$$

For an infinitesimal time-interval τ, we try the following hypothesis (generalizing (4.89) for a single step, $t_0 \to t_0 + \tau$)

$$f(q, t_0 + \tau) = \mathcal{N} \int_{-\infty}^{+\infty} \exp\left\{ -\frac{1}{2Q\tau} (q - q' - \tau K(q'))^2 \right\} f(q', t_0)\, dq'. \tag{6.150}$$

Readers, who are not so much interested in mathematical details may skip the following section and proceed directly to formula (6.162). We will expand (6.150) into a power series of τ, hereby proving, that (6.150) is equivalent to (6.147) up to and including the order τ, i.e., we want to show that the rhs of (6.150) may be transformed into

$$f + \tau\left(-\frac{d}{dq} Kf + \tfrac{1}{2} Q \frac{d^2 f}{dq^2} \right), \text{ where } f = f(q, t_0). \tag{6.151}$$

To this end, we introduce a new integration variable ξ by

$$q' = q + \xi. \tag{6.152}$$

We simultaneously evaluate the curly bracket in the exponent

$$f(q, t_0 + \tau) = \mathcal{N} \int_{-\infty}^{+\infty} \exp\left\{ -\frac{1}{2Q\tau} (\xi^2 + 2\tau\xi K(q + \xi) + \tau^2 K(q + \xi)^2) \right\} \cdot f(q + \xi, t_0)\, d\xi. \tag{6.153}$$

The basic idea is now this: the Gaussian distribution

$$\exp\left(-\frac{1}{2Q\tau} \xi^2 \right) \tag{6.154}$$

gets very sharply peaked as $\tau \to 0$. More precisely, only those terms under the integral are important for which $|\xi| < \sqrt{\tau}\sqrt{Q}$. This suggests to expand all factors of (6.154) in (6.153) into a power series of ξ and τ. Keeping the leading terms we obtain after some rearrangements

$$(6.153) = \mathcal{N} \int_{-\infty}^{+\infty} \exp\left\{ -\frac{1}{2Q\tau} \xi^2 \right\} [\cdots]\, d\xi, \tag{6.155}$$

where

$$[\cdots] = \left(1 - \frac{1}{Q} \xi^2 K' + \frac{1}{2} \frac{K^2}{Q^2} \xi^2 - \frac{\tau K^2}{2Q} \right) f + \xi^2 \left(-\frac{1}{Q} K \right) f' + \frac{1}{2} \xi^2 f''. \tag{6.156}$$

Terms odd in ξ have been dropped, because they vanish after integration over ξ. Here

$$K = K(q),\ K' = \frac{dK(q)}{dq}, \quad \text{and } f = f(q),\ f' = \frac{df}{dq},\ f'' = \frac{d^2 f}{dq^2}. \tag{6.157}$$

To evaluate (6.155) further, we integrate over ξ. Using well-known formulas

$$\int_{-\infty}^{+\infty} \exp\left\{ -\frac{1}{2Q\tau} \xi^2 \right\} d\xi = (2Q\pi\tau)^{1/2}, \tag{6.158}$$

$$\int_{-\infty}^{+\infty} \xi^2 \exp\left\{ -\frac{1}{2Q\tau} \xi^2 \right\} d\xi = Q\tau(2Q\pi\tau)^{1/2}, \tag{6.159}$$

we obtain (6.155) in the form

$$(6.155) = \mathcal{N}(2Q\tau\pi)^{1/2} \left\{ f - \tau K' f - \tau K f' + \tau \frac{Q}{2} f'' \right\}. \tag{6.160}$$

Choosing the constant $\mathcal{N}$,

$$\mathcal{N} = (2Q\tau\pi)^{-1/2}, \tag{6.161}$$

we may establish the required identity with (6.147). We may now repeat the whole procedure at times $t_2 = t_0 + 2\tau, \ldots, t_N = t_2 + N\tau$ and eventually pass over to the limit $N \to \infty$; $N\tau = t - t_0$ fixed. This yields an N-dimensional integral ($t_N \equiv t, t_0 = 0$)

$$f(q, t) = \lim_{\substack{N \to \infty \\ N\tau = t}} \int_{-\infty}^{+\infty} \cdots \int \mathscr{D}q \exp(-\tfrac{1}{2}O) f(q', t_0), \tag{6.162}$$

where

$$\begin{aligned} \mathscr{D}q &= (2Q\tau\pi)^{-N/2}\, dq_0 \cdots dq_{N-1}, \\ O &= \textstyle\sum_{\nu=1}^{N} \tau\{(q_\nu - q_{\nu-1})/\tau - K(q_{\nu-1})\}^2 Q^{-1}, \\ q_0 &\equiv q';\ q_N \equiv q. \end{aligned} \tag{6.163}$$

We leave it to the reader to compare (6.162) with the path integral (4.85) and to interpret (6.162) in the spirit of Chapter 4. The most probable path of the particle is that for which O has a minimum, i.e.,

$$\frac{1}{\tau}(q_\nu - q_{\nu-1}) = K(q_{\nu-1}). \tag{6.164}$$

Letting $\tau \to 0$ this equation is just the equation of motion under the force K. In the literature one often puts $(1/\tau)(q_\nu - q_{\nu-1}) = \dot{q}$ and replaces $K(q_\nu)$ by $K(q)$. This is correct, as long as this is only a shorthand for (6.163). Often different translations have been used leading to misunderstandings and even to errors.

2) In the *n-dimensional case* the path integral solution of (6.105) reads

$$f(\boldsymbol{q}, t) = \lim_{\substack{N \to \infty \\ N\tau = t}} \int_{-\infty}^{+\infty} \cdots \int \mathscr{D} \exp(-\tfrac{1}{2}\, O) f(\boldsymbol{q}', t_0), \tag{6.165}$$

where

$$\mathscr{D} = \textstyle\prod_{\mu=0}^{N-1} \{(2\pi\tau)^{-n/2} (\det Q)^{-1/2}\} (dq_1 \cdots dq_n)_\mu. \tag{6.166}$$

$$\begin{aligned} &\boldsymbol{q}_N = \boldsymbol{q};\ \boldsymbol{q}_0 = \boldsymbol{q}',\ \det = \text{determinant}, \\ &O = +\tau \textstyle\sum_{\nu=1}^{N} (\dot{\boldsymbol{q}}_\nu^T - \boldsymbol{K}_{\nu-1}^T) Q^{-1} (\dot{\boldsymbol{q}}_\nu - \boldsymbol{K}_{\nu-1}), \end{aligned} \tag{6.167}$$

where $\dot{\boldsymbol{q}}_\nu = (\boldsymbol{q}_\nu - \boldsymbol{q}_{\nu-1})/\tau$, $\boldsymbol{K}_{\nu-1} = \boldsymbol{K}(\boldsymbol{q}_{\nu-1})$, and T denotes the transposed vector. Q is the diffusion matrix occurring in (6.105). O may be called a generalized Onsager-Machlup function because these authors had determined O for the special case of K's *linear* in $\boldsymbol{q}$.

Exercise on 6.6

Verify that (6.165) is solution to (6.105).

Hint: Proceed as in the one-dimensional case and use (2.65), (2.66) with $\boldsymbol{m} \to \boldsymbol{K}$
What is the most probable path?
Hint: Q and thus Q^{-1} are positive definite.

6.7 Phase Transition Analogy

In the introductory Chapter 1, we mentioned a few examples of phase transitions of physical systems, for example that of the ferromagnet. A ferromagnet consists of very many atomistic elementary magnets. At a temperature T greater than a "critical" temperature $T_c, T > T_c$, these magnets point in random directions (see Fig. 1.7). When T is lowered, suddenly at $T = T_c$, a macroscopic number of these elementary magnets become aligned. The ferromagnet has now a spontaneous magnetization. Our considerations on the solutions of Fokker-Planck equations in the foregoing chapter will allow us to draw some very close analogies between phase transitions, occurring in thermal equilibrium, and certain disorder-order transitions in nonequilibrium systems. As we will demonstrate in later chapters, such systems may belong to physics, chemistry, biology, and other disciplines. To put these analogies on a solid basis we first consider the free energy of a physical system (in thermal equilibrium) (compare Sect. 3.3 and 3.4). The free energy, $\mathscr{F}$, depends on temperature, T, and possibly on further parameters, e.g., the volume. In the present case, we seek the minimum of the free energy under an additional constraint. Its significance can best be explained in the case of a ferromagnet. When M_1 elementary magnets point upwards and M_2 elementary magnets point downwards, the "magnetization" is given by

$$M = (M_\uparrow - M_\downarrow)m, \tag{6.168}$$

where m is the magnetic moment of a single elementary magnet. Our additional constraint requires that the average magnetization M equals a given value, or, in the notation of Section 3.3

$$f_k = M. \tag{6.169}$$

Because we also want to treat other systems, we shall replace M by a general coordinate q

$$M \to q, \text{ and } f_k = q. \tag{6.170}$$

In the following we assume that $\mathscr{F}$ is the minimum of the free energy for a fixed value of q. To proceed further we expand $\mathscr{F}$ into a power series of q,

$$\mathscr{F}(q, T) = \mathscr{F}(0, T) + \mathscr{F}'(0, T)q + \cdots + \frac{1}{4!}\mathscr{F}''''(0, T)q^4 + \cdots \tag{6.171}$$

and discuss $\mathscr{F}$ as a function of q. In a number of cases

$$\mathscr{F}' = \mathscr{F}''' = 0 \tag{6.172}$$

due to inversion symmetry (cf. Sect. 5.1). Let us discuss this case first. We write $\mathcal{F}$ in the form

$$\mathcal{F}(q, T) = \mathcal{F}(0, T) + \frac{\alpha}{2} q^2 + \frac{\beta}{4} q^4. \tag{6.173}$$

Following Landau we call q "order parameter". To establish a first connection of Section 6.4 with the Landau theory of phase transitions we identify (6.173) with the potential introduced in Section 6.4. As is shown in thermodynamics, (cf. Sections 3.3, 3.4)

$$f = \mathcal{N} \exp\{-\mathcal{F}/k_B T\} \tag{6.174}$$

gives the probability distribution if $\mathcal{F}$ is considered as a function of the order parameter, q. Therefore the most probable order parameter is determined by the requirement $\mathcal{F}$ = min! Apparently the minima of (6.173) can be discussed exactly as in the case of the potential $V(q)$. We investigate the position of those minima as function of the coefficient α. In the Landau theory this coefficient is assumed in the form

$$\alpha = a(T - T_c) \quad (a > 0), \tag{6.175}$$

i.e., it changes its sign at the critical temperature $T = T_c$. We therefore distinguish between the two regions $T > T_c$ and $T < T_c$ (compare Table 6.1). For $T > T_c$, $\alpha > 0$, and the minimum of $\mathcal{F}$ (or V) lies at $q = q_0 = 0$. As the entropy is connected with the free energy by the formula (cf. (3.93))

$$S = -\frac{\partial \mathcal{F}(q, T)}{\partial T}, \tag{6.176}$$

we obtain in this region, $T > T_c$,

$$S = S_0 = -\frac{\partial \mathcal{F}(0, T)}{\partial T}. \tag{6.176a}$$

The second derivative of $\mathcal{F}$ with respect to temperature yields the specific heat (besides the factor T):

$$c = T\left(\frac{\partial S}{\partial T}\right), \tag{6.177}$$

which, using (6.176a), gives

$$c = T\left(\frac{\partial S_0}{\partial T}\right). \tag{6.177a}$$

Now we perform the same procedure for the ordered phase $T < T_c$, i.e., $\alpha < 0$. This yields a new equilibrium value q_1 and a new entropy as exhibited in Table 6.1.

Table 6.1. Landau theory of second-order phase transition

Distribution function $f = \mathcal{N} \exp(-\mathcal{F}/k_B T)$,
free energy $\mathcal{F}(q, T) = \mathcal{F}(0, T) + (\alpha/2)q^2 + (\beta/4)q^4$,
$\alpha \equiv \alpha(T) = a(T - T_c)$.

State	Disordered	Ordered
Temperature	$T > T_c$	$T < T_c$
"External" parameter	$\alpha > 0$	$\alpha < 0$
Most probable order parameter q_0: $f(q) = \max!$, $\mathcal{F} = \min!$	$q_0 = 0$	$q_0 = \pm(-\alpha/\beta)^{1/2}$ Broken symmetry
Entropy $S = -\{\partial\mathcal{F}(q_0, T)/\partial T\}$	$S_0 = -\{\partial\mathcal{F}(0, T)/\partial T\}$	$S_0 + (a^2/2\beta)(T - T_c)$
	Continuous at $T = T_c$ ($\alpha = 0$)	
Specific heat $c = T(\partial S/\partial T)$	$T(\partial S_0/\partial T)$	$T(\partial S_0/\partial T) + (a^2/2\beta)T$
	Discontinuous at $T = T_c$ ($\alpha = 0$)	

One may readily check that S is continuous at $T = T_c$ for $\alpha = 0$. However (consider the last row of Table 6.1), when we calculate the specific heat we obtain two different expressions above and below the critical temperature and thus a discontinuity at $T = T_c$.

This phenomenon is called a phase transition of second order because the second derivative of the free energy is discontinuous. On the other hand, the entropy itself is continuous so that this transition is also referred to as a continuous phase transition. In statistical physics one also investigates the temporal change of the order parameter. Usually, in a more or less phenomenological manner, one assumes that q obeys an equation of the form

$$\dot{q} = -\frac{\partial\mathcal{F}}{\partial q} \tag{6.178}$$

which in the case of the potential (6.173) takes the explicit form

$$\dot{q} = -\alpha q - \beta q^3, \tag{6.179}$$

which we have met in various examples in our book in different context. For simplicity, we have omitted a constant factor in front of $\partial\mathcal{F}/\partial q$. For $\alpha \to 0$ we observe a phenomenon called critical slowing down, because the "particle" with coordinate q falls down the potential slope more and more slowly. Furthermore symmetry breaking occurs, which we have already encountered in Section 5.1. The critical slowing down is associated with a soft mode (cf. Sect. 5.1). In statistical mechanics one often includes a fluctuating force in (6.179) in analogy to Sect. 6.1. For $\alpha \to 0$ critical fluctuations arise: since the restoring force acts only via higher powers of q, the fluctuations of $q(t)$ become considerable.

We now turn to the case where the free energy has the form

$$\mathcal{F}(q, T) = \alpha\frac{q^2}{2} + \gamma\frac{q^3}{3} + \beta\frac{q^4}{4} \tag{6.180}$$

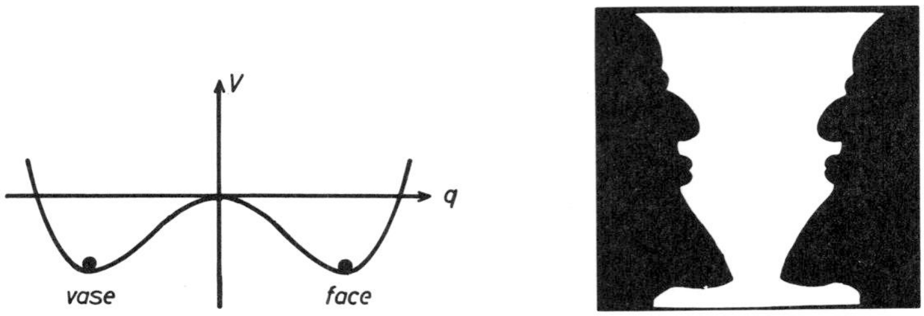

Fig. 6.12. Broken symmetry in visual perception. When focussing the attention to the center and interpreting it as foreground of a picture, a vase is seen, otherwise two faces

(β and γ positive but α may change its sign according to (6.175)). When we change the temperature T, i.e., the parameter α, we pass through a sequence of potential curves exhibited in Figs. 6.7a–d. Here we find the following situation.

When lowering temperature, the local minimum first remains at $q = 0$. When lowering temperature further we obtain the potential curve of Fig. 6.7d, i.e., now the "particle" may fall down from $q = 0$ to the new (global) minimum of $\mathscr{F}$ at q_1. The entropies of the two states, q_0 and q_1, differ. This phenomenon is called "first order phase transition" because the first derivative of $\mathscr{F}$ is discontinuous. Since the entropy S is discontinuous this transition is also referred to as a discontinuous phase transition. When we now increase the temperature we pass through the figures in the sequence $d \rightarrow a$. It is apparent that the system stays at q_1 longer than it had been before when going in the inverse direction of temperature. Quite obviously hysteresis is present. Fig. 6.13. (Consider also Figs. 6.7a–d).

Our above considerations must be taken with a grain of salt. We certainly can and will use the nomenclature connected with phase transitions as indicated above. Indeed these considerations apply to many cases of nonequilibrium systems as we shall substantiate in our later chapters. However, it must clearly be stated that the Landau theory of phase transitions (Table 6.1) was found inadequate for phase transitions of systems in thermal equilibrium. Here specific singularities of the specific heat, etc., occur at the phase transition point which are described by so-called critical exponents. The experimentally observed critical exponents do in general not agree with those predicted by the Landau theory. A main reason for this consists in an inadequate treatment of fluctuations as will transpire in the following part. These phenomena are nowadays successfully treated by Wilson's renormalization group techniques. We shall not enter into this discussion but rather interpret several *nonequilibrium* transitions in the sense of the Landau theory in regions where it is applicable.

In the rest of this chapter[1] we want to elucidate especially what the discontinuous transition of the specific heat means for nonequilibrium systems. As we will show by explicit examples in subsequent chapters, the action of a macroscopic system can be described by its order parameter q (or a set of them). In many cases a measure for that action is q^2. Adopting (6.174) as probability distribution, the average value of q^2 is defined by

[1] This part is somewhat technical and can be skipped when reading this book for the first time.

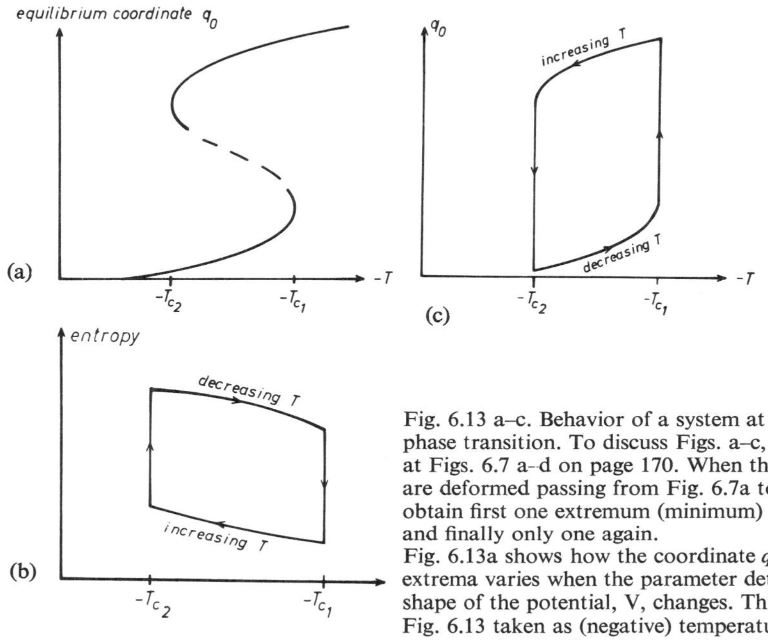

Fig. 6.13 a–c. Behavior of a system at a first order phase transition. To discuss Figs. a–c, first have a look at Figs. 6.7 a–d on page 170. When the potential curves are deformed passing from Fig. 6.7a to Fig. 6.7d, we obtain first one extremum (minimum) of V, then three, and finally only one again.
Fig. 6.13a shows how the coordinate q_0 of these extrema varies when the parameter determining the shape of the potential, V, changes. This parameter is in Fig. 6.13 taken as (negative) temperature. Evidently, when passing in Fig. 6.7 from a) to d), the system stays at $q_0 \approx 0$ until the situation d) is reached, where the system jumps to a new equilibrium value at the absolute minimum of V. On the other hand, when now passing from 6.7d to 6.7a the system remains first at $q_0 \neq 0$ and jumps only in the situation 6.7b again to $q_0 \approx 0$. These jumps are evident in Fig. 6.13c, where we have plotted the coordinate q_0 of the actually realized state of the system. Fig. 6.13b represents the corresponding variation of the entropy.

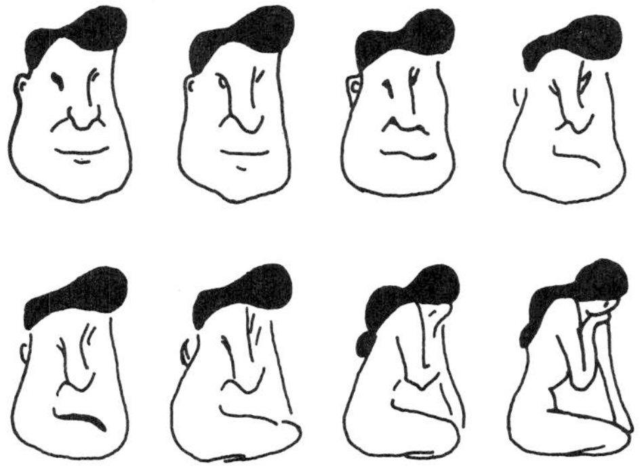

Fig. 6.14. Hysteresis effect in visual perception. Look at the picture first from upper left to lower right and then in the opposite direction. Note that perception switches at different points depending on the direction

$$\langle q^2 \rangle = \frac{\int q^2 \exp(-\hat{V})\, dq}{\int \exp(-\hat{V})\, dq}, \quad \mathscr{F}/k_B T \equiv \hat{V} = 2V/Q, \tag{6.181}$$

where we have now absorbed Q into $\hat{V}$, (cf. (6.116)). We assume $\hat{V}$ in the form

$$\hat{V} = (\alpha - \alpha_c) q^2 + \beta q^4. \tag{6.182}$$

Table 6.2. Phase transition analogy

Physical system in thermal equilibrium	Synergetic system with stationary distribution function $f(q)$
Order parameters $\boldsymbol{q}$	Order parameters $\boldsymbol{q}$
Distribution function $f = \mathcal{N} \exp(-\mathcal{F}/k_B T)$	Distribution function $f(\boldsymbol{q}) = \exp(-\hat{V})$, where $\hat{V}$ defined by $\hat{V} = -\ln f$
Temperature	External parameters, e.g., power input
Entropy	Action (e.g., power output)
Specific heat	Change of action with change of external parameter: efficiency

Apparently we can write q^2 in the form

$$q^2 = \frac{\partial \hat{V}}{\partial \alpha} \tag{6.183}$$

which allows us to write (6.181) in the form

$$\langle q^2 \rangle = \frac{\displaystyle\int \frac{\partial \hat{V}}{\partial \alpha} \exp(-\hat{V})\, dq}{\displaystyle\int \exp(-\hat{V})\, dq} \tag{6.184}$$

It is a simple matter to check that an equivalent form for (6.184) is

$$\langle q^2 \rangle = -\frac{\partial}{\partial \alpha} \ln\left(\int \exp(-\hat{V})\, dq \right). \tag{6.185}$$

For the evaluation of the integral of (6.185) we assume that $\exp(-\hat{V})$ is strongly peaked at $q = q_0$. If there are several peaks (corresponding to different minima of $\hat{V}(q)$) we assume that only one state $q = q_0$ is occupied. This is *an ad hoc assumption* to take into account symmetry breaking in the "second-order phase transition", or to select one of the two local minima of different depth in the "first-order phase transition". In both cases the assumption of a strongly peaked distribution implies that we are still "sufficiently far" away from the phase transition point, $\alpha = \alpha_c$. We expand the exponent around that minimum value keeping only terms of second order

$$\langle q^2 \rangle = -\frac{\partial}{\partial \alpha} \ln \left\{ \exp(-\hat{V}(q_0)) \int \exp(-\hat{V}''(q_0)(q - q_0)^2)\, dq \right\}. \tag{6.186}$$

Splitting the logarithm into two factors and carrying out the derivative with respect to α in the first factor we obtain

$$\langle q^2 \rangle = \frac{\partial \hat{V}(q_0)}{\partial \alpha} - \frac{\partial}{\partial \alpha} \ln \int \exp(-\hat{V}''(q_0)(q - q_0)^2)\, dq. \tag{6.187}$$

The last integral may be easily performed so that our final result reads

$$\langle q^2 \rangle = \frac{\partial \hat{V}(q_0)}{\partial \alpha} - \frac{1}{2}\frac{\partial}{\partial \alpha} \ln (\pi/\hat{V}''(q_0)). \tag{6.188}$$

A comparison of (6.188) with (6.176) by means of the correspondence

$$T \leftrightarrow -\alpha$$

$$\hat{V} \leftrightarrow \mathscr{F}$$

reveals that the first term on the rhs of (6.188) is proportional to the rhs of (6.176). Therefore the entropy S may be put in parallel with the output activity $\langle q^2 \rangle$. The discontinuity of (6.177) indicates a pronounced change of slope. (compare Fig. 1.12). The second part in (6.188) stems from fluctuations. They are significant in the neighborhood of the transition point $\alpha = \alpha_c$. Thus the Landau theory can be interpreted as a theory in which mean values have been replaced by the most probable values. It should be noted that at the transition point the behavior of $\langle q^2 \rangle$ is not appropriately described by (6.187), i.e., the divergence inherent in the second part of (6.188) does not really occur, but is a consequence of the method of evaluating (6.185). For an illustration the reader should compare the behavior of the specific heat which is practically identical with the laser output below and above threshold.

The Landau theory of second order phase transitions is very suggestive for finding an approximate (or sometimes even exact) stationary solution of a Fokker-Planck equation of one or several variables $\boldsymbol{q} = (q_1 \ldots q_n)$. We assume that for a parameter $\alpha > 0$ the maximum of the stationary solution $f(\boldsymbol{q})$ lies at $\boldsymbol{q} = 0$. To study the behavior of $f(\boldsymbol{q})$ at a critical value of α, we proceed as follows:

1) We write $f(\boldsymbol{q})$ in the form

$$f(\boldsymbol{q}) = \mathscr{N} \exp (-\hat{V}(\boldsymbol{q})),$$

($\mathscr{N}$: normalization factor)

2) Following Landau, we expand $\hat{V}(\boldsymbol{q})$ into a Taylor series around $\boldsymbol{q} = 0$ up to fourth order

$$\hat{V}(\boldsymbol{q}) = \hat{V}(0) + \textstyle\sum_\mu \hat{V}_\mu q_\mu + \frac{1}{2!}\sum_{\mu\nu} \hat{V}_{\mu\nu} q_\mu q_\nu$$

$$+ \frac{1}{3!}\textstyle\sum_{\mu\nu\lambda} \hat{V}_{\mu\nu\lambda} q_\mu q_\nu q_\lambda + \frac{1}{4!}\sum_{\mu\nu\lambda\kappa} \hat{V}_{\mu\nu\lambda\kappa} q_\mu q_\nu q_\lambda q_\kappa. \tag{6.189}$$

The subscripts of $\hat{V}$ indicate differentiation of $\hat{V}$ with respect to $q_\mu, q_\nu, \ldots$ at $\boldsymbol{q} = 0$.

3) It is required that $\hat{V}(\boldsymbol{q})$ is invariant under all transformations of $\boldsymbol{q}$ which leave the physical problem invariant. By this requirement, relations among the coefficients $\hat{V}_\mu$, $\hat{V}_{\mu\nu}, \ldots$, can be established so that the number of these expansion parameters can be considerably reduced. (The adequate methods to deal with this

problem are provided by group theory). If the stationary solution of the Fokker-Planck equation is unique, this symmetry requirement with respect to $f(\boldsymbol{q})$ (or $\hat{V}(\boldsymbol{q})$) can be proven rigorously.

An example:

Let a problem be invariant with respect to the inversion $\boldsymbol{q} \to -\boldsymbol{q}$. Then L in $\dot{f} = Lf$ is invariant and also $f(\boldsymbol{q})$ (with $\dot{f} = 0$). Inserting the postulate $f(\boldsymbol{q}) = f(-\boldsymbol{q})$ and thus $\hat{V}(\boldsymbol{q}) = \hat{V}(-\boldsymbol{q})$ into (6.189) yields $V_\mu = 0$ and $V_{\mu\nu\lambda} = 0$.

6.8 Phase Transition Analogy in Continuous Media: Space-Dependent Order Parameter

Let us again use the ferromagnet as example. In the foregoing section we introduced its total magnetization M. Now, we subdivide the magnet into regions (or "cells") still containing many elementary magnets so that we can still speak of a "macroscopic" magnetization in each cell. On the other hand, we choose the cells small compared to macroscopic dimensions, say 1 cm. Denoting the center coordinate of a cell by $\boldsymbol{x}$, we thus are led to introduce a space dependent magnetization, $M(\boldsymbol{x})$. Generalizing this idea we introduce a space dependent order parameter $q(\boldsymbol{x})$ and let the free energy $\mathscr{F}$ depend on the q's at all positions. In analogy to (6.171) we may expand $\mathscr{F}(\{q(\boldsymbol{x})\}, T)$ into a power series of all $q(\boldsymbol{x})$'s. Confining our present considerations to inversion symmetric problems we retain only even powers of $q(\boldsymbol{x})$.

We discuss the form of this expansion. In first approximation we assume that the cells, $\boldsymbol{x}$, do not influence each other. Thus $\mathscr{F}$ can be decomposed into a sum (or an integral in a continuum approximation) of contributions of each cell. In a second step we take the coupling between neighboring cells into account by a term describing an increase of free energy if the magnetizations $M(\boldsymbol{x})$, or, generally, $q(\boldsymbol{x})$ in neighboring cells differ from each other. This is achieved by a term $\gamma(\nabla q(\boldsymbol{x}))^2$. Thus we represent $\mathscr{F}$ in the form of the famous *Ginzburg-Landau functional*:

$$\mathscr{F}(\{q(\boldsymbol{x})\}; T) = \mathscr{F}_0(0, T) + \int d^n x \left\{ \frac{\alpha}{2} q(\boldsymbol{x})^2 + \frac{\beta}{4} q(\boldsymbol{x})^4 + \frac{\gamma}{2} (\nabla q(\boldsymbol{x}))^2 \right\}. \tag{6.190}$$

In the frame of a *phenomenological* approach the relations (6.174) and (6.178) are generalized as follows:

Distribution function:

$$f(\{q(\boldsymbol{x})\}) = \mathscr{N} \exp[-\mathscr{F}/k_B T] \tag{6.191}$$

with $\mathscr{F}$ defined by (6.190).

Equation for the relaxation of $q(\boldsymbol{x})$

$$\dot{q}(\boldsymbol{x}) = -\frac{\delta \mathscr{F}}{\delta q(\boldsymbol{x})}, \tag{6.192}$$

where q is now treated as function of space, $\boldsymbol{x}$, *and* time, t. (Again, a constant factor on the rhs is omitted). Inserting (6.190) into (6.192) yields the *time-dependent Ginzburg-Landau equation*

$$\dot{q} = -\alpha q - \beta q^3 + \gamma \Delta q + (F). \tag{6.193}$$

The typical features of it are:
a linear term, $-\alpha q$, where the coefficient α changes its sign at a certain "threshold", $T = T_c$,
a nonlinear term, $-\beta q^3$, which serves for a stabilization of the system,
a "diffusion term", $\gamma \Delta q$, where Δ is the Laplace operator.
Finally, to take fluctuations into account, a fluctuating force $F(\boldsymbol{x}, t)$ is added ad hoc. In Section 7.6–7.8 we will develop a theory yielding equations of the type (6.193) or generalizations of it. If the fluctuating forces F are Gaussian and Markovian with zero mean and

$$\langle F(\boldsymbol{x}', t')F(\boldsymbol{x}, t)\rangle = Q\delta(\boldsymbol{x} - \boldsymbol{x}')\delta(t - t'), \tag{6.194}$$

the Langevin equation (6.193) is equivalent to the functional Fokker-Planck equation:

$$f^{\cdot} = \int d^n x \left\{ \frac{\delta}{\delta q(\boldsymbol{x})} (\alpha q(\boldsymbol{x}) + \beta q(\boldsymbol{x})^3 - \gamma \Delta q(\boldsymbol{x})) + \frac{Q}{2} \frac{\delta^2}{\delta q(\boldsymbol{x})^2} \right\} f. \tag{6.195}$$

Its stationary solution is given (generalizing (6.116)) by

$$f = \mathcal{N} \exp\left[-\frac{2}{Q} \int d^n x \left\{ \frac{\alpha}{2} q(\boldsymbol{x})^2 + \frac{\beta}{4} q(\boldsymbol{x})^4 + \frac{\gamma}{2} (\nabla q(\boldsymbol{x}))^2 \right\} \right]. \tag{6.196}$$

A direct solution of the *nonlinear* equation (6.193) or of the *time-dependent* Fokker-Planck equation (6.195) appears rather hopeless, since computers are necessary for the solution of the corresponding equations even with an $\boldsymbol{x}$-independent q. Thus we first study for $\alpha > 0$ the linearized equation (6.193), i.e.,

$$\dot{q} = -\alpha q + \gamma \Delta q + F. \tag{6.197}$$

It can be easily solved by Fourier analyzing $q(\boldsymbol{x}, t)$ and $F(\boldsymbol{x}, t)$. Adopting the correlation function (6.194), we may calculate the two-point correlation function

$$\langle q(\boldsymbol{x}', t')q(\boldsymbol{x}, t)\rangle. \tag{6.198}$$

We quote one case: one dimension, equal times, $t' = t$, but different space points:

$$\langle q(x', t)q(x, t)\rangle = Q/(\alpha\gamma)^{1/2} \exp\left(-(\alpha/\gamma)^{1/2}|x' - x|\right). \tag{6.199}$$

The factor of $|x' - x|$ in the exponential function has the meaning of $(\text{length})^{-1}$.

We therefore put $l_c = (\alpha/\gamma)^{-1/2}$. Since (6.199) describes the *correlation* between two space points, l_c is called correlation length. Apparently

$$l_c \to \infty \text{ as } \alpha \to 0,$$

at least in the linearized theory. The exponent μ in $l_c \propto \alpha^\mu$ is called critical exponent. In our case, $\mu = -1/2$. The correlation function $\langle q(x', t)q(x, t)\rangle$ has been evaluated for the nonlinear case by a computer calculation (see end of this chapter). In many practical cases the order parameter $q(\boldsymbol{x})$ is a complex quantity. We denote it by $\xi(\boldsymbol{x})$. The former equations must then be replaced by the following:

Langevin equation

$$\dot{\xi} = -\alpha\xi - \beta|\xi|^2\xi + \gamma\Delta\xi + F. \tag{6.200}$$

Correlation function of fluctuating forces

$$\langle FF\rangle = 0, \langle F^*F^*\rangle = 0, \tag{6.201}$$

$$\langle F^*(\boldsymbol{x}', t')F(\boldsymbol{x}, t)\rangle = Q\delta(\boldsymbol{x} - \boldsymbol{x}')\delta(t - t'). \tag{6.202}$$

Fokker-Planck equation

$$\dot{f} = \int d^n x \left\{\frac{\delta}{\delta\xi(\boldsymbol{x})}(\alpha\xi(\boldsymbol{x}) + \beta|\xi(\boldsymbol{x})|^2\xi(\boldsymbol{x}) - \gamma\Delta\xi(\boldsymbol{x})) + c.c. + \frac{Q}{2}\frac{\delta^2}{\delta\xi(\boldsymbol{x})\delta\xi^*(\boldsymbol{x})}\right\} f. \tag{6.203}$$

Stationary solution of Fokker-Planck equation

$$f = \mathcal{N} \exp\left[-\frac{2}{Q}\int d^n x \left\{\alpha|\xi(\boldsymbol{x})|^2 + \frac{\beta}{2}|\xi(\boldsymbol{x})|^4 + \gamma|\nabla\xi(\boldsymbol{x})|^2\right\}\right]. \tag{6.204}$$

A typical correlation function reads e.g.,

$$\langle \xi^*(\boldsymbol{x}', t')\xi(\boldsymbol{x}, t)\rangle. \tag{6.205}$$

For equal times, $t = t'$, the correlation functions (6.198) and (6.205) have been determined for the *nonlinear* one-dimensional case (i.e., $\beta \neq 0$) by using path-integrals and a computer calculation. To treat the real q and the complex ξ in the same way, we put

$$\left.\begin{matrix} q(x) \\ \xi(x) \end{matrix}\right\} \equiv \Psi(x).$$

It can be shown that the amplitude correlation functions (6.198) and (6.205) can be written in a good approximation, as

$$\langle \Psi^*(x', t)\Psi(x, t)\rangle = \langle|\Psi|\rangle^2 \exp(-l_1^{-1}|x - x'|). \tag{6.206}$$

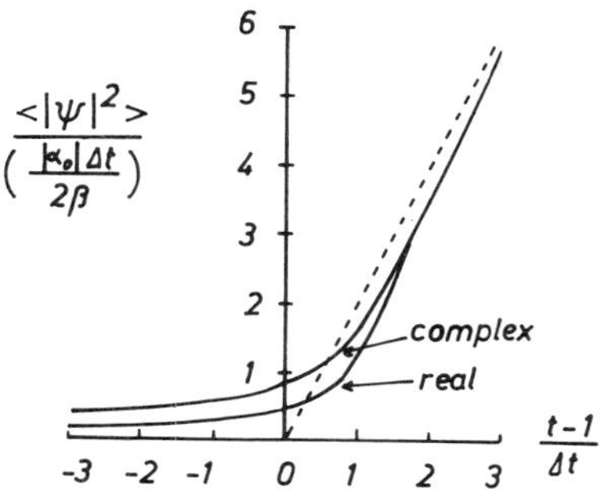

Fig. 6.15. $\langle|\Psi|^2\rangle$ versus "pump parameter" $(t-1)/\Delta t$. (Compare text) $\Delta t=2(\beta Q/2l_0\alpha_0^2)^{2/3}$, $l_0=(2\gamma/\alpha_0)^{1/2}$ (Redrawn after Scalapino et al.)

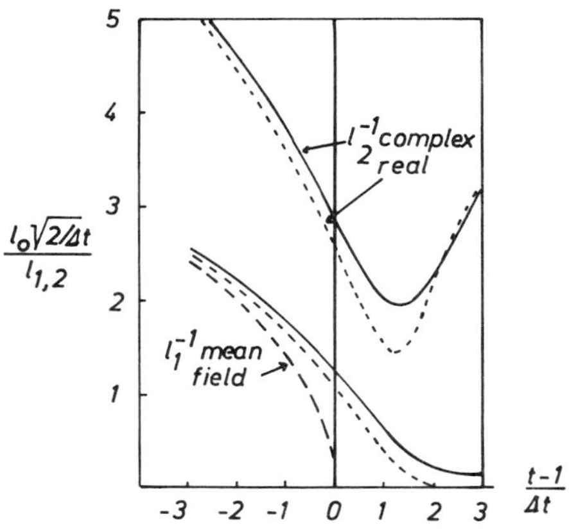

Fig. 6.16. Inverse correlation lengths l_1 and l_2 for real (dashed) and complex (solid) fields. Dashed curve: linearized theory ("mean field theory"). (Redrawn after Scalapino et al.)

i.e., they can again be expressed by a single correlation length, l_1. $\langle|\Psi|^2\rangle$ is the average of Ψ over the steady state distribution. For reasons which will transpire later we put $a = \alpha_0(1 - t)$. (For illustration we mention a few examples. In superconductors, $t = T/T_c$, where T: absolute temperature, T_c: critical temperature, $\alpha_0 < 0$; in lasers, $t = D/D_c$, where D: (unsaturated) atomic inversion, D_c: critical inversion, $\alpha_0 > 0$; in chemical reactions: $t = b/b_c$ where b concentration of a certain chemical, b_c critical concentration, etc.) We further define the length $l_0 = (2\gamma/\alpha_0)^{1/2}$. Numerical results near threshold are exhibited in Figs. 6.15 and 6.16 for $\langle|\Psi|^2\rangle$ and l_1^{-1}, l_2^{-1} respectively. The intensity correlation function can be (again approximately) expressed as

$$\langle|\Psi(x', t)|^2|\Psi(x, t)|^2\rangle - \langle|\Psi(x', t)|^2\rangle\langle|\Psi(x, t)|^2\rangle = \{\langle|\Psi|^4\rangle - \langle|\Psi|^2\rangle^2\} \exp(-l_2^{-1}|x - x'|). \tag{6.207}$$

Numerical results for l_2^{-1} near threshold are exhibited in Fig. 6.16.

7. Self-Organization

Long-Living Systems Slave Short-Living Systems

In this chapter we come to our central topic, namely, organization and self-organization. Before we enter into the mathematical treatment, let us briefly discuss what we understand by these two words in ordinary life.

a) *Organization*

Consider, for example, a group of workers. We then speak of organization or, more exactly, of organized behavior if each worker acts in a well-defined way on given external orders, i.e., by the boss. It is understood that the thus-regulated behavior results in a joint action to produce some product.

b) *Self-Organization*

We would call the same process as being self-organized if there are no external orders given but the workers work together by some kind of mutual understanding, each one doing his job so as to produce a product.

Let us now try to cast this rather vague description of what we understand by organization or self-organization into rigorous mathematical terms. We have to keep in mind that we have to develop a theory applicable to a large class of different systems comprising not only the above-mentioned case of sociological systems but, still more, physical, chemical, and biological systems.

7.1 Organization

The above-mentioned orders of the boss are the cause for the subsequent action of the workers. Therefore we have to express causes and actions in mathematical terms. Consider to this end an example from mechanics: Skiers on a skilift pulled uphill by the lift. Here the causes are the forces acting on the skiers. The action consists in a motion of the skiers. Quite another example comes from chemistry: Consider a set of vessels into which different chemicals are poured continuously. This input causes a reaction, i.e., the output of new chemicals. At least in these examples we are able to express causes and actions quantitatively, for example by the velocity of skiers or the concentrations of the produced chemicals.

Let us discuss what kind of equations we will have for the relations between causes and actions (effects). We confine our analysis to the case where the action (effect), which we describe by a quantity q, changes in a small time interval Δt by an amount proportional to Δt and to the size F of the cause. Therefore, mathematically, we consider only equations of the type

$$\dot{q}(t) = F_0(q(t); t).$$

Furthermore we require that without external forces, there is no action or no output. In other words, we wish $q = 0$ in the absence of external forces. Furthermore, we require that the system come back to the state $q = 0$ when the force is switched off. Thus we require that the system is stable and damped for $F = 0$. The simplest equation of this type is

$$\dot{q} = -\gamma q, \tag{7.1}$$

where γ is a damping constant. When an external "force" F is added, we obtain the simple equation

$$\dot{q} = -\gamma q + F(t). \tag{7.2}$$

In a chemical reaction, F will be a function of the concentration of chemical reactants. In the case of population dynamics, F could be the food supply, etc. The solution of (7.2) can be written in the form

$$q(t) = \int_0^t e^{-\gamma(t-\tau)} F(\tau)\, d\tau, \tag{7.3}$$

where we shall neglect here and lateron transient effects. (7.3) is a simple example for the following relation: The quantity q represents the response of the system with respect to the applied force $F(\tau)$. Apparently the value of q at time t depends not only on the "orders" given at time t but also given in the past. In the following we wish to consider a case in which the system reacts instantaneously i.e., in which $q(t)$ depends only on $F(t)$. For further discussion we put for example

$$F(t) = a e^{-\delta t}. \tag{7.4}$$

The integral (7.3) with (7.4) can immediately be performed yielding

$$q(t) = \frac{a}{\gamma - \delta} (e^{-\delta t} - e^{-\gamma t}). \tag{7.5}$$

With help of formula (7.5) we can quantitatively express the condition under which q acts instantaneously. This is the case if $\gamma \gg \delta$, namely

$$q(t) \approx \frac{a}{\gamma} e^{-\delta t} \equiv \frac{1}{\gamma} F(t), \tag{7.6}$$

or, in other words, the time constant $t_0 = 1/\gamma$ inherent in the system must be much shorter than the time constant $t' = 1/\delta$ inherent in the orders. We shall refer to this assumption, which turns out to be of fundamental importance for what follows, as "adiabatic approximation". We would have found the same result (7.6) if we had put $\dot{q} = 0$ in eq. (7.2) from the very beginning, i.e., if we had solved the equation

$$0 = -\gamma q + F(t). \tag{7.7}$$

We generalize our considerations in a way which is applicable to quite a number of practical systems. We consider a set of subsystems distinguished by an index μ. Each subsystem may be described by a whole set of variables $q_{\mu,1} \ldots q_{\mu,2}$. Furthermore a whole set of "forces" $F_1 \ldots F_m$ is admitted. We allow for a coupling between the q's and for coupling coefficients depending on the external forces F_j. Finally the forces may occur in an inhomogeneous term as in (7.2), where this term may be a complicated nonlinear function of the F_j's. Thus, written in matrix form, our equations are

$$\dot{\boldsymbol{q}}_\mu = A\boldsymbol{q}_\mu + B(\boldsymbol{F})\boldsymbol{q}_\mu + \boldsymbol{C}(\boldsymbol{F}), \tag{7.8}$$

where A and B are matrices independent of $\boldsymbol{q}_\mu$. We require that all matrix elements of B which are linear or nonlinear functions of the $\boldsymbol{F}$'s vanish when $\boldsymbol{F}$ tends to zero. The same is assumed for $\boldsymbol{C}$. To secure that the system (7.8) is damped in the absence of external forces (or, in other words, that the system is stable) we require that the eigenvalues of the matrix A have all negative real parts

$$\operatorname{Re} \lambda < 0. \tag{7.9}$$

Incidentally this guarantees the existence of the inverse of A, i.e., the determinant is unequal zero

$$\det A \neq 0. \tag{7.10}$$

Furthermore, on account of our assumptions on B and C, we are assured that the determinant

$$\det |A + B(\boldsymbol{F})| \tag{7.11}$$

does not vanish, provided that the $\boldsymbol{F}$'s are small enough. Though the set of equations (7.8) is linear in $\boldsymbol{q}_\mu$, a general solution is still a formidable task. However, within an adiabatic elimination technique we can immediately provide an explicit and unique solution of (7.8). To this end we assume that the $\boldsymbol{F}$'s change much more slowly than the free system $\boldsymbol{q}_\mu$. This allows us in exactly the same way as discussed before to put

$$\dot{\boldsymbol{q}}_\mu \approx 0 \tag{7.12}$$

so that the differential equations (7.8) reduce to simple algebraic equations which are solved by

$$\boldsymbol{q}_\mu = -(A + B(\boldsymbol{F}))^{-1}\boldsymbol{C}(\boldsymbol{F}). \tag{7.13}$$

Note that A and B are matrices whereas $\boldsymbol{C}$ is a vector. For practical applications an important generalization of (7.8) is given by allowing for different quantities A, B and $\boldsymbol{C}$ for different subsystems, i.e., we make the replacements

$$\left.\begin{aligned} A &\to A^{(\mu)} \\ B &\to B^{(\mu)} \\ \boldsymbol{C} &\to \boldsymbol{C}^{(\mu)} \end{aligned}\right\} \tag{7.14}$$

in (7.13). The response $\boldsymbol{q}_\mu$ to the $\boldsymbol{F}$'s is unique and instantaneous and is in general a nonlinear function of the $\boldsymbol{F}$'s.

Let us consider some further generalizations of (7.8) and discuss its usefulness. a) (7.8) could be replaced by equations containing higher order derivatives of $\boldsymbol{q}_\mu$ with respect to time. Since equations of higher order can always be reduced to a set of equations of lower order e.g. of first order, this case is already contained in (7.8). b) B and $\boldsymbol{C}$ could depend on $\boldsymbol{F}$'s of earlier times. This case presents no difficulty here provided we still stick to the adiabatic elimination technique, but it leads to an enormous difficulty when we treat the case of selforganization below. c) The right hand side of (7.8) could be a nonlinear function of $\boldsymbol{q}_\mu$. This case may appear in practical applications. However, we can exclude it if we assume that the systems μ are heavily damped. In this case the $\boldsymbol{q}$'s are relatively small and the rhs of (7.8) can be expanded in powers of $\boldsymbol{q}$ where in many cases one is allowed to keep only linear terms. d) A very important generalization must be included later. (7.8) is a completely causal equation, i.e., there are no fluctuations allowed for. In many practical systems fluctuations play an important rôle, however.

In summary we can state the following: To describe organization quantitatively we will use (7.8) in the adiabatic approximation. They describe a fairly wide class of responses of physical, chemical and biological and, as we will see later on, sociological systems to external causes. A note should be added for physicists. A good deal of present days physics is based on the analysis of response functions in the nonadiabatic domain. Certainly further progress in the theory of selforganization (see below) will be made when such time-lag effects are taken into account.

7.2 Self-Organization

A rather obvious step to describe self-organization consists in including the external forces as parts of the whole system. In contrast to the above-described cases, however, we must not treat the external forces as given fixed quantities but rather as obeying by themselves equations of motion. In the simplest case we have only one force and one subsystem. Identifying now $\boldsymbol{F}$ with q_1 and the former variable $\boldsymbol{q}$ with q_2, an explicit example of such equations is

$$\dot{q}_1 = -\gamma_1 q_1 - a q_1 q_2, \tag{7.15}$$

$$\dot{q}_2 = -\gamma_2 q_2 + b q_1^2. \tag{7.16}$$

Again we assume that the system (7.16) is damped in the absence of the system

(7.15), which requires $\gamma_2 > 0$. To establish the connection between the present case and the former one we want to secure the validity of the adiabatic technique. To this end we require

$$\gamma_2 \gg \gamma_1. \tag{7.17}$$

Though γ_1 appears in (7.15) with the minus sign we shall later allow for both $\gamma_1 \gtrless 0$. On account of (7.17) we may solve (7.16) approximately by putting $\dot{q}_2 = 0$ which results in

$$q_2(t) \approx \gamma_2^{-1} b q_1^2(t). \tag{7.18}$$

Because (7.18) tells us that the system (7.16) follows immediately the system (7.15) the system (7.16) is said to be slaved by the system (7.15). However, the slaved system reacts on the system (7.15). We can substitute q_2 (7.18) in (7.15). Thus we obtain the equation

$$\dot{q}_1 = -\gamma_1 q_1 - \frac{ab}{\gamma_2} q_1^3, \tag{7.19}$$

which we have encountered before in Section 5.1. There we have seen that two completely different kinds of solutions occur depending on whether $\gamma_1 > 0$ or $\gamma_1 < 0$. For $\gamma_1 > 0$, $q_1 = 0$, and thus also $q_2 = 0$, i.e., no action occurs at all. However, if $\gamma_1 < 0$, the steady state solution of (7.19) reads

$$q_1 = \pm(|\gamma_1|\gamma_2/ab)^{1/2} \tag{7.20}$$

and consequently $q_2 \neq 0$ according to (7.18). Thus the system, consisting of the two subsystems (7.15) and (7.16) has internally decided to produce a finite quantity q_2, i.e., nonvanishing action occurs. Since $q_1 \neq 0$ or $q_1 = 0$ are a measure if action or if no action occurs, we could call q_1 an action parameter.

For reasons which will become obvious below when dealing with complex systems, q_1 describes the degree of order. This is the reason why we shall refer to q_1 as "order parameter". In general we shall call variables, or, more physically spoken, modes "order parameters" if they slave subsystems. This example lends itself to the following generalization: We deal with a whole set of subsystems which again are described by several variables. For all these variables we now use a single kind of suffices running from 1 to n. For the moment being we assume these equations in the form

$$\begin{aligned}
\dot{q}_1 &= -\gamma_1 q_1 + g_1(q_1, \ldots, q_n),\\
\dot{q}_2 &= -\gamma_2 q_2 + g_2(q_1, \ldots, q_n),\\
&\vdots\\
\dot{q}_n &= -\gamma_n q_n + g_n(q_1, \ldots, q_n).
\end{aligned} \tag{7.21}$$

To proceed further, we imagine that we have arranged the indices in such a way that there are now two distinct groups in which $i = 1, \ldots, m$ refers to modes with small damping which can even become unstable modes (i.e., $\gamma \lesssim 0$) and another group with $s = m + 1, \ldots, n$ referring to stable modes. It is understood that the functions g_j are *nonlinear* functions of $q_1, \ldots, q_n$ (with no constant or linear terms) so that in a first approximation these functions can be neglected compared to the linear terms on the right hand side of equations (7.21). Because

$$\gamma_i \to 0, \text{ but } \gamma_s > 0 \text{ and finite,} \qquad i = 1, \ldots, m; \; s = m + 1, \ldots, n \tag{7.22}$$

holds we may again invoke the adiabatic approximation principle putting $\dot{q}_s = 0$. Furthermore we assume that the $|q_s|$'s are much smaller than the $|q_i|$'s which is motivated by the size of the γ_s (but which must be checked explicitly in each practical case). As a consequence we may put all $q_s = 0$ in g_s. This allows us to solve the (7.21) for $s = m + 1, \ldots, n$ with $q_1 \ldots q_m$ as given quantities:

$$\gamma_s q_s = g_s(q_1, \ldots, q_n), \; s = m + 1, \ldots, n. \tag{7.23}$$

where $q_{m+1}, \ldots, q_n$ must be put equal zero in g_s. Reinserting (7.23) into the first m equations of (7.21) leads us to nonlinear equations for the q_i's alone.

$$\dot{q}_i = -\gamma_i q_i + g_i(q_1, \ldots, q_m; q_{m+1}(q_i), \ldots). \tag{7.24}$$

The solutions of these equations then determine whether nonzero action of the subsystems is possible or not. The simplest example of (7.24) leads us back to an equation of the type (7.19) or of e.g., the type

$$\dot{q}_1 = -\gamma_1 q_1 + a q_1^2 + b q_1^3. \tag{7.25}$$

The equations (7.21) are characterized by the fact that we could group them into two clearly distinct groups with respect to their damping, or, in other words, into stable and (virtually) unstable variables (or "modes"). We now want to show that self-organized behavior need not be subject to such a restriction. We rather start now with a system in which from the very beginning the q's need not belong to two distinct groups. We rather consider the following system of equations

$$\dot{q}_j = h_j(q_1, \ldots, q_n), \tag{7.26}$$

where h_j are in general nonlinear functions of the q's. We assume that the system (7.26) is such that it allows for a time independent solution denoted by q_j^0. To understand better what follows, let us have a look at the simpler system (7.21). There, the rhs depends on a certain set of parameters, namely, the γ's. Thus in a more general case we shall also allow that the h_j's on the rhs of (7.26) depend on parameters called $\sigma_1, \ldots, \sigma_l$. Let us first assume that these parameters are chosen in such a way that the q^0's represent stable values. By a shift of the origin of the coordinate system of the q's we can put the q^0's equal to zero. This state will be referred to as the quiescent state in which no action occurs. In the following we put

$$q_j(t) = q_j^0 + u_j(t) \quad \text{or} \quad \boldsymbol{q}(t) = \boldsymbol{q}^0 + \boldsymbol{u}(t) \tag{7.27}$$

and perform the same steps as in the stability analysis (compare Sect. 5.3 where $\xi(t)$ is now called $u(t)$). We insert (7.27) into (7.26). Since the system is stable we may assume that the u_j's remain very small so that we can linearize the equations (7.26) (under suitable assumptions about the h_j's, which we don't formulate here explicitly). The linearized equations are written in the form

$$\dot{u}_j = \sum_{j'} L_{jj'} u_{j'}, \tag{7.28}$$

where the matrix element $L_{jj'}$ depends on $\boldsymbol{q}^0$ and simultaneously on the parameters $\sigma_1, \sigma_2, \ldots$. In short we write instead of (7.28)

$$\dot{\boldsymbol{u}} = \boldsymbol{L}\boldsymbol{u}. \tag{7.29}$$

(7.28) or (7.29) represent a set of first order differential equations with constant coefficients. Solutions can be found as in 5.3 in the form

$$\boldsymbol{u} = \boldsymbol{u}^{(\mu)}(0) \exp(\lambda_\mu t), \tag{7.30}$$

where λ_μ's are the eigenvalues of the problem

$$\lambda_\mu \boldsymbol{u}^{(\mu)}(0) = \boldsymbol{L}\boldsymbol{u}^{(\mu)}(0) \tag{7.31}$$

and $\boldsymbol{u}^{(\mu)}(0)$ are the right hand eigenvectors. The most general solution of (7.28) or (7.29) is obtained by a superposition of (7.30)

$$\boldsymbol{u} = \sum_\mu \xi_\mu \exp(\lambda_\mu t) \boldsymbol{u}^{(\mu)}(0) \tag{7.32}$$

with arbitrary constant coefficients ξ_μ. We introduce left-hand eigenvectors $\boldsymbol{v}^{(\mu)}$ which obey the equation

$$\lambda_\mu \boldsymbol{v}^{(\mu)} = \boldsymbol{v}^{(\mu)} \boldsymbol{L}. \tag{7.33}$$

Because we had assumed that the system is stable the real part of all eigenvalues λ_μ is negative. We now require that the decomposition (7.27) satisfies the original nonlinear equations (7.26) so that $\boldsymbol{u}(t)$ is a function still to be determined

$$\dot{\boldsymbol{u}} = \boldsymbol{L}\boldsymbol{u} + \boldsymbol{N}(\boldsymbol{u}). \tag{7.34}$$

We have encountered the linear part $\boldsymbol{L}\boldsymbol{u}$ in (7.28), whereas $\boldsymbol{N}$ stems from the residual nonlinear contributions. We represent the vector $\boldsymbol{u}(t)$ as a superposition of the right-hand eigenvectors (7.30) in the form (7.32) where, however, the ξ's are now unknown *functions of time*, and $\exp(\lambda_\mu t)$ is dropped. To find appropriate equations for the time dependent amplitudes $\xi_\mu(t)$ we multiply (7.34) from the left by

$$\boldsymbol{v}^{(\mu)}(0) \tag{7.35}$$

and observe the orthogonality relation

$$\langle \boldsymbol{v}^{(\mu)} \boldsymbol{u}^{(\mu')} \rangle = \delta_{\mu\mu'} \tag{7.36}$$

known from linear algebra. (7.34) is thus transformed into

$$\dot{\xi}_\mu = \lambda_\mu \xi_\mu + g_\mu(\xi_1, \xi_2, \ldots), \tag{7.37}$$

where

$$g_\mu = \langle \boldsymbol{v}^{(\mu)}, \boldsymbol{N}(\textstyle\sum_\mu \xi_\mu \boldsymbol{u}^{(\mu)}) \rangle. \tag{7.38}$$

The lhs (7.37) stems from the corresponding lhs of (7.34). The first term on the rhs of (7.37) comes from the first expression on the rhs of (7.34). Similarly, $\boldsymbol{N}$ is transformed into $\boldsymbol{g}$. Note that g_μ is a nonlinear function in the ξ's. When we now identify ξ_μ with q_j and λ_μ with $-\gamma_j$ we notice that the equations (7.37) have exactly the form of (7.21) so that we can immediately apply our previous analysis. To do this we change the parameters $\sigma_1, \sigma_2, \ldots$ in such a way that the system (7.28) becomes unstable, or in other words, that one or several of the λ_μ's acquire a vanishing or positive real part while the other λ's are still connected with damped modes. The modes ξ_μ with $\text{Re}\,\lambda_\mu \geqq 0$ then play the role of the order parameters which slave all the other modes. In this procedure we assume explicitly that there are two *distinct* groups of λ_μ's for a given set of parameter values of $\sigma_1, \ldots, \sigma_n$.

In many practical applications (see Sect. 8.2) only one or very few modes ξ_μ become unstable, i.e., $\text{Re}\,\lambda_\mu \geqq 0$. If all the other modes ξ_μ remain damped, which again is satisfied in many practical applications, we can safely apply the adiabatic elimination procedure. The important consequence lies in the following: Since all the damped modes follow the order parameters adiabatically, the behavior of the whole system is determined by the behavior of very few order parameters. Thus even very complex systems may show a well regulated behavior. Furthermore we have seen in previous chapters that order parameter equations may allow for bifurcation. Consequently, complex system can operate in different "modes", which are well defined by the behavior of the order parameters (Fig. 7.1). The above sections of the present chapter are admittedly somewhat abstract. We therefore strongly advise students to repeat all the individual steps by an explicit example presented in the exercise. In practice, we often deal with a hierarchical structure, in which the relaxation constants can be grouped so that

$$\gamma^{(1)} \gg \gamma^{(2)} \gg \gamma^{(3)} \quad \ldots$$

In this case one can apply the adiabatic elimination procedure first to the variables

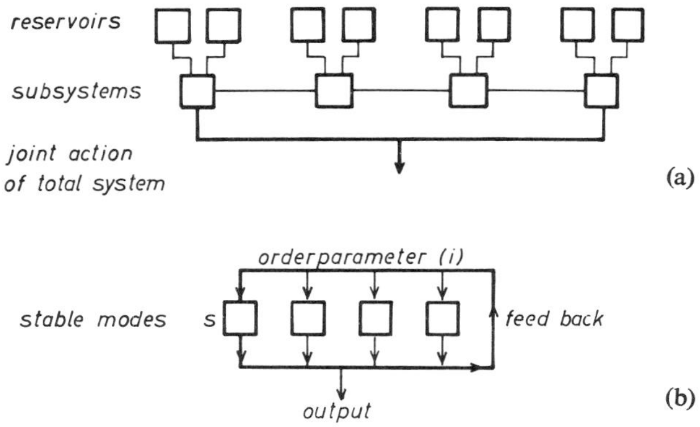

Fig. 7.1. Typical example of a system composed of interacting subsystems each coupled to reservoirs (a). The reservoirs contain many degrees of freedom and we have only very limited knowledge. They are treated by information theory (or in physics by thermodynamics or statistical physics). After elimination of the reservoir variables, the "unstable" modes of the subsystems are determined and serve as order parameters. The stable modes generate in many cases a feedback loop selecting and stabilizing certain configurations of the order parameters (b)

connected with $\gamma^{(1)}$, leaving us with the other variables. Then we can apply this methods to the variables connected with $\gamma^{(2)}$ and so on.

There are three generalizations of our present treatment which are absolutely necessary in quite a number of important cases of practical interest.

1) The equations (7.26) or equivalently (7.37) suffer from a principle draw back. Assume that we have first parameters σ_1, σ_2 for the stable regime. Then all $u = 0$ in the stationary case, or, equivalently all $\xi = 0$ in that case. When we now go over to the unstable regime, $\xi = 0$ remains a solution and the system will never go to the new bifurcated states. Therefore to understand the onset of self-organization, additional considerations are necessary. They stem from the fact that in practically all systems fluctuations are present which push the system away from the unstable points to new stable points with $\xi \neq 0$ (compare Sect. 7.3).

2) The adiabatic elimination technique must be applied with care if the λ's have imaginary parts. In that case our above procedure can be applied only if also the imaginary parts of $\lambda_{\text{unstable}}$ are much smaller than the real parts of λ_{stable}.

3) So far we have been considering only discrete systems, i.e., variables q depending on discrete indices j. In continuously extended media, such as fluids, or in continuous models of neuron networks, the variables q depend on space-points $\boldsymbol{x}$ in a continuous fashion.

The points 1–3 will be fully taken into account by the method described in Sections 7.7 to 7.11.

In our above considerations we have implicitely assumed that after the adiabatic elimination of stable modes we obtain equations for *order parameters* which are *now stabilized* (cf. (7.19)). This is a self-consistency requirement which must be checked in the individual case. In Section 7.8 we shall present an extension of the present procedure which may yield stabilized order-parameters in higher order of a certain iteration scheme. Finally there may be certain exceptional cases where the damping constants γ_s of the damped modes are not large enough. Consider for example the equation for a damped mode

$$\dot{q}_s = -\gamma_s q_s + q_i q_s + b q_i^2.$$

If the order parameter q_i can become too large so that $q_i - \gamma_s > 0$, the adiabatic elimination procedure breaks down, because $\gamma_{\text{eff}} \equiv -\gamma_s + q_i > 0$. In such a case, very interesting new phenomena occur, which are used e.g., in electronics to construct the so-called "universal circuit". We shall come back to this question in Section 12.4.

Exercise on 7.2

Treat the equations

$$\dot{q}_1 = -q_1 + \beta q_2 - a(q_1^2 - q_2^2) \equiv h_1(q_1, q_2), \tag{E.1}$$

$$\dot{q}_2 = \beta q_1 - q_2 + b(q_1 + q_2)^2 \equiv h_2(q_1, q_2), \tag{E.2}$$

using the steps (7.27) till (7.38). β is to be considered as a parameter ≥ 0, starting from $\beta = 0$. Determine $\beta = \beta_c$, so that $\lambda_1 = 0$.

7.3 The Role of Fluctuations: Reliability or Adaptibility? Switching

The typical equations for self-organizing systems are intrinsically homogeneous, i.e., $\boldsymbol{q} = 0$ must be a solution (except for a trivial displacement of the origin of $\boldsymbol{q}$). However, if the originally inactive system is described by $\boldsymbol{q} = 0$, it remains at $\boldsymbol{q} = 0$ forever and no self-organization takes place. Thus we must provide a certain initial push or randomly repeated pushes. This is achieved by random forces which we came across in chapter 6. In all explicit examples of natural systems such fluctuations occur. We quote just a few: laser: spontaneous emission of light, hydrodynamics: hydrodynamic fluctuations, evolution: mutations.

Once self-organization has occurred and the system is in a certain state $\boldsymbol{q}^{(1)}$, fluctuations drive the system to explore new states. Consider Fig. 7.2 where the system is described by a single order parameter q. Without fluctuations the system would

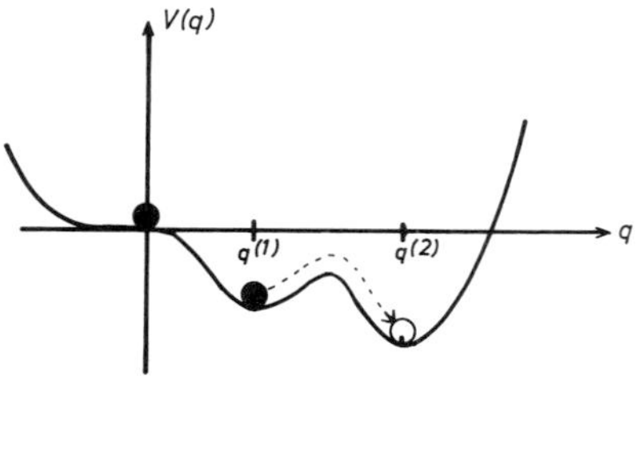

Fig. 7.2. (compare text)

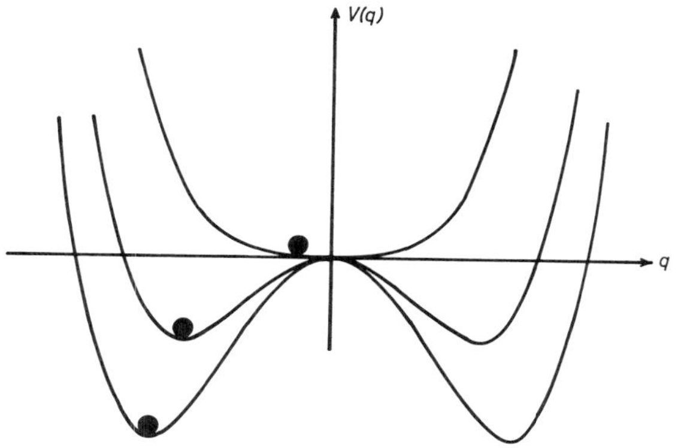

Fig. 7.3. Switching of a device by deformation of potential $V(q)$ (after Landauer)

never realize that at $q = q^{(2)}$ there is a still more stable state. However, fluctuations can drive the system from $q^{(1)}$ to $q^{(2)}$ by some diffusion process. Since q describes the macroscopic performance of a system, among the new states may be those in which the system is better adapted to its surroundings. When we allow for an ensemble of such systems and admit competition, selection will set in (compare Sect. 10.3). Thus the interplay between fluctuations and selection leads to an evolution of systems.

On the other hand certain devices, e.g., tunnel diodes in electronics operate in states described by an order parameter q with an effective potential curve e.g. that of Fig. 7.3. By external means we can bring the tunnel diode (or other systems) into the state $q^{(1)}$ and thus store information at $q^{(1)}$. With this state a certain macroscopic feature of the system is connected (e.g., a certain electrical current). Thus we can measure the state of the system from the outside, and the device serves as a *memory*. However, due to fluctuations, the system may diffuse to state $q^{(2)}$ thus losing its memory. Therefore the *reliability* of the device is lowered due to fluctuations.

To process information we must be able to *switch* a device, thus bringing it from $q^{(1)}$ to $q^{(2)}$. This can be done by letting the potential barrier become first lower and lower (cf. Fig. 7.3). Diffusion then drives the system from $q^{(1)}$ to $q^{(2)}$. When the potential barrier is increased again, the system is now trapped at $q^{(2)}$. It may be that nature supports evolution by changing external parameters so that the just-described switching process becomes effective in developing new species. Therefore the size of the order parameter fluctuations is crucial for the performance of a system, acting in two opposite ways: adaptibility and ease of switching require large fluctuations and flat potential curves, whereas reliability requires small fluctuations and deep potential valleys. How can we control the size of fluctuations? In self-organizing systems which contain several (identical) subsystems this can be achieved by the number of components. For a fixed, (i.e., nonfluctuating) order parameter each subsystem, s, has a deterministic output $q_d^{(s)}$ and a randomly fluctuating output $q_r^{(s)}$. Let us assume that the total output $q^{(s)}$ is additive

$$q^{(s)} = q_d^{(s)} + q_r^{(s)}. \tag{7.39}$$

Let us further assume that $q^{(s)}$ are independent stochastic variables. The total output $q_{\text{total}} = \sum_s q^{(s)}$ can then be treated by the central limit theorem (cf. Sect. 2.15). The total output increases with the number N of subsystems, while the fluctuations increase only with $\sqrt{N}$. Thus we may control the reliability and adaptibility just by the number of subsystems. Note that this estimate is based on a linear analysis (7.39). In reality, the feedback mechanism of nonlinear equations leads to a still more pronounced suppression of noise as we will demonstrate most explicitly in the laser case (Chapter 8).

In conclusion, we discuss reliability of a system in the sense of stability against malfunctioning of some of its subsystems. To illustrate a solution that self-organizing systems offer let us consider the laser (or the neuron network). Let us assume that laser light emission occurs in a regular fashion by atoms all emitting light at the same frequency ω_0. Now assume that a number of atoms get a different transition frequency ω_1. While in a usual lamp both lines ω_0 and ω_1 appear—indicating a malfunction—the laser (due to nonlinearities) keeps emitting *only* ω_0. This is a consequence of the competition between different order parameters (compare Sect. 5.4). The same behavior can be expected for neurons, when some neurons would try to fire at a different rate. In these cases the output signal becomes merely somewhat weaker, but it retains its characteristic features. If we allow for fluctuations of the subsystems, the fluctuations remain small and are outweighed by the "correct" macroscopic order parameter.

7.4* Adiabatic Elimination of Fast Relaxing Variables from the Fokker-Planck Equation

In the foregoing Sections 7.1 and 7.2 we described methods of eliminating fast variables from the equations of motion. In several cases, e.g., chemical reaction dynamics, a Fokker-Planck equation is more directly available than the corresponding Langevin equations. Therefore it is sometimes desirable to perform the adiabatic elimination technique with the Fokker-Planck equation. We explain the main ideas by means of the example (7.15) (7.16) where now fluctuations are included. To distinguish explicitly between the "unstable" and "stable" modes, we replace the indices

1 by u ("unstable")

2 by s ("stable")

The Fokker-Planck equation corresponding to (7.15) and (7.16) reads

$$f(q_u, q_s)^{\cdot} = \Big\{\frac{\partial}{\partial q_u}\underbrace{(\gamma_u q_u + a q_u q_s)}_{-F_u} + \frac{\partial}{\partial q_s}\underbrace{(\gamma_s q_s - b q_u^2)}_{-F_s}\Big\} f(q_u, q_s)$$
$$+ \tfrac{1}{2}\left(Q_u \frac{\partial^2}{\partial q_u^2} + Q_s \frac{\partial^2}{\partial q_s^2}\right) f(q_u, q_s). \tag{7.40}$$

We write the joint distribution function $f(q_u, q_s)$ in the form

$$f(q_u, q_s) = h(q_s \mid q_u) g(q_u), \tag{7.41}$$

where we impose the normalization conditions

$$\int h(q_s \mid q_u)\, dq_s = 1 \tag{7.42}$$

and

$$\int g(q_u)\, dq_u = 1. \tag{7.43}$$

Obviously, $h(q_s \mid q_u)$ can be interpreted as conditional probability to find q_s under the condition that the value of q_u is given. It is our goal to obtain an equation for $g(q_u)$, i.e., to eliminate q_s. Inserting (7.41) into (7.40) yields

$$\begin{aligned} \underline{\dot{g}}h + g\dot{h} = &-\underline{\frac{\partial}{\partial q_u}(F_u g h)} + \underline{\tfrac{1}{2} Q_u \frac{\partial^2 g}{\partial q_u^2} h} - \underline{g \frac{\partial}{\partial q_s}(F_s h)} + \underline{\tfrac{1}{2} Q_s g \frac{\partial^2 h}{\partial q_s^2}} \\ &+ Q_u \left(\frac{\partial g}{\partial q_u} \frac{\partial h}{\partial q_u} + \tfrac{1}{2} g \frac{\partial^2 h}{\partial q_u^2} \right). \end{aligned} \tag{7.44}$$

To obtain an equation for $g(q_u)$ alone we integrate (7.44) over q_s. Using (7.42) we obtain

$$\dot{g}(q_u) = -\frac{\partial}{\partial q_u} \underbrace{\int F_u h\, dq_s}_{=\hat{F}} g(q_u) + \tfrac{1}{2} Q_u \frac{\partial^2 g}{\partial q_u^2}. \tag{7.45}$$

As indicated we use the abbreviation

$$\hat{F}(q_u) = \int F_u(q_u, q_s) h(q_s \mid q_u)\, dq_s. \tag{7.46}$$

Apparently, (7.45) contains the still unknown function $h(q_s \mid q_u)$. A closer inspection of (7.44) reveals that it is a reasonable assumption to determine h by the equation

$$\dot{h} = -\frac{\partial}{\partial q_s} F_s h + \tfrac{1}{2} Q_s \frac{\partial^2 h}{\partial q_s^2} \equiv L_s h. \tag{7.47}$$

This implies that we assume that h varies much more slowly as a function of q_u than as a function of q_s, or, in other words, derivates of first (second) order of h with respect to q_u can be neglected compared to the corresponding ones with respect to q_s. As we will see explicitly, this requirement is fulfilled if γ_s is sufficiently large. Furthermore we require $\dot{h} = 0$ so that h has to fulfil the equation

$$L_s h = 0, \tag{7.48}$$

where L_s is the differential operator of the rhs of (7.47). As in our special example

$$F_s = -\gamma_s q_s + \varphi(q_u), \tag{7.49}$$

the solution of (7.48) can be found explicitly with aid of (6.129). (Note that q_u is here treated as a constant parameter)

$$h(q_s \mid q_u) = \mathcal{N} \exp\{-Q_s^{-1}\gamma_s(q_s - \varphi(q_u)/\gamma_s)^2\}. \tag{7.50}$$

Since in practically all applications F_u is composed of powers of q_s, the integral (7.46) can be explicitly performed. In our present example we obtain

$$\hat{F} = -\gamma_u q_u - \frac{1}{\gamma_s}\varphi(q_u)\cdot q_u. \tag{7.51}$$

Provided the adiabatic condition holds, our procedure can be applied to general functions F_u and F_s (instead of those in (7.40)). In one dimension, (7.48) can then be solved explicitly (cf. Section 6.4). But also in several dimensions (see exercise) (7.48) can be solved in a number of cases, e.g., if the potential conditions (6.114,115) hold. In view of Sections 7.6 and 7.7 we mention that this procedure applies also to functional Fokker-Planck equations.

Exercise on 7.4

Generalize the above procedure to a set of q_u's and q_s's, $u = 1, \ldots, k$, $s = 1, \ldots, l$.

7.5* Adiabatic Elimination of Fast Relaxing Variables from the Master Equation

In a number of applications one may identify slow and fast (stable) quantities, or, more precisely, random variables X_s, X_u, the probability distribution of which obeys a master equation. In that case we may eliminate the stable variables from the master equation in a way closely analogous to our previous procedure applied to the Fokker-Planck equation. We denote the values of the unstable (stable) variables by $\boldsymbol{m}_u(\boldsymbol{m}_s)$. The probability distribution P obeys the master equation (4.111)

$$\dot{P}(\boldsymbol{m}_s,\boldsymbol{m}_u; t) = \sum_{\boldsymbol{m}_s',\boldsymbol{m}_u'} w(\boldsymbol{m}_s,\boldsymbol{m}_u; \boldsymbol{m}_s',\boldsymbol{m}_u')P(\boldsymbol{m}_s',\boldsymbol{m}_u'; t) - P(\boldsymbol{m}_s,\boldsymbol{m}_u; t) \sum_{\boldsymbol{m}_s',\boldsymbol{m}_u'} w(\boldsymbol{m}_s',\boldsymbol{m}_u'; \boldsymbol{m}_s,\boldsymbol{m}_u). \tag{7.52}$$

We put

$$P(\boldsymbol{m}_s, \boldsymbol{m}_u, t) = G(\boldsymbol{m}_u)H(\boldsymbol{m}_s \mid \boldsymbol{m}_u), \tag{7.53}$$

and require

$$\sum_{m_s} H(m_s \mid m_u) = 1, \tag{7.54}$$

$$\sum_{m_u} G(m_u) = 1. \tag{7.55}$$

Inserting (7.53) into (7.52) and summing up over m_s on both sides yields the still exact equation

$$\dot{G}(m_u) = \sum_{m_u'} \tilde{w}(m_u; m_u')G(m_u') - G(m_u) \sum_{m_u'} \tilde{w}(m_u'; m_u), \tag{7.56}$$

where a little calculation shows that

$$\tilde{w}(m_u; m_u') = \sum_{m_s, m_s'} w(m_s, m_u; m_s', m_u') H(m_s' \mid m_u'). \tag{7.57}$$

To derive an equation for the conditional probability $H(m_s \mid m_u)$ we invoke the adiabatic hypothesis namely that X_u changes much more slowly than X_s. Consequently we determine H from that part of (7.52), in which transitions between m_s' and m_s occur for fixed $m_u = m_u'$: Requiring furthermore $\dot{H} = 0$, we thus obtain

$$\sum_{m_s'} w(m_s, m_u; m_s', m_u) H(m_s' \mid m_u) - H(m_s \mid m_u) \sum_{m_s'} w(m_s', m_u; m_s, m_u) = 0. \tag{7.58}$$

The equations (7.58) and (7.56) can be solved explicitly in a number of cases, e.g., if detailed balance holds. The quality of this procedure can be checked by inserting (7.53) with G and H determined from (7.56) and (7.58) into (7.52) and making an estimate of the residual expressions. If the adiabatic hypothesis is valid, these terms can be now taken into account as a small perturbation. Inserting the solution H of (7.58) into (7.57) yields an explicit expression for $\tilde{w}$, which can now be used in (7.56). This last equation determines the probability distribution G of the order parameters.

7.6 Self-Organization in Continuously Extended Media. An Outline of the Mathematical Approach

In this and the following sections we deal with equations of motion of continuously extended systems containing fluctuations. We first assume external parameters permitting only stable solutions and then linearize the equations, which define a set of modes. When external parameters are changed, the modes becoming unstable are taken as order parameters. Since their relaxation time tends to infinity, the damped modes can be eliminated adiabatically leaving us with a set of nonlinear coupled order-parameter equations. In two and three dimensions they allow for example for hexagonal spatial structures. Our procedure has numerous practical applications (see Chap. 8).

To explain our procedure we look at the general form of the equations which are used in hydrodynamics, lasers, nonlinear optics, chemical reaction models and related problems. To be concrete we consider macroscopic variables, though in many cases our procedure is also applicable to microscopic quantities. We denote the physical quantities by $\boldsymbol{U} = (U_1, U_2, \ldots)$ mentioning the following examples. In lasers, $\boldsymbol{U}$ stands for the electric field strength, for the polarization of the medium, and for the inversion density of laser active atoms. In nonlinear optics, $\boldsymbol{U}$ stands for the field strengths of several interacting modes. In hydrodynamics, $\boldsymbol{U}$ stands e.g., for the components of the velocity field, for the density, and for the temperature. In chemical reactions, $\boldsymbol{U}$ stands for the numbers (or densities) of molecules participating in the chemical reaction. In all these cases, $\boldsymbol{U}$ obeys equations of the following type

$$\frac{\partial}{\partial t} U_\mu = G_\mu(\nabla, U) + D_\mu \nabla^2 U_\mu + F_\mu(t); \quad \mu = 1, 2, \ldots, n. \tag{7.59}$$

In it, G_μ are nonlinear functions of $\boldsymbol{U}$ and perhaps of a gradient. In most applications like lasers or hydrodynamics, G is a linear or bilinear function of $\boldsymbol{U}$, though in certain cases (especially in chemical reaction models) a cubic coupling term may equally well occur. The next term in (7.59) describes diffusion (D real) or wave type propagation (D imaginary). In this latter case the second order time derivative of the wave equation has been replaced by the first derivative by use of the "slowly varying amplitude approximation". The $F_\mu(t)$'s are fluctuating forces which are caused by external reservoirs and internal dissipation and which are connected with the damping terms occurring in (7.59).

We shall not be concerned with the derivation of equations (7.59). Rather, our goal will it be to derive from (7.59) equations for the undamped modes which acquire a macroscopic size and determine the dynamics of the system in the vicinity of the instability point. These modes form a mode skeleton which grows out from fluctuations above the instability and thus describes the "embryonic" state of the evolving spatio-temporal structure.

7.7* Generalized Ginzburg-Landau Equations for Nonequilibrium Phase Transitions

We now start with a treatment of (7.59). We assume that the functions G_μ in (7.59) depend on external parameters $\sigma_1, \sigma_2, \ldots$ (e.g., energy pumped into the system). First we consider such values of σ so that $\boldsymbol{U} = \boldsymbol{U}_0$ is a stable solution of (7.59). At higher instabilities, $\boldsymbol{U}_0$ may be space- and time-dependent in the form

$$\boldsymbol{U}_l(\boldsymbol{x}) = \textstyle\sum_m T_{lm} \tilde{\boldsymbol{U}}_m \exp(i\boldsymbol{k}_m\boldsymbol{x} - i\omega_m t).$$

In a number of cases the dependence of $\boldsymbol{U}_l$ on $\boldsymbol{x}$ and t can be transformed away so that again a new space- and time-independent $\boldsymbol{U}_0$ (or $\tilde{\boldsymbol{U}}_0$) results. We then decompose $\boldsymbol{U}$

$$\boldsymbol{U} = \boldsymbol{U}_0 + \boldsymbol{q} \tag{7.60}$$

with

$$q = \begin{pmatrix} q_1(x, t) \\ \vdots \\ q_n(x, t) \end{pmatrix}. \tag{7.61}$$

Splitting the rhs of (7.59) into a linear part, Kq, and a nonlinear part q we obtain (7.59) in the form

$$\left(\frac{\partial}{\partial t} - K(\nabla^2)\right)q = g(q) + F(t). \tag{7.62}$$

In it the matrix

$$K = (\hat{K}_{\mu\nu}) \tag{7.63}$$

has the form

$$\hat{K}_{\mu\nu} = K_{\mu\nu} + \delta_{\mu\nu} D_\mu \nabla_\mu^2, \text{ where } K_{\mu\nu} = \left.\frac{\partial G_\mu}{\partial U_\nu}\right|_{U_{\nu,0}} \tag{7.64}$$

Our whole procedure applies, however, to a matrix K which depends in a general way on ∇. g is assumed in the form

$$g_i(q) = \sum_{\mu\nu} q_\mu g^{(2)}_{i\mu\nu}(\nabla) q_\nu + \sum_{\mu\nu\kappa} g^{(3)}_{i\mu\nu\kappa} q_\mu q_\nu q_\kappa. \tag{7.65}$$

$g^{(2)}$ may or may not depend on ∇, and equally $g^{(3)}$ may or may not depend on ∇ (If $g^{(3)}$ depends on ∇, the sequence of g and q must be chosen properly). We first deal with K and introduce operators

$$O^{(j)} = O^{(j)}(\nabla) \tag{7.66}$$

which still depend on ∇ and which are defined as eigenvectors satisfying the equation

$$K(\nabla) O^{(j)} = \lambda_j(\nabla) O^{(j)} \tag{7.67}$$

When ∇ is replaced by $i \cdot k$, (7.67) becomes a linear algebraic equation so that (7.67) can easily be solved without resorting to any operator techniques. We furthermore introduce eigenfunctions of the wave equation

$$\nabla^2 \chi_k(x) = -k^2 \chi_k(x), \tag{7.68}$$

provided K depends on ∇^2. χ_k can be chosen lateron adequately. We shall use a notation which suggests that the χ_k's are plane-wave solutions, but it may be advantageous to use other representations as well, e.g., Bessel functions or spherical wave functions, depending on the problem. If K depends on ∇ in odd powers (and in even powers), $\chi_{\boldsymbol{k}}$'s are taken as complex plane wave solutions of (7.68). Because in general K is not selfadjoint, we introduce the solutions $\overline{\boldsymbol{O}}$ of the adjoint equation

$$\overline{\boldsymbol{O}}^{(j)}K(\nabla) = \lambda_j(\nabla)\overline{\boldsymbol{O}}^{(j)}, \tag{7.69}$$

where

$$\overline{\boldsymbol{O}}^{(j)} = (\overline{O}_1^{(j)}, \ldots, \overline{O}_n^{(j)}) = \overline{\boldsymbol{O}}^{(j)}(\nabla), \tag{7.70}$$

and we may choose $\overline{\boldsymbol{O}}$ always so that

$$\overline{\boldsymbol{O}}^{(j)}\boldsymbol{O}^{(j')} = \delta_{jj'}. \tag{7.71}$$

The requirements (7.67), (7.69), (7.71) fix the $\overline{\boldsymbol{O}}$'s and $\boldsymbol{O}$'s besides "scaling" factors $S_j(\nabla)$: $\boldsymbol{O}^{(j)} \to \boldsymbol{O}^{(j)}S_j(\nabla)$, $\overline{\boldsymbol{O}}^{(j)} \to \overline{\boldsymbol{O}}^{(j)}S_j(\nabla)^{-1}$. This can be used in the mode expansion (7.72) to introduce suitable units for the $\xi_{\boldsymbol{k},j}$'s. Since ξ and O occur jointly, this kind of scaling does not influence the convergence of our adiabatic elimination procedure used below. We represent $\boldsymbol{q}(\boldsymbol{x})$ as superposition

$$\boldsymbol{q}(\boldsymbol{x}, t) = \textstyle\sum_{\boldsymbol{k},j} \boldsymbol{O}^{(j)}\xi_{\boldsymbol{k},j}\chi_{\boldsymbol{k}}(\boldsymbol{x}). \tag{7.72}$$

Now we take a decisive step for what follows. It will be shown later by means of explicit examples like laser or hydrodynamic instabilities that there exist small band excitations of unstable modes. This suggests to build up wave packets (as in quantum mechanics) so that we take sums over small regions of $\boldsymbol{k}$ together. We thus obtain carrier modes with discrete wave numbers $\boldsymbol{k}$ and slowly varying amplitudes $\xi_{\boldsymbol{k},j}(\boldsymbol{x})$. Our first goal is to derive a general set of equations for the mode amplitudes ξ. To do so we insert (7.72) into (7.62), multiply from the left hand side by $\chi_{\boldsymbol{k}'}^*(\boldsymbol{x})\overline{\boldsymbol{O}}^{(j')}$ and integrate over a region which contains many oscillations of $\chi_{\boldsymbol{k}}$ but in which ξ changes very little. The resulting expressions on the left hand side of (7.62) can be evaluated as follows:

$$\int \chi_{\boldsymbol{k}'}^*(\boldsymbol{x})\xi_{\boldsymbol{k},j}(\boldsymbol{x})\chi_{\boldsymbol{k}}(\boldsymbol{x})\, d^3x \approx \xi_{\boldsymbol{k},j}(\boldsymbol{x})\delta_{\boldsymbol{k}\boldsymbol{k}'}, \tag{7.73}$$

and

$$\int \chi_{\boldsymbol{k}'}^*(\boldsymbol{x})\lambda_j(\nabla)\xi_{\boldsymbol{k},j}(\boldsymbol{x})\chi_{\boldsymbol{k}}(\boldsymbol{x})\, d^3x \approx \hat{\lambda}_j(\nabla, \boldsymbol{k})\xi_{\boldsymbol{k},j}(\boldsymbol{x})\delta_{\boldsymbol{k}\boldsymbol{k}'}, \tag{7.74}$$

where

$$[\lambda_j(\nabla), \boldsymbol{O}^{(j')}] = 0 \text{ and } [a, b] \equiv ab - ba \tag{7.75}$$

has been used. Because $\xi_{\boldsymbol{k}j}(\boldsymbol{x})$ are slowly varying functions of $\boldsymbol{x}$, the application of ∇ yields a small quantity. We are therefore permitted to expand $\hat{\lambda}_j$ into a power series of ∇, and to confine ourselves to the leading term. In an isotropic medium we thus obtain

$$\hat{\lambda}_j(\nabla, \boldsymbol{k}) = \hat{\lambda}_j(0, \boldsymbol{k}) + \bar{\gamma}_{j,\boldsymbol{k}}\nabla + \gamma_{j,\boldsymbol{k}}\nabla^2. \tag{7.76}$$

On the rhs of (7.62) we have to insert (7.72) into g (see (7.62) and (7.65)), and to do the multiplication and integration described above. Denoting the resulting function of ξ by $\hat{g}$, we have to evaluate

$$\int \chi_{\boldsymbol{k}'}^{*}\overline{\boldsymbol{O}}^{(j')}\hat{\boldsymbol{g}}(\boldsymbol{O}^{(i)}\xi_{\boldsymbol{k},i}\chi_{\boldsymbol{k}}(\boldsymbol{x}))\, d^3x. \tag{7.77}$$

The explicit result will be given below, (7.87) and (7.88). Here we mention only a few basic facts. As we shall see later the dependence of $\xi_{\boldsymbol{k},i}(\boldsymbol{x})$ on $\boldsymbol{x}$ is important only for the unstable modes and here again only in the term connected with $\hat{\lambda}$. This means that we make the following replacements in (7.77):

$$\overline{\boldsymbol{O}}^{(j)}(\nabla) = \overline{\boldsymbol{O}}^{(j)}(\boldsymbol{k}'), \tag{7.78}$$

and

$$\boldsymbol{O}^{(i)}(\nabla) = \boldsymbol{O}^{(i)}(\boldsymbol{k}). \tag{7.79}$$

Since g (or $\hat{g}$) contains only quadratic and cubic terms in $\boldsymbol{q}$, the evaluation of (7.77) amounts to an evaluation of integrals of the form

$$\int \chi_{\boldsymbol{k}'}^{*}\chi_{\boldsymbol{k}}\chi_{\boldsymbol{k}''}\xi_{\boldsymbol{k},i}(\boldsymbol{x})\xi_{\boldsymbol{k}'',i''}(\boldsymbol{x})\, d^3x \approx \xi_{\boldsymbol{k},i}(\boldsymbol{x})\xi_{\boldsymbol{k}'',i''}(\boldsymbol{x})I_{\boldsymbol{k}';\boldsymbol{k}\boldsymbol{k}''}, \tag{7.80}$$

where

$$I_{\boldsymbol{k}';\boldsymbol{k}\boldsymbol{k}''} = \int \chi_{\boldsymbol{k}'}^{*}\chi_{\boldsymbol{k}}\chi_{\boldsymbol{k}''}\, d^3x, \tag{7.81}$$

and

$$\int \chi_{\boldsymbol{k}'}^{*}\chi_{\boldsymbol{k}}\chi_{\boldsymbol{k}''}\chi_{\boldsymbol{k}'''}\xi_{\boldsymbol{k},i}\xi_{\boldsymbol{k}'',i''}\xi_{\boldsymbol{k}''',i'''}\, d^3x = \xi_{\boldsymbol{k},i}(\boldsymbol{x})\xi_{\boldsymbol{k}'',i''}(\boldsymbol{x})\xi_{\boldsymbol{k}''',i'''}(\boldsymbol{x})J_{\boldsymbol{k}',\boldsymbol{k}\boldsymbol{k}''\boldsymbol{k}'''}, \tag{7.82}$$

where

$$J_{\boldsymbol{k}'\boldsymbol{k}\boldsymbol{k}''\boldsymbol{k}'''} = \int \chi_{\boldsymbol{k}'}^{*}\chi_{\boldsymbol{k}}\chi_{\boldsymbol{k}''}\chi_{\boldsymbol{k}'''}\, d^3x. \tag{7.83}$$

Finally, the fluctuating forces F_μ give rise to new fluctuating forces in the form

$$\hat{F}_{\boldsymbol{k},j} = \int d^3x\, \chi_{\boldsymbol{k}}^{*}(\boldsymbol{x})\overline{\boldsymbol{O}}^{(j)}\boldsymbol{F}(\boldsymbol{x}, t). \tag{7.84}$$

After these intermediate steps the basic set of equations reads as follows:

$$\dot{\xi}_{k,j} - \hat{\lambda}_j(\nabla, k)\xi_{k,j} = H_{k,j}(\{\xi(x)\}) + \hat{F}_{k,j}, \tag{7.85}$$

where

$$H_{k,j}(\{\xi(x)\}) \equiv \sum_{\substack{k'k'' \\ j'j''}} a_{kk'k'',j,j',j''} I_{kk'k''} \xi_{k'j'} \xi_{k''j''} + \sum_{\substack{k'k''k''' \\ j'j''j'''}} b_{kk'k''k''',j,j',j'',j'''} \cdot J_{kk'k''k'''} \xi_{k'j'} \xi_{k''j''} \xi_{k'''j'''}. \tag{7.86}$$

The coefficients a and b are given by

$$a_{kk'k''jj'j''} = (\tfrac{1}{2}) \sum_{\mu\nu\nu'} \bar{O}_{\mu}^{(j)}(k) O_{\nu}^{(j')}(k') O_{\nu'}^{(j'')}(k'') \{g_{\mu\nu\nu'}^{(2)}(k'') + g_{\mu\nu'\nu}^{(2)}(k')\}, \tag{7.87}$$

and

$$b_{kk'k''k'''jj'j''j'''} = \sum_{\mu\nu'\nu''\nu'''} g_{\mu\nu'\nu''\nu'''}^{(3)} \bar{O}_{\mu}^{(j)}(k) O_{\nu'}^{(j')}(k') O_{\nu''}^{(j'')}(k'') O_{\nu'''}^{(j''')}(k'''), \tag{7.88}$$

respectively. So far practically no approximations have been made, but to cast (7.85) into a practicable form we have to *eliminate* the unwanted or uninteresting modes which are the *damped modes.* Accordingly we put

$$j = u \text{ unstable if } \operatorname{Re} \hat{\lambda}_u(0, k) \gtrsim 0, \tag{7.89}$$

and

$$j = s \text{ stable if } \operatorname{Re} \hat{\lambda}_s(0, k) < 0. \tag{7.90}$$

An important point should be observed at this stage: Though ξ bears two indices, $\boldsymbol{k}$ and u, these indices are not independent of each other. Indeed the instability usually occurs only in a small region of $k = k_c$ (compare Fig. 7.4a–c). Thus we must carefully distinguish between the $\boldsymbol{k}$ values at which (7.95) and (7.96) (see below) are evaluated. Because of the "wave packets" introduced earlier, $\boldsymbol{k}$ runs over a set of discrete values with $|\boldsymbol{k}| = k_c$. The prime at the sums in the following formulas indicates this restriction of the summation over $\boldsymbol{k}$. Fig. 7.4c provides us with an example of a situation where branch j of $\xi_{k,j}$ can become unstable at two different values of $\boldsymbol{k}$. If $k = 0$ and $k = k_c$ are connected with a hard and soft mode, respectively, parametrically modulated spatio-temporal mode patterns appear.

The basic idea of our further procedure is this. Because the undamped modes may grow unlimited provided the nonlinear terms are neglected, we expect that the amplitudes of the undamped modes are considerably bigger than those of the damped modes. Since, on the other hand, close to the "phase transition" point the relaxation time of the undamped modes tends to infinity, i.e., the real part of $\hat{\lambda}$ tends to zero, the damped modes must adiabatically follow the undamped modes. Though the amplitudes of the damped modes are small, they must not be neglected completely. This neglect would lead to a catastrophe if in (7.86), (7.87), (7.88) the

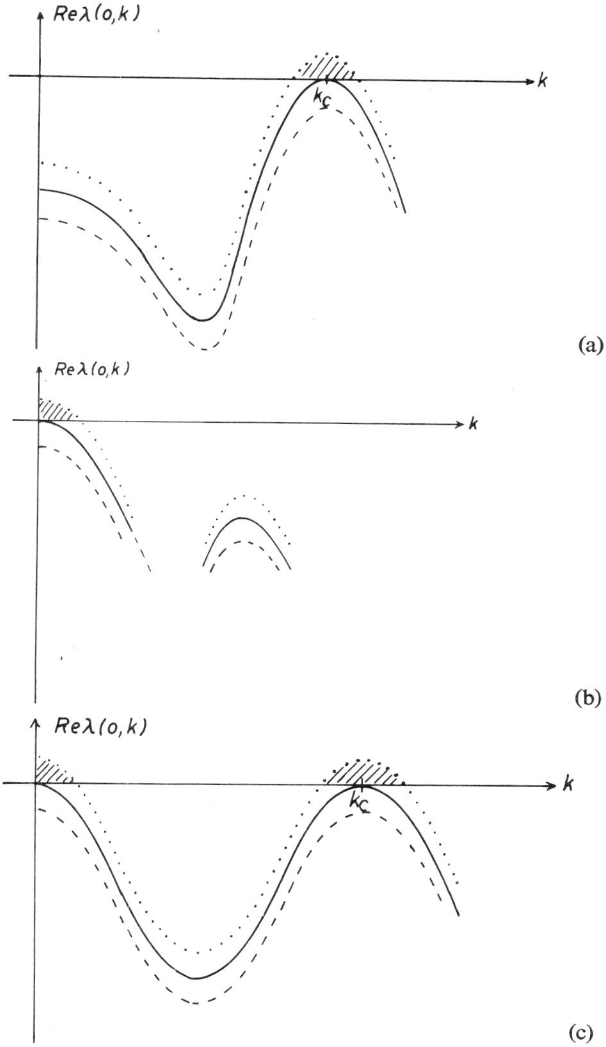

Fig. 7.4. (a) Example of an eigenvalue λ leading to an instability at k_c. This dependence of λ and k corresponds qualitatively to that of the Brusselator model of chemical reactions. (cf. Section 9.4) The dashed curve corresponds to the stable situation, the solid curve to the marginal, and the dotted curve leads to an instability for a mode continuum near k_c
(b) The situation is similar to that of Fig. 7.4a the instability occurring now at $k=0$. In the case of the Brusselator model, this instability is connected with a hard mode
(c) Example of two simultaneously occurring instabilities. If the mode at $k=0$ is a hard mode and that at $k=k_c$, a soft mode, spatial and temporal oscillations can occur

cubic terms are lacking. As one convinces oneself very quickly, quadratic terms can never lead to a globally stable situation. Thus cubic terms are necessary for a stabilization. Such cubic terms are introduced even in the absence of those in the original equations by the elimination of the damped modes. To exhibit the main features of our elimination procedure more clearly, we put for the moment

$$\xi_{\boldsymbol{k},j} \to (\boldsymbol{k}, j), \tag{7.91}$$

and drop all coefficients in (7.85). We assume $|\xi_s| \ll |\xi_u|$ and, in a selfconsistent way, $\xi_s \propto \xi_u^2$. Keeping in (7.85) only terms up to third order in ξ_u, we obtain

$$\begin{aligned}\left(\frac{d}{dt} - \hat{\lambda}_u\right)(\boldsymbol{k}, u) &= \sum_{\boldsymbol{k}'\boldsymbol{k}''u's} (\boldsymbol{k}', u')\cdot(\boldsymbol{k}'', s) + \sum_{\substack{\boldsymbol{k}'\boldsymbol{k}''\\ u'u''}} (\boldsymbol{k}', u')(\boldsymbol{k}'', u'') \\ &+ \sum_{\substack{\boldsymbol{k}'\boldsymbol{k}''\boldsymbol{k}'''\\ u'u''u'''}} (\boldsymbol{k}', u')(\boldsymbol{k}'', u'')(\boldsymbol{k}''', u''') + F_{\boldsymbol{k},u}.\end{aligned} \tag{7.92}$$

Consider now the corresponding equation for $j = s$. In it we keep only terms necessary to obtain an equation for the unstable modes up to third order

$$\left(\frac{d}{dt} - \hat{\lambda}_s\right)(\boldsymbol{k}, s) = \sum_{\substack{\boldsymbol{k}'\boldsymbol{k}''\\ u'u''}} (\boldsymbol{k}', u')(\boldsymbol{k}'', u'') + \cdots. \tag{7.93}$$

If we adopt an iteration scheme using the inequality $|\xi_s| \ll |\xi_u|$ one readily convinces oneself that ξ_s is at least proportional to ξ_u^2 so that the only relevant terms in (7.93) are those exhibited explicitly. We now use our second hypothesis, namely, that the stable modes are damped much more quickly than the unstable ones which is well fulfilled for the soft mode instability. In the case of a hard mode, we must be careful to remove the oscillatory part of $(\boldsymbol{k}', u')(\boldsymbol{k}'', u'')$ in (7.93). This is achieved by keeping the time derivative in (7.93). We therefore write the solution of (7.93) in the form

$$(\boldsymbol{k}, s) = \left(\frac{d}{dt} - \hat{\lambda}_s\right)^{-1} \sum_{\substack{\boldsymbol{k}'\boldsymbol{k}''\\ u'u''}} (\boldsymbol{k}', u')(\boldsymbol{k}'', u''). \tag{7.94}$$

The prescription to evaluate d/dt is this: in the soft mode case it can be neglected, whereas in the case $(\boldsymbol{k}, u) \propto \exp(i\omega_{\boldsymbol{k}}t)$, d/dt is to be replaced by $i(\omega_{\boldsymbol{k}'} + \omega_{\boldsymbol{k}''})$. The equations of type (7.93) can now be readily solved. If time and space derivatives are neglected, the solution of (7.93) is a purely algebraic problem so that one could keep also higher order terms in (7.93) without difficulties, at least in principle. Inserting the result into (7.92) we obtain the fundamental set of equations for the order parameters

$$\begin{aligned}\left(\frac{d}{dt} - \hat{\lambda}_u(\nabla, \boldsymbol{k})\right)\xi_{\boldsymbol{k},u} &= \sum_{\substack{\boldsymbol{k}'\boldsymbol{k}''\\ u'u''}} a_{\boldsymbol{k}\boldsymbol{k}'\boldsymbol{k}'';uu'u''} I_{\boldsymbol{k}\boldsymbol{k}'\boldsymbol{k}''} \xi_{\boldsymbol{k}'u'}\xi_{\boldsymbol{k}''u''} \\ &+ \sum_{\substack{\boldsymbol{k}'\boldsymbol{k}''\boldsymbol{k}'''\\ u'u''u'''}} \xi_{\boldsymbol{k}'u'} C_{\boldsymbol{k}\boldsymbol{k}'\boldsymbol{k}''\boldsymbol{k}''', uu'u''u'''} \xi_{\boldsymbol{k}''u''}\xi_{\boldsymbol{k}'''u'''} + \tilde{F}_{\boldsymbol{k},u} \\ &\equiv H_{\boldsymbol{k},u}^{(r)}(\{\xi_u(\boldsymbol{x})\}) + \tilde{F}_{\boldsymbol{k},u},\end{aligned} \tag{7.95}$$

where we have used the abbreviation

$$C_{kk'k''k''',uu'u''u'''} = b_{kk'k''k'''uu'u''u'''}J_{kk'k''k'''} + 2\sum'_{\hat{k}s} a_{kk'\hat{k},uu's}I_{kk'\hat{k}} \left\{\frac{d}{dt} - \hat{\lambda}_s(0, \hat{k})\right\}^{-1} a_{\hat{k}k''k''',su''u'''}I_{\hat{k}k''k'''}. \quad (7.96)$$

$\tilde{F}_{k,u}$ is defined by

$$\tilde{F}_{k,u}(x, t) = \hat{F}_{k,u}(x, t) + 2\sum_{\substack{k'u'\\ \hat{k}\, s}} a_{kk'\hat{k}uu's}I_{kk'\hat{k}}\xi_{k'u'}(x) \cdot \cdot \left\{\frac{d}{dt} - \hat{\lambda}_s(0, \hat{k})\right\}^{-1} \hat{F}_{\hat{k},s}(x, t). \quad (7.97)$$

7.8* Higher-Order Contributions to Generalized Ginzburg-Landau Equations

It is the purpose of this section to show how the first steps of the procedure of the foregoing chapter can be extended to a systematic approach which allows us to calculate all higher correction terms to generalized Ginzburg-Landau equations by explicit construction. We first cast our basic equations (7.85) into a more concise form. We introduce the vector

$$\boldsymbol{l} = (\xi_{k_1 l_1}, \xi_{k_2 l_2}, \ldots, \ldots), \quad (7.98a)$$

and

$$\hat{\boldsymbol{F}}_{\boldsymbol{m}} = \begin{pmatrix} \hat{F}_{k_1,m_1} \\ \hat{F}_{k_2,m_2} \\ \vdots \end{pmatrix}. \quad (7.98b)$$

The ξ's are the expansion coefficients introduced in (7.72) which may be slowly space dependent and also functions of time. The $\hat{F}$'s are fluctuating forces defined in (7.84). $\boldsymbol{k}_j$ are wave vectors, whereas $l_1, l_2, \ldots$ or $m_1, m_2, \ldots$ distinguish between stable or unstable modes i.e.

$$l_j = u \text{ or } s. \quad (7.99)$$

Note that $\boldsymbol{k}$ and l_j are not independent variables because in some regions of $\boldsymbol{k}$ the modes may be unstable while they remain stable in other areas of $\boldsymbol{k}$. We now introduce the abbreviation

$$A_{s,l,l'}\colon \boldsymbol{l}\colon \boldsymbol{l}' = \sum_{\substack{k',k''\\ l,\, l'}} a_{kk'k'';s,l,l'}I_{kk'k''}\xi_{k'l}\xi_{k''l'}, \quad (7.100)$$

which defines the lhs of that equation. The coefficients a and I have been given in

(7.87), (7.81). In a similar way we introduce the notation

$$B_{sll'l''}: \boldsymbol{l}: \boldsymbol{l}': \boldsymbol{l}'' = \sum_{\substack{k'k''k''' \\ l\, l'\, l''}} b_{kk'k''k''',s,l,l',l''} J_{kk'k''k'''} \cdot \xi_{k'l}\xi_{k''l'}\xi_{k'''l''}, \tag{7.101}$$

where b and J can be found in (7.88), (7.83). We further introduce the matrix

$$\Lambda_{\boldsymbol{m}} = \begin{pmatrix} \hat{\lambda}_{m_1}(\nabla, \boldsymbol{k}_1) & 0 & \cdot & 0 \\ 0 & \hat{\lambda}_{m_2}(\nabla, \boldsymbol{k}_2) & & 0 \\ \cdot & \cdot & \cdot & \\ 0 & 0 & & \end{pmatrix}, \tag{7.102}$$

where the $\hat{\lambda}$'s are the eigenvalues occurring in (7.85). With the abbreviations (7.98), (7.100) to (7.102) the (7.85), if specialized to the stable modes, can be written in the form

$$\left(\frac{d}{dt} - \Lambda_s\right)\boldsymbol{s} = A_{suu}: \boldsymbol{u}: \boldsymbol{u} + 2A_{sus}: \boldsymbol{u}: \boldsymbol{s} + A_{sss}: \boldsymbol{s}: \boldsymbol{s} + B_{suuu}: \boldsymbol{u}: \boldsymbol{u}: \boldsymbol{u}$$
$$+ 3B_{suus}: \boldsymbol{u}: \boldsymbol{u}: \boldsymbol{s} + 3B_{suss}: \boldsymbol{u}: \boldsymbol{s}: \boldsymbol{s} + B_{ssss}: \boldsymbol{s}: \boldsymbol{s}: \boldsymbol{s} + \boldsymbol{F}_s. \tag{7.103}$$

In the spirit of our previous approach we assume that the vector $\boldsymbol{s}$ is completely determined by this equation. Because (7.103) is a nonlinear equation, we must devise an iteration procedure. To this end we make the ansatz

$$\boldsymbol{s} = \sum_{n=2}^{\infty} \boldsymbol{C}^{(n)}(\boldsymbol{u}), \tag{7.104}$$

where $\boldsymbol{C}^{(n)}$ contains the components of $\boldsymbol{u}$ exactly n times. We insert (7.104) into (7.103) and compare terms which contain exactly the same numbers of factors of $\boldsymbol{u}$. Since all stable modes are damped while the unstable modes are undamped, we are safe that the operator $d/dt - \Lambda_s$ possesses an inverse. By use of this we find the following relations

$$\boldsymbol{C}^{(2)}(\boldsymbol{u}) = \left(\frac{d}{dt} - \Lambda_s\right)^{-1} \{A_{suu}: \boldsymbol{u}: \boldsymbol{u} + \hat{\boldsymbol{F}}_s\}, \tag{7.105}$$

and for $n \geq 3$

$$\boldsymbol{C}^{(n)}(\boldsymbol{u}) = \left(\frac{d}{dt} - \Lambda_s\right)^{-1} \{\cdots\}, \tag{7.106}$$

where the bracket is an abbreviation for

$$\begin{aligned}\{\cdots\} = {} & 2A_{sus}: \boldsymbol{u}: \boldsymbol{C}^{(n-1)} + (1 - \delta_{n,3}) \sum_{m=2}^{n-2} A_{sss}: \boldsymbol{C}^{(m)}: \boldsymbol{C}^{(n-m)} \\ & + \delta_{n,3} B_{suuu}: \boldsymbol{u}: \boldsymbol{u}: \boldsymbol{u} + 3(1 - \delta_{n,3}) B_{suus}: \boldsymbol{u}: \boldsymbol{u}: \boldsymbol{C}^{(n-2)} \\ & + 3(1 - \delta_{n,3})(1 - \delta_{n,4}) B_{suss}: \boldsymbol{u}: \boldsymbol{C}^{(m)}: \boldsymbol{C}^{(n-1-m)} \\ & + (1 - \delta_{n,3})(1 - \delta_{n,4})(1 - \delta_{n,5}) \sum_{\substack{m_1, m_2, m_3 \geq 2 \\ m_1 + m_2 + m_3 = n}} B_{ssss}: \boldsymbol{C}^{(m_1)}: \boldsymbol{C}^{(m_2)}: \boldsymbol{C}^{(m_3)}.\end{aligned} \tag{7.107}$$

This procedure allows us to calculate all $C^{(n)}$'s consecutively so that the C's are determined uniquely. Because the modes are damped, we can neglect solutions of the homogeneous equations provided we are interested in the stationary state or in slowly varying states which are not affected by initial distributions of the damped modes. Note that $\hat{F}$ is formally treated as being of the same order as $\boldsymbol{u}:\boldsymbol{u}$. This is only a formal trick. In applications the procedure must possibly be altered to pick out the correct orders of $\boldsymbol{u}$ and $\hat{F}$ in the final equations. Note also that $\hat{\lambda}$ and thus Λ may still be differential operators with respect to $\boldsymbol{x}$ but must not depend on time t. We now direct our attention to the equations of motion for the unstable modes. In our present notation they read

$$\left(\frac{d}{dt} - \Lambda_u\right)\boldsymbol{u} = A_{\boldsymbol{uuu}}:\boldsymbol{u}:\boldsymbol{u} + 2A_{\boldsymbol{uus}}:\boldsymbol{u}:\boldsymbol{s} + A_{\boldsymbol{uss}}:\boldsymbol{s}:\boldsymbol{s}$$
$$+ B_{\boldsymbol{uuuu}}:\boldsymbol{u}:\boldsymbol{u}:\boldsymbol{u} + 3B_{\boldsymbol{uuus}}:\boldsymbol{u}:\boldsymbol{u}:\boldsymbol{s}$$
$$+ 3B_{\boldsymbol{uuss}}:\boldsymbol{u}:\boldsymbol{s}:\boldsymbol{s} + B_{\boldsymbol{usss}}:\boldsymbol{s}:\boldsymbol{s}:\boldsymbol{s} + \hat{\boldsymbol{F}}_{\boldsymbol{u}}. \tag{7.108}$$

As mentioned above our procedure consists in first evaluating the stable modes as a functional or function of the unstable modes. We now insert the expansion (7.104) where the C's are consecutively determined by (7.105) and (7.106) into (7.108). Using the definition

$$\boldsymbol{C}^{(0)} = \boldsymbol{C}^{(1)} = 0 \tag{7.109}$$

and collecting terms containing the same total number of $\boldsymbol{u}$'s, our final equation reads

$$\left(\frac{d}{dt} - \Lambda_u\right)\boldsymbol{u} = A_{\boldsymbol{uuu}}:\boldsymbol{u}:\boldsymbol{u} + 2A_{\boldsymbol{uus}}:\boldsymbol{u}:\sum_{\nu=0}^{n-1}\boldsymbol{C}^{(\nu)} + \sum_{\substack{\nu_1,\nu_2=0\\ \nu_1+\nu_2\le n}} A_{\boldsymbol{uss}}:\boldsymbol{C}^{(\nu_1)}:\boldsymbol{C}^{(\nu_2)}$$
$$+ B_{\boldsymbol{uuuu}}:\boldsymbol{u}:\boldsymbol{u}:\boldsymbol{u} + 3B_{\boldsymbol{uuus}}:\boldsymbol{u}:\sum_{\nu=0}^{n-2}\boldsymbol{C}^{(\nu)}$$
$$+ 3B_{\boldsymbol{uuss}}:\boldsymbol{u}:\sum_{\substack{\nu_1,\nu_2=0\\ \nu_1+\nu_2\le n-1}}\boldsymbol{C}^{(\nu_1)}:\boldsymbol{C}^{(\nu_2)}$$
$$+ \sum_{\substack{\nu_1,\nu_2,\nu_3=0\\ \nu_1+\nu_2+\nu_3\le n}} B_{\boldsymbol{usss}}:\boldsymbol{C}^{(\nu_1)}:\boldsymbol{C}^{(\nu_2)}:\boldsymbol{C}^{(\nu_3)} + \hat{\boldsymbol{F}}_{\boldsymbol{u}}. \tag{7.110}$$

In the general case the solution of (7.110) may be a formidable task. In practical applications, however, in a number of cases solutions can be found (cf. Chap. 8).

7.9* Scaling Theory of Continuously Extended Nonequilibrium Systems

In this section we want to develop a general scaling theory applicable to large classes of phase transition-like phenomena in nonequilibrium systems. We treat an N-component system and take fluctuations fully into account. Our present approach is in general only applicable to the one-dimensional case, though there may be cases in which it applies also to two and three dimensions. The basic idea of our procedure is as follows:

We start from a situation in which the external parameters permit only stable solutions. In this case the equations can be linearized around their steady state values and the resulting equations allow for damped modes only. When changing an external parameter, we eventually reach a marginal situation with one or several modes becoming unstable. We now expand the nonlinear equations around the marginal point with respect to powers of ε, where ε^2 denotes the deviation of the actual parameter from the parameter value at the marginal point. In the framework of perturbation theory applied to the linearized equations one readily establishes that the complex frequency depends on ε^2. This leads to the idea to scale the time with ε^2. On the other hand, if the linearized equations contain spatial derivatives, one may show rather simply that the corresponding changes are so that the space coordinate $\boldsymbol{r}$ goes with ε. In the following we shall therefore use these two scalings and expand the solutions of the nonlinear equations into a superposition of solutions of the linearized equations at the marginal point. Including terms up to third order in ε, we then find a self-consistent equation for the amplitude of the marginal solution. The resulting equation is strongly reminiscent of time-dependent Ginzburg-Landau equations with fluctuating forces (compare Sect. 7.6, 7.7, 7.8.). As we treat essentially the same system as in Section 7.6 we start with (7.62) which we write as

$$\Gamma \boldsymbol{q} = \boldsymbol{g}(\boldsymbol{q}) + \boldsymbol{F}(\boldsymbol{x}, t) \tag{7.111}$$

using the abbreviation

$$\Gamma_{ik} = \left(\frac{\partial}{\partial t} - D_i \nabla_x^2\right)\delta_{ik} - K_{ik}. \tag{7.112}$$

In a perturbation approach one could easily solve (7.111) by taking the inverse of Γ. However, some care must be exercised because the determinant may vanish in actual cases. We therefore write (7.111) in the form

$$\det|\Gamma|\boldsymbol{q} = \tilde{\Gamma}[\boldsymbol{g}(\boldsymbol{q}) + \boldsymbol{F}(\boldsymbol{x}, t)], \tag{7.113}$$

where $\tilde{\Gamma}_{ik}$ are the subdeterminants of order $(N - 1)$ belonging to the element Γ_{ki} of the matrix (7.112).

a) *The Homogeneous Problem*

The linear homogeneous problem (7.111) defines a set of right-hand eigenvectors (modes)

$$\hat{K} \cdot \boldsymbol{u}_n = \lambda_n \boldsymbol{u}_n. \tag{7.114}$$

with $\hat{K}$ defined by (7.64). Because Γ is in general not self-adjoint, we have to define corresponding left-hand eigenvectors by means of

$$\boldsymbol{v}_n \hat{K} = \lambda_n \cdot \boldsymbol{v}_n. \tag{7.115}$$

Note that $\boldsymbol{u}_n$ and $\boldsymbol{v}_n$ are time and space dependent functions. They may be represented by plane wave solutions. The index n distinguishes both, the $\boldsymbol{k}$-vector and the corresponding eigenvalue of Γ: $n = (\boldsymbol{k}, m)$. The vectors (7.114) and (7.115) form a biorthogonal set with the property

$$(\boldsymbol{v}_n \mid \boldsymbol{u}_{n'}) = \delta_{nn'}, \tag{7.116}$$

where (|) denotes the scalar product.

b) Scaling

We assume that one or several external parameters are changed so that one goes beyond the marginal point. The size of this change is assumed to be proportional to ε^2 which is a smallness parameter for our following purposes. In general the change of an external parameter causes a change of the coefficients of the diffusion matrix and of the matrix K

$$\left.\begin{aligned} D_i &= D_i^{(0)} + D_i^{(2)}\varepsilon^2 \\ K_{ij} &= K_{ij}^{(0)} + K_{ij}^{(2)}\varepsilon^2 \end{aligned}\right| . \tag{7.117}$$

Correspondingly we expand the solution $\boldsymbol{q}$ into a power series of ε

$$\boldsymbol{q} = \varepsilon \boldsymbol{q}^{(1)} + \varepsilon^2 \boldsymbol{q}^{(2)} + \cdots = \textstyle\sum_{m=1}^{\infty} \varepsilon^m \boldsymbol{q}^{(m)}. \tag{7.118}$$

To take into account finite band width effects, space and time are scaled simultaneously by

$$\begin{aligned} \varepsilon \boldsymbol{x} &= \boldsymbol{R}, \\ \varepsilon^2 t &= T. \end{aligned} \tag{7.119}$$

c) Perturbation Theory

Inserting the expansions (7.117), (7.118) and the scaling laws (7.119) in (7.113) we obtain

$$\begin{aligned} &(\textstyle\sum_{l=0}^{\infty} \varepsilon^l \det |\Gamma|^{(l)}) \cdot (\sum_{n=1}^{\infty} \varepsilon^n \boldsymbol{q}^{(n)}) \\ &\quad = (\textstyle\sum_{l=0}^{\infty} \varepsilon^l \tilde{\Gamma}^{(l)})(\sum_{m=1}^{\infty} \varepsilon^m \boldsymbol{g}^{(m)} + \sum_{m=1}^{\infty} \varepsilon^m \boldsymbol{F}^{(m)}(\boldsymbol{x}, t)). \end{aligned} \tag{7.120}$$

In order ε^1 we come back to the homogeneous problem (7.114). In the general order ε^l (7.120) reduces to

$$\textstyle\sum_{r=0}^{l-1} \det |\Gamma|^{(r)} \boldsymbol{q}^{(l-r)} = \sum_{r=0}^{l-1} \tilde{\Gamma}^{(r)}(\boldsymbol{g}^{(l-r)}(\boldsymbol{q}) + \boldsymbol{F}^{(l-r)}(\boldsymbol{x}, t)). \tag{7.121}$$

We extract $\boldsymbol{q}^{(l)}$ from the lhs and bring the rest to the rhs. This yields

$$\begin{aligned} \det |\Gamma|^{(0)} \boldsymbol{q}^{(l)} &= -\textstyle\sum_{r=1}^{l-1} \det |\Gamma|^{(r)} \boldsymbol{q}^{(l-r)} \\ &\quad + \textstyle\sum_{r=0}^{l-1} \tilde{\Gamma}^{(r)}[\boldsymbol{g}^{(l-r)}(\boldsymbol{q}) + \boldsymbol{F}^{(l-r)}(\boldsymbol{x}, t)]. \end{aligned} \tag{7.122}$$

We will explicitly treat two different situations which may arise at the critical point. These situations are distinguished by a different behavior of the complex frequencies λ.

1) one soft mode, that is for one and only one mode $\lambda_1 = 0$,
2) the hard mode case in which a pair of complex conjugate modes becomes unstable so that Re $\lambda_1 = 0$ but Im $\lambda_1 \neq 0$.

To write the perturbation theory in the same form for both cases, we introduce the concept of modified Green's functions. To this end we rewrite the homogeneous problem (7.114)

$$\hat{K}\boldsymbol{u}_n = (K^{(0)} + D^{(0)}\nabla_x^2)\boldsymbol{u}_n = \lambda_n \boldsymbol{u}_n. \tag{7.123}$$

We now define Ω by

$$\Omega = K^{(0)} + D^{(0)}\nabla_x^2 - \lambda I, \tag{7.124}$$

where I is the unity matrix. The Green's function belonging to (7.124) reads

$$G^{\mathrm{mod}} = \sum_{n\neq 1} \frac{1}{\lambda_n - \lambda} |\boldsymbol{u}_n)(\boldsymbol{v}_n|. \tag{7.125}$$

In connection with the modified Green's function we have the orthogonality condition

$$(\boldsymbol{v}_1 \mid -\textstyle\sum_{r=1}^{l-1} \det |\Gamma|^{(r)} \boldsymbol{q}^{(l-r)} + \sum_{r=0}^{l-1} \tilde{\Gamma}^{(r)} [\boldsymbol{g}^{(l-r)}(\boldsymbol{q}) + \boldsymbol{F}^{(l-r)}(\boldsymbol{x}, t)]) = 0. \tag{7.126}$$

The concept of modified Green's functions is inevitable to treat the soft mode case $\lambda_1 = 0$, where the index 1 means ($k = k_c$, $m = 1$). By means of the Green's function the solution can be written in the form

$$\boldsymbol{q}^{(l)}(\lambda) = \sum_{n\neq 1} \frac{1}{\lambda_n - \lambda} |\boldsymbol{u}_n)(\boldsymbol{v}_n| - \textstyle\sum_{r=1}^{l-1} \Gamma^{(r)} \boldsymbol{q}^{(l-r)} + \boldsymbol{g}^{(l)}(\boldsymbol{q}) + \boldsymbol{F}^{(l)}(\boldsymbol{x}, t)). \tag{7.127}$$

Obviously the same procedure is applicable to the left hand eigenvectors. If the critical mode is degenerate, it is straightforward to define an appropriate, modified Green's function. Instead of (7.126), one now gets a set of h orthogonality conditions. (h is the degree of degeneracy). Each one of these conditions must be fulfilled separately.

7.10* Soft-Mode Instability

We assume that the time and space dependence of $\boldsymbol{u}_n$ can be represented by plane waves

$$\boldsymbol{u}_n \propto \exp[i\omega_m t - ik\boldsymbol{x}]. \tag{7.128}$$

As mentioned before, the critical point is fixed by

$$\lambda_1 = i\omega_1 = 0. \tag{7.129}$$

From (7.129) we get

$$\det |\Gamma^{(0)}(\lambda_1 = 0)| = 0. \tag{7.130}$$

Equation (7.130) is a condition for the coefficients D_i, K_{ik} as functions of k^2 at the marginal point. k_c is fixed by the condition

$$\frac{d\lambda_1}{d(k^2)} = 0 \tag{7.131}$$

which has the consequence

$$\frac{d}{d(k^2)} \det |\Gamma^{(0)}(\lambda_1 = 0)| = 0. \tag{7.132}$$

From (7.132) we get

$$\textstyle\sum_i D_i^{(0)} \tilde{\Gamma}_{ii}^{(0)}(\lambda_1 = 0; k^2) \equiv \mathrm{tr}\,(D\tilde{\Gamma}(\lambda_1 = 0, k^2)) = 0, \tag{7.133}$$

where tr means "trace". The quantities $\tilde{\Gamma}_{ii}$ are the adjoint subdeterminants of order $(N - 1)$ to the elements Γ_{ii} of the matrix Γ. Now one may construct the right-hand and left-hand eigenvectors of the critical mode. We represent the right-hand eigenvector as

$$\boldsymbol{u}_1 = \mathrm{e}^{-ik_cx} \begin{pmatrix} a_1(k_c, 1) \\ a_2(k_c, 1) \\ \cdots \\ a_N(k_c, 1) \end{pmatrix}. \tag{7.134}$$

$a_i(k_c, 1)$ are the components of the right-hand eigenvector to the eigenvalue $\lambda_1 = 0$ in the vector space $(q_1, q_2, \ldots, q_N)$. Here it is assumed that there exists only one eigenvector to λ_1. (In other words, the problem is nondegenerate). The left-hand eigenvector is given by

$$\boldsymbol{v}_1 = \mathrm{e}^{ik_cx}(\bar{a}_1(k_c,1), \bar{a}_2(k_c,1), \ldots \bar{a}_N(k_c,1)). \tag{7.135}$$

The scalar product between left-hand and right-hand eigenvectors is defined as follows:

$$\frac{1}{L}\int_{-L/2}^{+L/2} dx \cdot \mathrm{e}^{i(k-k')x} \textstyle\sum_{n=1}^{N} \bar{a}_n(k, i) a_n(k', j) = \delta_{kk'}\delta_{ji}. \tag{7.136}$$

Because of the critical mode which is dominating all other modes at the marginal point the solution of (7.122) in order ε^1 is given by

$$\boldsymbol{q}^{(1)} = \xi(R, T)\mathrm{e}^{-ik_cx}\begin{pmatrix} a_1(k_c, 1) \\ a_2(k_c, 1) \\ \cdots \\ a_N(k_c, 1) \end{pmatrix}. \tag{7.137}$$

The common factor $\xi(R, T)$ is written as a slowly varying amplitude in space and time. In order ε^2 we find the following equation

$$\det|\Gamma|^{(0)}\boldsymbol{q}^{(2)} + \det|\Gamma|^{(1)}\boldsymbol{q}^{(1)} = \tilde{\Gamma}^{(0)}\boldsymbol{g}^{(2)}(\boldsymbol{q}). \tag{7.138}$$

Because $\det|\Gamma|^{(1)}$ contains differential operators, one may readily show that on account of (7.133)

$$\det|\Gamma|^{(1)}\boldsymbol{q}^{(1)} = 2(\nabla_x\nabla_R)\operatorname{tr}(D^{(0)}\tilde{\Gamma}^{(0)}(\lambda = 0; k_c))\boldsymbol{q}^{(1)} = 0. \tag{7.139}$$

To write (7.138) more explicitly, we represent $\boldsymbol{g}^{(2)}(\boldsymbol{q})$ as

$$g_i^{(2)} = g_{ikl}^{(2)}q_k^{(1)}q_l^{(1)} + g_{ikl}^{(2)}q_k^{(1)*}q_l^{(1)} + c.c. \tag{7.140}$$

(summation is taken over dummy indices). $q_k^{(1)}$ may be approximated by the solution $\boldsymbol{q}^{(1)}$ of (7.137). The solvability condition (7.126) is automatically fulfilled. Therefore the solution $\boldsymbol{q}^{(2)}$ reads

$$q_j^{(2)} = \xi^2\cdot\sum_m \frac{1}{\lambda_m(2k_c)}a_j(2k_c, m)\bar{a}_i(2k_c, m)g_{ikl}^{(2)}a_k(k_c, 1)a_l(k_c, 1)\,\mathrm{e}^{-2ik_cx}$$
$$+ |\xi|^2\sum_m \frac{1}{\lambda_m(0)}a_j(0, m)\bar{a}_i(0, m)g_{ikl}^{(2)}a_k^*(k_c, 1)a_l(k_c, 1). \tag{7.141}$$

We finally discuss the balance equation of third order. The corresponding solvability condition

$$(\boldsymbol{v}_{1,k_c}\,|\,\det|\Gamma|^{(2)}\boldsymbol{q}^{(1)} + \det|\Gamma|^{(1)}\boldsymbol{q}^{(2)} + \det|\Gamma^{(0)}|\boldsymbol{q}^{(3)})$$
$$= (\boldsymbol{v}_{1,k_c}\,|\,\tilde{\Gamma}^{(0)}[\boldsymbol{g}^{(3)}(\boldsymbol{q}) + \boldsymbol{F}(x, t)] + \tilde{\Gamma}^{(1)}\boldsymbol{g}^{(2)}(\boldsymbol{q}) + \tilde{\Gamma}^{(0)}\boldsymbol{g}'^{(2)}\nabla_R) \quad (7.142)^1$$

reduces on account of

$$(\boldsymbol{v}_{1,k_c}\,|\,\det|\Gamma|^{(0)}\boldsymbol{q}^{(3)}) = 0 \tag{7.143}$$

and

$$(\boldsymbol{v}_{1,k_c}\,|\,\det|\Gamma|^{(1)}\boldsymbol{q}^{(2)}) = (\boldsymbol{v}_{1,k_c}\,|\,\tilde{\Gamma}^{(1)}\boldsymbol{g}^{(2)}(\boldsymbol{q})) = (\boldsymbol{v}_{1,k_c}\,|\,\tilde{\Gamma}^{(0)}\boldsymbol{g}^{(2)\prime}\nabla_R) = 0 \tag{7.144}$$

[1] We have split the original $g^{(2)}$ into a ∇-independent part, $g^{(2)}$, and a part with the scaled ∇, $g'^{(2)}$.

to

$$(\boldsymbol{v}_{1,k_c} \mid \det|\Gamma|^{(2)}\boldsymbol{q}^{(1)}) = (\boldsymbol{v}_{1,k_c} \mid \tilde{\Gamma}^{(0)}[\boldsymbol{g}^{(3)}(\boldsymbol{q}) + \boldsymbol{F}(x, t)]). \tag{7.145}$$

To evaluate (7.145) we calculate the single terms separately. For $\det|\Gamma|^{(2)}$ we find

$$\det|\Gamma|^{(2)} = \sum_{i,j}\left[\left(\frac{\partial}{\partial T} - D_i^{(0)}\nabla_R^2\right)\delta_{ij} - K_{ij}^{(2)} + D_i^{(2)}k_c^2\delta_{ij}\right]\tilde{\Gamma}_{ij}^{(0)}$$
$$- 2\sum_{i,j} D_i^{(0)}D_j^{(0)}\gamma_{ii,jj}\nabla_R^2 k_c^2. \tag{7.146}$$

$\gamma_{ii,jj}$ are the subdeterminants of order $(N - 2)$ to Γ. Because of (7.146) the structure of the lhs of (7.145) is given by

$$A\cdot\frac{\partial}{\partial T}\xi(R, T) - B\nabla_R^2\xi(R, T) + C\cdot\xi(R, T) \tag{7.147}$$

with the coefficients

$$A = \sum_i \tilde{\Gamma}_{ii}^{(0)}(\lambda_1, k_c) \tag{7.148a}$$
$$B = \sum_i D_i^{(0)}\tilde{\Gamma}_{ii}^{(0)}(\lambda_1, k_c) + 2\sum_{i,j} D_i^{(0)}D_j^{(0)}\gamma_{ii,jj}k_c^2 \tag{7.148b}$$
$$C = \sum_{ij}(-K_{ij}^{(2)} + D_i^{(2)}k_c^2\delta_{ij})\tilde{\Gamma}_{ij}^{(0)}(\lambda_1, k_c). \tag{7.148c}$$

We now evaluate the right-hand side of (7.145). Inserting the explicit form of $\boldsymbol{v}_1$, $\boldsymbol{g}^{(2)}$ and $\boldsymbol{q}^{(2)}$ we obtain

$$\bar{a}_m(k_c, 1)\{\tilde{\Gamma}_{mi}(\lambda_1, k_c)[g_{ijkl}^{(3)} + g_{ikjl}^{(3)} + g_{ilkj}^{(3)}]a_j^*(k_c, 1)a_k(k_c, 1)a_l(k_c, 1)$$
$$+ \tilde{\Gamma}_{mi}(\lambda_1, k_c)(g_{ijk}^{(2)} + g_{ikj}^{(2)})[a_k(k_c, 1)\sum_l \frac{1}{\lambda_l(0)} a_j(0, l)\bar{a}_p(0, l)g_{pqr}a_q^*(k_c, 1)a_r(k_c, 1)$$
$$+ a_k^*(k_c, 1)\sum_l \frac{1}{\lambda_l(2k_c)} a_j(2k_c, l)\bar{a}_p(2k_c, l)g_{pqr}^{(2)}a_q(k_c, 1)a_r(k_c, 1)]\}$$
$$\times|\xi(R, T)|^2\xi(R, T). \tag{7.149}$$

This can be written in the form

$$M|\xi(R, T)|^2\xi(R, T). \tag{7.150}$$

Finally we discuss the fluctuating forces. We thereby assume that they are slowly varying in time and contribute in order ε^3. For the fluctuating force we get

$$(\boldsymbol{v}_{1,k_c} \mid \tilde{\Gamma}^{(0)}\boldsymbol{F}(x, t)) = F(R, T). \tag{7.151}$$

$F(R, T)$ can be written as

$$F(R, T) = \sum_k c_{1k}\cdot\tilde{F}_k(R, t) \tag{7.152}$$

with

$$\tilde{F}(R, T) = e^{ik_c x} F(R, t) \tag{7.153}$$

and

$$c_{1k} = \bar{a}_j(k_c, 1)\tilde{\Gamma}^{(0)}_{jk}(\lambda_1, k_c). \tag{7.154}$$

Again the fluctuating forces are Gaussian. Combining the expressions (7.147), (7.150), (7.151) we can write the final equation for the slowly varying amplitude $\xi(R, T)$ which reads

$$A \cdot \frac{\partial \xi(R, T)}{\partial T} - B \cdot \nabla_R^2 \xi(R, T) = (-C + M \mid \xi(R, T)|^2)\xi(R, T) + F(R, T) \tag{7.155}$$

with A, B, C defined in (7.148). Equation (7.155) is analogous to the time dependent Ginzburg-Landau equation in superconductivity and the equation for the continuous mode laser. The stochastically equivalent Fokker-Planck equation has a potential solution which is formally the same as the Ginzburg-Landau free energy (cf. Sect. 6.8).

7.11* Hard-Mode Instability

In this section we assume that the instability occurs at $k_c = 0$ but we assume that the eigenvalue is complex allowing for two conjugate solutions. Basically the procedure is analogous to that of Section 7.10 so that we mainly exhibit the differences. When looking at the homogeneous problem, we may again define a biorthogonal set of eigenvectors. However, we have now to define a scalar product with respect to time by

$$(\boldsymbol{v}_{n,p} \mid \boldsymbol{u}_{n',p'}) = \frac{1}{T}\int_{-T/2}^{+T/2} e^{-i\omega(p-p')t} \textstyle\sum_{j=1}^{N} \bar{a}_j(p, n) a_j(p', n') dt = \delta_{nn'}\delta_{pp'} \tag{7.156}$$

where

$$T = \frac{2\pi}{\omega_0}, \ p \text{ and } p' \text{ are integers.} \tag{7.157}$$

We assume that $\operatorname{Im} \lambda_1 \neq 0$ is practically fixed, whereas the real part of λ_1 goes with ε^2 which again serves as small expansion parameter. While the ε^1 balance had defined the homogeneous problem, the ε^2 balance yields the equation

$$\det |\Gamma|^{(0)} \boldsymbol{q}^{(2)} + \det |\Gamma|^{(1)} \boldsymbol{q}^{(1)} = \tilde{\Gamma}^{(0)} \boldsymbol{g}^{(2)}. \tag{7.158}$$

On account of

$$\det |\Gamma|^{(1)}\boldsymbol{q}^{(1)} = 0 \tag{7.159}$$

(7.158) has the same structure as (7.138) of the preceding paragraph. Thus the solvability condition is fulfilled. Taking the Green's function

$$G = \sum_p \sum_n' \frac{1}{\lambda_n - ip\omega_0} \mathrm{e}^{ip\omega_0(t-t')} |\boldsymbol{u}_{n,p})(\boldsymbol{v}_{n,p}|, \tag{7.160}$$

the solution of the corresponding equation reads

$$\begin{aligned} q_j^{(2)} &= \xi^2 \sum_n \frac{1}{\lambda_n - 2i\omega_0} a_j(2, n)\bar{a}_j(2, n) g_{ikl}^{(2)} a_k(1, 1) a_l(1, 1) \mathrm{e}^{2i\omega_0 t} \\ &\quad + |\xi|^2 \sum_n \frac{1}{\lambda_n} a_j(0, n)\, \bar{a}_i(0, n) g_{ikl}^{(2)} a_k^*(1, 1) a_l(1, 1). \end{aligned} \tag{7.161}$$

Inserting (7.161) into the ε^3 balance equation, leads on the rhs to terms of the form

$$M|\xi|^2\xi(R, T) + F(R, T), \tag{7.162}$$

where $F(R, T)$ is the projection of the fluctuations on the unstable modes. On the lhs we obtain expressions from $\det |\Gamma|^{(2)}$. These terms read

$$\sum_{i,j} \left[\left(\frac{\partial}{\partial T} - D_i \nabla_R^2 \right) \delta_{ij} - K_{ij}^{(2)} \right] \tilde{\Gamma}_{ij}^{(0)}(\omega_0), \tag{7.163}$$

where $\tilde{\Gamma}_{ij}$ is the subdeterminant belonging to the element Γ_{ji}. As a result we find expressions having exactly the same structure as (7.147). The only difference rests in complex coefficients which are now given by

$$A = \operatorname{tr} \tilde{\Gamma}^{(0)}(\omega_0), \tag{7.164a}$$

$$B = \operatorname{tr} (D^{(0)} \tilde{\Gamma}^{(0)}(\omega_0)), \tag{7.164b}$$

$$C = -\sum_{ij} K_{ij}^{(2)} \tilde{\Gamma}_{ij}^{(0)}(\omega_0). \tag{7.164c}$$

M has the same structure as in the soft mode case

$$\begin{aligned} M = \bar{a}_m(1, 1)\{&\tilde{\Gamma}_{mi}^{(0)}(\omega_0)[g_{ijkl}^{(3)} + g_{ikjl}^{(3)} + g_{ilkj}^{(3)}] \\ &+ \tilde{\Gamma}_{mi}^{(0)}(\omega_0)[g_{ijk}^{(2)} + g_{ikj}^{(2)}][a_k(1, 1) \sum_n \frac{1}{\lambda_n} a_j(1, n)\bar{a}_l(1, n) g_{lpq}^{(2)} a_p^*(1, 1) a_q(1, 1) \\ &+ a_k^*(1, 1) \sum_n \frac{1}{\lambda_n - 2i\omega_0} a_j(2, n)\bar{a}_l(2, n) g_{lpq}^{(2)} a_p(1, 1) a_q(1, 1)\}. \end{aligned} \tag{7.165}$$

8. Physical Systems

8.1 Cooperative Effects in the Laser: Self-Organization and Phase Transition

The laser is nowadays one of the best understood many-body problems. It is a system far from thermal equilibrium and it allows us to study cooperative effects in great detail. We take as an example the solid-state laser which consists of a set of laser-active atoms embedded in a solid state matrix (cf. Fig. 1.9). As usual, we assume that the laser end faces act as mirrors serving two purposes: They select modes in axial direction and with discrete cavity frequencies. In our model we shall treat atoms with two energy levels. In thermal equilibrium the levels are occupied according to the Boltzmann distribution function. By exciting the atoms, we create an inverted population which may be described by a negative temperature. The excited atoms now start to emit light which is eventually absorbed by the surroundings, whose temperature is much smaller than $\hbar\omega/k_B$ (where ω is the light frequency of the atomic transition and k_B is Boltzmann's constant) so that we may put this temperature ≈ 0. From a thermodynamic point of view the laser is a system (composed of the atoms and the field) which is coupled to reservoirs at different temperatures. Thus the laser is a system far from thermal equilibrium.

The essential feature to be understood in the laser is this. If the laser atoms are pumped (excited) only weakly by external sources, the laser acts as an ordinary lamp. The atoms independently of each other emit wavetracks with random phases. The coherence time of about 10^{-11} sec is evidently on a microscopic scale. The atoms, visualized as oscillating dipoles, are oscillating completely at random. If the pump is further increased, suddenly within a very sharp transition region the linewidth of the laser light may become of the order of one cycle per second so that the phase of the field remains unchanged on the macroscopic scale of 1 sec. Thus the laser is evidently in a new, highly ordered state on a macroscopic scale. The atomic dipoles now oscillate in phase, though they are excited by the pump completely at random. Thus the atoms show the phenomenon of self-organization. The extraordinary coherence of laser light is brought about by the cooperation of the atomic dipoles. When studying the transition-region lamp laser, we shall find that the laser shows features of a second-order phase transition.

8.2 The Laser Equations in the Mode Picture

In the laser, the light field is produced by the excited atoms. We describe the field by its electric field strength, E, which depends on space and time. We consider only a single direction of polarization. We expand $E = E(\mathbf{x}, t)$ into cavity modes

$$E(\mathbf{x}, t) = i \sum_\lambda \{(2\pi\hbar\omega_\lambda/V)^{1/2} \exp(i\mathbf{k}_\lambda \mathbf{x}) b_\lambda - c.c.\}, \tag{8.1}$$

where we assume, for simplicity, running waves. λ is an index distinguishing the different modes. ω_λ is the mode frequency, V the volume of the cavity, $\mathbf{k}_\lambda$ the wave vector, b_λ and b_λ^* are time-dependent complex amplitudes. The factor $(2\pi\hbar\omega_\lambda/V)^{1/2}$ makes b_λ, b_λ^* dimensionless (its precise form stems from quantum theory). We distinguish the atoms by an index μ. Of course, it is necessary to treat the atoms by quantum theory. When we confine our treatment to two laser-active atomic energy levels, the treatment can be considerably simplified. As shown in laser theory, we may describe the physical properties of atom μ by its complex dipole moment α_μ and its inversion σ_μ. We use α_μ in dimensionless units. σ_μ is the difference of the occupation numbers N_2 and N_1 of the upper and lower atomic energy level:

$$\sigma_\mu = (N_2 - N_1)_\mu.$$

As is also shown in laser theory with the help of "quantum-classical-correspondence", the amplitudes b_λ, the dipole moments α_μ and the inversion σ_μ can be treated as classical quantities, obeying the following sets of equations:

a) Field Equations

$$\dot{b}_\lambda = (-i\omega_\lambda - \kappa_\lambda) b_\lambda - i \sum_\mu g_{\mu\lambda} \alpha_\mu + F_\lambda(t). \tag{8.2}$$

κ_λ is the decay constant of mode λ if left alone in the cavity without laser action. κ_λ takes into account field losses due to semi-transparent mirrors, to scattering centers etc. $g_{\mu\lambda}$ is a coupling constant describing the interaction between mode λ and atom μ. $g_{\mu\lambda}$ is proportional to the atomic dipole matrix element. F_λ is a stochastic force which occurs necessarily due to the unavoidable fluctuations when dissipation is present. Eq. (8.2) describes the temporal change of the mode amplitude b_λ due to different causes: the free oscillation of the field in the cavity ($\sim \omega_\lambda$), the damping ($-\kappa_\lambda$), the generation by oscillating dipole moments ($-ig_{\mu\lambda}\alpha_\mu$), due to fluctuations (e.g., in the mirrors), ($\sim F$). On the other hand, the field modes influence the atoms. This is described by the

b) Matter Equations

1) *Equations for the Atomic Dipole Moments*

$$\dot{\alpha}_\mu = (-i\nu - \gamma)\alpha_\mu + i \sum_\lambda g_{\mu\lambda}^* b_\lambda \sigma_\mu + \Gamma_\mu(t). \tag{8.3}$$

ν is the central frequency of the atom, γ its linewidth caused by the decay of the atomic dipole moment. $\Gamma_\mu(t)$ is the fluctuating force connected with the damping constant γ. According to (8.3), α_μ changes due to the free oscillation of the atomic dipole moment, $(-i\nu)$, due to its damping $(-\gamma)$ and due to the field amplitudes $(\sim b_\lambda)$. The factor σ_μ serves to establish the correct phase-relation between field and dipole moment depending on whether light is absorbed $(\sigma_\mu \equiv (N_2 - N_1)_\mu < 0)$ or emitted $(\sigma_\mu > 0)$. Finally, the inversion also changes when light is emitted or absorbed.

2) *Equation for the Atomic Inversion*

$$\dot{\sigma}_\mu = \gamma_{\parallel}(d_0 - \sigma_\mu) + 2i \textstyle\sum_\lambda (g_{\mu\lambda}\alpha_\mu b_\lambda^+ - \text{c.c.}) + \Gamma_{\sigma,\mu}(t). \tag{8.4}$$

d_0 is an equilibrium inversion which is caused by the pumping process and incoherent decay processes if no laser action takes place, $\gamma_{\parallel}$ is the relaxation time after which the inversion comes to an equilibrium. In (8.3) and (8.4) the Γ's are again fluctuating forces.

Let us first consider the character of the equations (8.2) to (8.4) from a mathematical view point. They are coupled, first-order differential equations for many variables. Even if we confine ourselves to modes within an atomic line width, there may be in between dozens to thousands of modes. Furthermore there are typically 10^{14} laser atoms or still many more, so that the number of variables of the system (8.2) to (8.4) is enormous. Furthermore the system is nonlinear because of the terms $b\sigma$ in (8.3) and αb^+, $\alpha^+ b$ in (8.4). We shall see in a moment that these nonlinearities play a crucial role and must not be neglected. Last but not least, the equations contain stochastic forces. Therefore, at a first sight, the solution of our problem seems rather hopeless. We want to show, however, that by the concepts and methods of Chapter 7 the solution is rather simple.

8.3 The Order Parameter Concept

A discussion of the physical content of (8.2) to (8.4) will help us to cut the problem down and solve it completely. Eq. (8.2) describes the temporal change of the mode amplitude under two forces: a driving force stemming from the oscillating dipole moments (α_μ) quite analogous to the classical theory of the Hertzian dipole, and a stochastic force F. Eqs. (8.3) and (8.4) describe the reaction of the field on the atoms. Let us first assume that in (8.3) the inversion σ_μ is kept constant. Then b acts as a driving force on the dipole moment. If the driving force has the correct phase and is near resonance, we expect a feedback between the field and the atoms, or, in other words, we obtain stimulated emission. This stimulation process has two opponents. On the one hand the damping constants κ and γ will tend to drive the field to zero and, furthermore, the fluctuating forces will disturb the total emission process by their stochastic action. Thus we expect a damped oscillation. As we shall see more explicitly below, if we increase σ_μ, suddenly the system becomes unstable with exponential growth of the field and correspondingly of the dipole moments. Usually it is just a single-field mode that first becomes undamped or, in other words, that becomes unstable. In this instability region its internal relaxation time is apparently very long. This makes us anticipate that the mode amplitudes, which virtually become undamped, may serve as the order parameters. These slowly varying amplitudes now slave the atomic system. The atoms have to obey the orders of the order parameters as described by the rhs of (8.3) and (8.4). If the atoms follow immediately the orders of the order parameter, we may eliminate the "atomic" variables α^+, α, σ, adiabatically, obtaining equations for the order parameters b_λ alone. These equations describe most explicitly the competition of the order parameters among each other. The atoms will then obey that order parameter which wins the competition. In order to learn more about this mechanism, we anticipate that one b_λ has won the competition and we first confine our analysis to this single-mode case.

8.4 The Single-Mode Laser

We drop the index λ in (8.2) to (8.4), assume exact resonance, $\omega = \nu$, and eliminate the main time dependence by the substitutions

$$b = \tilde{b}e^{-i\omega t}, \; \alpha_\mu = \tilde{\alpha}_\mu e^{-i\nu t}, \; F = \tilde{F}e^{-i\omega t}, \tag{8.5}$$

where we finally drop the tilde, $\sim$. The equations we consider are then

$$\dot{b} = -\kappa b - i\textstyle\sum_\mu g_\mu \alpha_\mu + F(t), \tag{8.6}$$

$$\dot{\alpha}_\mu = -\gamma\alpha_\mu + ig_\mu^* b\sigma_\mu + \Gamma_\mu(t), \tag{8.7}$$

$$\dot{\sigma}_\mu = \gamma_\parallel(d_0 - \sigma_\mu) + 2i(g_\mu \alpha_\mu b^+ - \text{c.c.}) + \Gamma_{\sigma,\mu}(t). \tag{8.8}$$

We note that for running waves (in a single direction) the coupling coefficients g_μ have the form

$$g_\mu = g e^{-ikx_\mu}. \tag{8.9}$$

g is assumed real. Note that the field-mode amplitude b is supported via a sum of dipole moments

$$\textstyle\sum_\mu \alpha_\mu e^{-ikx_\mu} = S_k. \tag{8.10}$$

We first determine the oscillating dipole moment from (8.7) which yields in an elementary way

$$\alpha_\mu = ig_\mu^* \int_{-\infty}^t e^{-\gamma(t-\tau)}(b\sigma_\mu)_\tau \, d\tau + \hat{\Gamma}_\mu(t) \tag{8.11}$$

with

$$\hat{\Gamma}_\mu(t) = \int_{-\infty}^t e^{-\gamma(t-\tau)}\Gamma_\mu(\tau)\, d\tau. \tag{8.12}$$

We now make a very important assumption which is quite typical for many cooperative systems (compare Section 7.2). We assume that the relaxation time of the atomic dipole moment α is much smaller than the relaxation time inherent in the order parameter b as well as in σ_μ. This allows us to take $b \cdot \sigma_\mu$ out of the integral in (8.11). By this *adiabatic approximation* we obtain

$$\alpha_\mu = \frac{ig_\mu^*}{\gamma} b\sigma_\mu + \hat{\Gamma}_\mu(t). \tag{8.13}$$

(8.13) tells us that the atoms obey instantaneously the order parameter. Inserting (8.13) into (8.6) yields

$$\dot{b} = -\kappa b + \frac{g^2}{\gamma} b \sum_\mu \sigma_\mu + \hat{F}(t), \tag{8.14}$$

where $\hat{F}$ is now composed of the field noise source, F, and the atomic noise sources, Γ,

$$\hat{F}(t) = F(t) - i \sum_\mu g_\mu \hat{\Gamma}_\mu(t). \tag{8.15}$$

In order to eliminate the dipole moments completely, we insert (8.13) into (8.8). A rather detailed analysis shows that one may safely neglect the fluctuating forces. We therefore obtain immediately

$$\dot{\sigma}_\mu = \gamma_\parallel(d_0 - \sigma_\mu) - 4\frac{g^2}{\gamma} b^+ b\sigma_\mu. \tag{8.16}$$

We now again assume that the atom obeys the field instantaneously, i.e., we put

$$\dot{\sigma}_\mu = 0 \tag{8.17}$$

so that the solution of (8.16) reads

$$\sigma_\mu = d_0(1 + 4(g^2/\gamma\gamma_\parallel)b^+b)^{-1}. \tag{8.18}$$

Because we shall later be mainly interested in the threshold region where the characteristic laser features emerge, and in that region b^+b is still a small quantity, we replace (8.18) by the expansion

$$\sigma_\mu = d_0 - 4(g^2/\gamma\gamma_\parallel)d_0 b^+ b. \tag{8.19}$$

As we shall see immediately, laser action will start at a certain value of the inversion d_0. Because in this case b^+b is a small quantity, we may replace d_0 by d_c in the second term of (8.19) to the same order of approximation. We introduce the total inversion

$$\sum_\mu \sigma_\mu = D \tag{8.20}$$

and correspondingly (N = number of laser atoms)

$$Nd_0 = D_0. \tag{8.21}$$

Inserting (8.19) into (8.14) we obtain (with $Nd_c = \kappa\gamma/g^2$)

$$b^{\cdot} = \left(-\kappa + \frac{g^2}{\gamma} D_0\right) b - 4\frac{g^2\kappa}{\gamma\gamma_{\parallel}} b^+ bb + \hat{F}(t). \tag{8.22}$$

If we treat for the moment being b as a real quantity, q, (8.22) is evidently identical with the overdamped anharmonic oscillator discussed in Sections 5.1 and 6.4, where we may identify

$$\left(-\kappa + \frac{g^2}{\gamma} D_0\right) = -\alpha, \quad 4\frac{g^2\kappa}{\gamma\gamma_{\parallel}} = \beta. \tag{8.22a}$$

(compare (6.118)).

Thus we may apply the results of that discussion in particular to the critical region, where the parameter α changes its sign. We find that the concepts of symmetry breaking instability, soft mode, critical fluctuations, critical slowing down, are immediately applicable to the single mode laser and reveal a pronounced analogy between the laser threshold and a (second order) phase transition. (cf. Sect. 6.7). While we may use the results and concepts exhibited in Sections 5.1, 6.4, 6.7 we may also interpret (8.22) in the terms of laser theory. If the inversion D_0 is small enough, the coefficient of the linear term of (8.22) is negative. We may safely neglect the nonlinearity, and the field is merely supported by stochastic processes (spontaneous emission noise). Because $\hat{F}$ is (approximately) Gaussian also b is Gaussian distributed (for the definition of a Gaussian process see Sect. 4.4). The inverse of the relaxation time of the field amplitude b may be interpreted as the optical line width. With increasing inversion D_0 the system becomes more and more undamped. Consequently the optical line width decreases, which is a well-observed phenomenon in laser experiments. When α (8.22a) passes through zero, b acquires a new equilibrium position with a stable amplitude. Because b is now to be interpreted as a field amplitude, this means that the laser light is completely coherent. This coherence is only disturbed by small superimposed amplitude fluctuations caused by $\hat{F}$ and by very small phase fluctuations.

If we consider (8.22) as an equation for a complex quantity b we may derive the right hand side from the potential

$$V(|b|) = -\left(-\kappa + \frac{g^2}{\gamma} D_0\right)|b|^2 + 2\frac{g^2\kappa}{\gamma\gamma_{\parallel}}|b|^4. \tag{8.23}$$

By methods described in Sections 6.3 and 6.4, the Fokker-Planck equation can be established and readily solved yielding

$$f(b) = \mathcal{N} \exp\left(-\frac{2V(|b|)}{Q}\right), \tag{8.24}$$

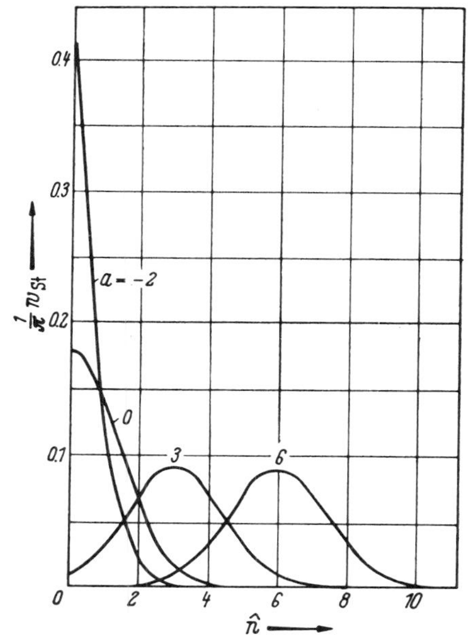

Fig. 8.1. The stationary distribution as a function of the normalized "intensity" $\hat{n}$. (After *H. Risken:* Z. Physik **186**, 85 (1965))

where Q (compare (6.91)) measures the strength of the fluctuating force. The function (8.24) (cf. Fig. 8.1) describes the photon distribution of laser light, and has been checked experimentally with great accuracy. So far we have seen that by the adiabatic principle the atoms are forced to obey immediately the order parameter. We must now discuss in detail why only one order parameter is dominant. If all parameters occurred simultaneously, the system could still be completely random.

Exercise on 8.4

Verify that

$$\dot{f} = \left[-\frac{\partial}{\partial b}(-\alpha b - \beta |b|^2 b) + \text{c.c.} + Q \frac{\partial^2}{\partial b \partial b^*} \right] f$$

is the Fokker-Planck equation belonging to (8.22).

8.5 The Multimode Laser

We now repeat the steps done before for the multimode case. We anticipate that the field mode with amplitude b_λ may be decomposed into a rapidly oscillating part with frequency Ω_λ and a slowly varying amplitude B_λ:

$$b_\lambda = B_\lambda e^{-i\Omega_\lambda t} \tag{8.25}$$

Inserting (8.25) into (8.3) yields after integration

$$\alpha_\mu = i \sum_\lambda g^*_{\mu\lambda} \int_{-\infty}^{t} e^{(-i\nu-\gamma)(t-\tau)} (b_\lambda \sigma_\mu)_\tau \, d\tau + \hat{\Gamma}_\mu, \tag{8.26}$$

or, if again the adiabatic approximation is made

$$\alpha_\mu = i \sum_\lambda g^*_{\mu\lambda} \{-i(\Omega_\lambda - \nu) + \gamma\}^{-1} b_\lambda \sigma_\mu + \hat{\Gamma}_\mu. \tag{8.27}$$

We insert (8.27) into (8.2) and use the abbreviation

$$\delta\omega_\lambda = \omega_\lambda - \Omega_\lambda. \tag{8.28}$$

We thus obtain

$$e^{-i\Omega_\lambda t} \dot{B}_\lambda = (-i\delta\omega_\lambda - \kappa_\lambda) b_\lambda + \sum_{\mu\lambda'} \frac{g_{\mu\lambda} g^*_{\mu\lambda'}}{-i(\Omega_{\lambda'} - \nu) + \gamma} b_{\lambda'} \sigma_\mu + \hat{F}_\lambda. \tag{8.29}$$

We now consider explicitly the case in which we have a discrete spectrum of modes and we assume further that we may average over the different mode phases, which in many cases is quite a good approximation. (It is also possible, however, to treat phaselocking which is of practical importance for the generation of ultra-short pulses). Multiplying (8.29) by b_λ and taking the phase average we have

$$\overline{B^+_\lambda B_{\lambda'}} = n_\lambda \delta_{\lambda\lambda'}, \tag{8.30}$$

where n_λ is the number of photons of the mode λ. If we neglect for the moment being the fluctuating forces in (8.29) we thus obtain

$$\dot{n}_\lambda = -2\kappa_\lambda n_\lambda + n_\lambda w_\lambda D \tag{8.31}$$

with

$$w_\lambda = \frac{2\gamma g^2}{(\Omega_\lambda - \nu)^2 + \gamma^2}, \tag{8.32}$$

$$|g_{\mu\lambda}|^2 = g^2. \tag{8.33}$$

In the same approximation we find

$$\dot{\sigma}_\mu = \gamma_\parallel (d_0 - \sigma_\mu) - 2 \sum_\lambda w_\lambda n_\lambda \sigma_\mu, \tag{8.34}$$

or, after solving (8.34) again adiabatically

$$D \equiv \sum_\mu \sigma_\mu \approx D_0 - \frac{2D_c}{\gamma_\parallel} \sum_\lambda w_\lambda n_\lambda, \tag{8.35}$$

where D_c is the critical inversion of all atoms at threshold. To show that (8.31)–

(8.35) lead to the selection of modes (or order parameters), consider as example just two modes treated in the exercise of Section 5.4. This analysis can be done quite rigorously also for many modes and shows that in the laser system only a single mode with smallest losses and closest to resonance survives. All the others die out. It is worth mentioning that equations of the type (8.31) to (8.35) have been proposed more recently in order to develop a mathematical model for evolution. We will come back to this point in Section 10.3.

As we have seen in Section 6.4, it is most desirable to establish the Fokker-Planck equation and its stationary solution because it gives us the overall picture of global and local stability and the size of fluctuations. The solution of the Fokker-Planck equation, which belongs to (8.29), with (8.35) can be found by the methods of Section 6.4 and reads

$$f(B_\lambda) = \mathcal{N} \exp\left(-\frac{2\Phi}{Q}\right), \tag{8.36}$$

where

$$2\Phi = \sum_\lambda |B_\lambda|^2(2\kappa_\lambda - w_\lambda D_0) + \frac{2D_c}{\gamma_\parallel} \sum_{\lambda\lambda'} w_\lambda w_{\lambda'} |B_\lambda|^2 |B_{\lambda'}|^2. \tag{8.37}$$

The local minima of Φ describe stable or metastable states. This allows us to study multimode configurations if some modes are degenerate.

8.6 Laser with Continuously Many Modes. Analogy with Superconductivity

The next example, which is slightly more involved, will allow us to make contact with the Ginzburg-Landau theory of superconductivity. Here we assume a *continuum of modes* all running in one direction. Similar to the case just considered, we expect that only modes near resonance will have a chance to participate at laser action; but because the modes are now continuously spaced, we must take into consideration a whole set of modes near the vicinity of resonance. Therefore we expect (which must be proven in a self-consistent way) that only modes with

$$|\Omega_\lambda - \nu| \ll \gamma \tag{8.38}$$

and

$$|\Omega_\lambda - \Omega_{\lambda'}| \ll \gamma_\parallel \tag{8.39}$$

are important near laser threshold. Inserting (8.27) into (8.4), we obtain

$$\dot{\sigma}_\mu = \gamma_\parallel(d_0 - \sigma_\mu) - 2\sigma_\mu \sum_{\lambda\lambda'} \left(\frac{g_{\mu\lambda} g^*_{\mu\lambda'}}{i(\Omega_\lambda - \nu) + \gamma} b^+_\lambda b_{\lambda'} + \text{c.c.} \right), \tag{8.40}$$

which under the just mentioned simplifications reduces to

$$\sigma_\mu \approx \left(d_0 - \frac{2d_c}{\gamma_\parallel} \sum_{\lambda\lambda'} \gamma^{-1} g_{\mu\lambda} g^*_{\mu\lambda'} b^+_\lambda b_{\lambda'} + \text{c.c.} \right). \tag{8.41}$$

Inserting this into (8.29) yields

$$\dot{b}_\lambda = \left(-i\omega_\lambda - \kappa_\lambda + D_0 \frac{g^2}{-i(\Omega_\lambda - \nu) + \gamma} \right) b_\lambda + \hat{F}_\lambda(t)$$
$$- \frac{4d_c}{\gamma_\parallel \gamma^2} \sum_{\mu\lambda'\lambda''\lambda'''} g_{\mu\lambda} g^*_{\mu\lambda'} g^*_{\mu\lambda''} g_{\mu\lambda'''} b_{\lambda'} b_{\lambda''} b^+_{\lambda'''}. \tag{8.42}$$

Using the form (8.9), one readily establishes

$$\sum_\mu g_{\mu\lambda} g^*_{\mu\lambda'} g^*_{\mu\lambda''} g_{\mu\lambda'''} = N g^4 \delta(k_\lambda - k_{\lambda''} - k_{\lambda'} + k_{\lambda'''}), \tag{8.43}$$

where N is the number of laser atoms. Note that we have again assumed (8.38) in the nonlinear part of (8.42). If

$$\omega_\lambda + \text{Im} \frac{D_0 g^2}{i(\Omega_\lambda - \nu) + \gamma} \tag{8.44}$$

possesses no dispersion, i.e., $\Omega_\lambda \propto k_\lambda$, the following exact solution of the corresponding Fokker-Planck equation holds

$$f(b) = \mathcal{N}_0 \exp\left(\frac{2\Phi}{Q} \right), \tag{8.45}$$

where

$$\Phi = \sum_\lambda |b_\lambda|^2 \left(D_0 \frac{\gamma g^2}{(\Omega_\lambda - \nu)^2 + \gamma^2} - \kappa_\lambda \right)$$
$$- \frac{2D_c}{\gamma_\parallel \gamma^2} g^4 \sum_{\lambda\lambda'\lambda''\lambda'''} \delta(k_\lambda - k_{\lambda''} - k_{\lambda'} + k_{\lambda'''}) b^+_{\lambda'} b^+_{\lambda''} b_\lambda b_{\lambda'''}. \tag{8.46}$$

We do not continue the discussion of this problem here in the mode picture but rather establish the announced analogy with the Ginzburg-Landau theory. To this end we assume

$$\omega_\lambda = c|k_\lambda|, \tag{8.47}$$

$$\Omega_\lambda = v|k_\lambda|, \tag{8.48}$$

$$\kappa_\lambda = \kappa. \tag{8.49}$$

Confining ourselves again to modes close to resonance we use the expansion

$$\frac{g^2}{i(\Omega_\lambda - \nu) + \gamma} = \frac{g^2}{\gamma} - i\frac{g^2}{\gamma^2}(\Omega_\lambda - \nu) - \frac{g^2}{\gamma^3}(\Omega_\lambda - \nu)^2. \tag{8.50}$$

We now replace the index λ by the wave number k and form the wave packet

$$\Psi(x, t) = \int_{-\infty}^{+\infty} B_k e^{+ikx - iv|k|t}\, dk. \tag{8.51}$$

The Fourier transformation of (8.42) is straightforward, and we obtain

$$\Psi(x, t)^{\cdot} = -\alpha\Psi(x, t) + c\left(iv\frac{d}{dx} + \nu\right)^2 \Psi(x, t) - 2\beta|\Psi(x, t)|^2\Psi(x, t) + F(x, t), \tag{8.52}$$

where in particular the coefficient α is given by

$$-\alpha = \left(-\kappa + \frac{g^2}{\gamma} D_0\right). \tag{8.53}$$

Eq. (8.52) is identical with the equation of the electron-pair wave function of the Ginzburg-Landau theory of superconductivity for the one-dimensional case if the following identifications are made:

Table 8.1

Superconductor	Laser
Ψ, pair wave function	Ψ, electric field strength
$\alpha \propto T\text{-}T_c$	$\alpha \propto D_c\text{-}D$
T temperature	D total inversion
T_c critical temperature	D_c critical inversion
$\nu \propto A_x$-component of vector potential	ν atomic frequency
$F(x, t)$ thermal fluctuations	fluctuations caused by spontaneous emission, etc.

Note, however, that our equation holds for systems far from thermal equilibrium where the fluctuating forces, in particular, have quite a different meaning. We may again establish the Fokker-Planck equation and translate the solution (8.45) (8.46) to the continuous case which yields

$$f = \mathcal{N}_0 \exp\left(\frac{2\Phi}{Q}\right) \tag{8.54}$$

with

$$\Phi = \int\left\{-\alpha|\Psi(x, t)|^2 - \beta|\Psi(x, t)|^4 - c\left|\left(iv\frac{d}{dx} - \nu\right)\Psi\right|^2\right\}dx. \tag{8.55}$$

Eq. (8.55) is identical with the expression for the distribution function of the Ginzburg-Landau theory of superconductivity if we identify (in addition to Table 8.1) Φ with the free energy $\mathcal{F}$ and Q with $2k_B T$. The analogy between systems away from thermal equilibrium and in thermal equilibrium is so evident that it needs no further discussion. As a consequence, however, methods originally developed for one-dimensional superconductors are now applicable to lasers and vice versa.

8.7 First-Order Phase Transitions of the Single-Mode Laser

So far we have found that at threshold the single-mode laser undergoes a transition which has features of a second-order phase transition. In this section we want to show that by some other physical conditions the character of the phase transition can be changed.

a) *Single Mode Laser with an Injected External Signal*

When we send a light wave on a laser, the external field interacts directly only with the atoms. Since the laser field is produced by the atoms, the injected field will indirectly influence the laser field. We assume the external field in the form of a plane wave

$$E = E_0 e^{i\varphi_0 + i\omega_0 t - ik_0 x} + \text{c.c.}, \tag{8.56}$$

where E_0 is the real amplitude and φ_0 a fixed phase. The frequency ω_0 is assumed in resonance with the atomic transition and also with the field mode. k_0 is the corresponding wave number. We assume that the field strength E_0 is so weak that it practically does not influence the atomic inversion. The only laser equation that E enters into is the equation for the atomic dipole moments. Since here the external field plays the same role as the laser field itself, we have to replace b by b + const $\cdot E$ in (8.7). This amounts to making the replacement

$$\Gamma_\mu \to \Gamma_\mu + \vartheta_\mu E_0 e^{-i\varphi_0}\sigma_\mu, \tag{8.57}$$

where

$$\vartheta_\mu = \vartheta e^{ik_0 x_\mu}. \tag{8.58}$$

and ϑ is proportional to the atomic dipole moment transition matrix element. In (8.57) we have taken into account only the resonant term. Since E_0 is anyway only a higher order correction term to the equations we will replace in the following σ_μ by d_0.

It is now a simple matter to repeat all the iteration steps performed after (8.11). We then see that the only effect of the change (8.57) is to make in (8.15) the following replacements

$$\hat{F} \to \hat{F} - i \textstyle\sum_\mu g_\mu \vartheta_\mu d_0 E_0 e^{-i\varphi_0}\gamma^{-1}. \tag{8.59}$$

Inserting (8.59) into (8.22) yields the basic equation

$$b^{\cdot} = \left(-\kappa + \frac{g^2}{\gamma} D_0\right) b - 4\frac{g^2\kappa}{\gamma\gamma_{\parallel}} b^+ bb + \hat{E}_0 e^{-i\phi_0} + \hat{F}(t), \tag{8.60}$$

where we have used the abbreviation

$$-\gamma^{-1}d_0 e^{-i\varphi_0}E_0 i\sum_\mu g_\mu \vartheta_\mu = \hat{E}_0 e^{-i\hat{\varphi}_0}. \tag{8.61}$$

Again it is a simple matter to write the rhs of (8.60) and of its conjugate complex as derivatives of a potential where we still add the fluctuating forces

$$db^+/dt = -\frac{\partial V}{\partial b} + \hat{F}^+, \tag{8.62}$$

$$db/dt = -\frac{\partial V}{\partial b^+} + \hat{F}. \tag{8.63}$$

Writing b in the form

$$b = re^{-i\varphi} \tag{8.64}$$

the potential reads

$$V = V_0(r) - 2\hat{E}_0 r\cos(\hat{\varphi}_0 - \varphi), \tag{8.65}$$

where V_0 is identical with the rotation symmetric potential of formula (8.23). The additional cosine term destroys the rotation symmetry. If $\hat{E}_0 = 0$, the potential is rotation symmetric and the phase can diffuse in an undamped manner giving rise to a finite linewidth of the laser mode. This phase diffusion is no more possible if $\hat{E}_0 \neq 0$, since the potential considered as a function of φ has now a minimum at

$$\varphi = \hat{\varphi}_0 + \pi. \tag{8.66}$$

The corresponding potential is exhibited in Fig. 8.2 which really exhibits a pinning

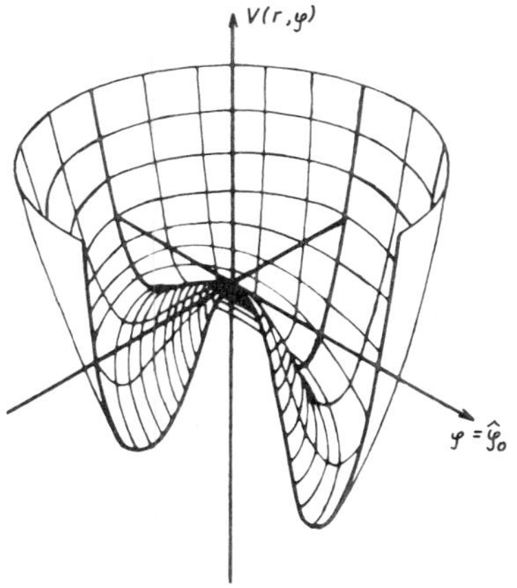

Fig. 8.2. The rotational symmetry of the potential is broken due to an injected signal. (After *W. W. Chow, M. O. Scully, E. W. van Stryland*: Opt. Commun. **15**, 6 (1975))

of φ. Of course, there are still fluctuating forces which let φ fluctuate around the value (8.66).

b) *Single Mode Laser in the Presence of a Saturable Absorber*

Here we treat an experimental setup in which a so-called saturable absorber is inserted between the laser material and one of the mirrors (compare Fig. 1.9). The saturable absorber is a material having the following property. If it is irradiated by light of weak intensity the saturable absorber absorbs light. On the other hand at high light intensities the saturable absorber becomes transparent. To take into account the effect of a saturable absorber in the basic equations we have to allow for intensity dependent losses. This could be done by replacing κ in (8.14) by

$$\kappa(b) \approx \kappa_0 + \kappa_s - |b|^2\kappa_s/I_s. \tag{8.67}$$

While (8.67) is a reasonable approximation for not too high intensities $|b|^2$, at very high fields (8.67) would lead to the wrong result that the loss becomes negative. Indeed one may show that the loss constant should be replaced by

$$\kappa \to \kappa(b) = \kappa_0 + \frac{\kappa_s}{1 + |b|^2/I_s}. \tag{8.68}$$

From it the meaning of I_s becomes clear. It is that field intensity where the decrease of the loss becomes less and less important. It is a simple matter to treat also the variation of the inversion more exactly which we take now from (8.18). Inserting this into (8.13) and the resulting expression into (8.6) we obtain, using as loss constant (8.68)

$$db^+/dt = -\left(\kappa_0 + \frac{\kappa_s}{1 + |b|^2/I_s}\right)b^+ + \frac{G}{1 + |b|^2/I_{as}}b^+ + \hat{F}^+(t) \tag{8.69}$$

Here we have used the abbreviations

$$G = d_0 \textstyle\sum_\mu |g_\mu|^2/\gamma \equiv \dfrac{g^2}{\gamma} D_0, \tag{8.70}$$

and

$$1/I_{as} = 4g^2/\gamma\gamma_{\|}. \tag{8.71}$$

Again (8.69) and its conjugate complex possess a potential which can be found by integration in the explicit form

$$-V(b) = -\kappa_0|b|^2 - I_s\kappa_s \ln(1 + |b|^2/I_s) + I_{as}G \ln(1 + |b|^2/I_{as}). \tag{8.72}$$

For a discussion of (8.72) let us consider the special case in which $I_s \ll I_{as}$. This allows us to expand the second logarithm for not too high $|b|^2$, but we must keep the first logarithm in (8.72). The resulting potential curve of

$$-V(b) = (G - \kappa_0)|b|^2 - I_s\kappa_s \ln(1 + |b|^2/I_s) - \tfrac{1}{2}G|b|^4/I_{as} \tag{8.73}$$

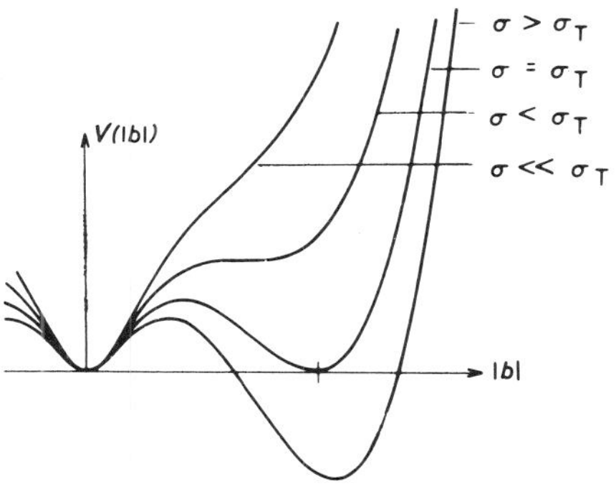

Fig. 8.3. Change of the potential curve for different pump parameters $\sigma = d_0$. (After *J. W. Scott, M. Sargent III, C. D. Cantrell*: Opt. Commun. **15**, 13 (1975))

is exhibited in Fig. 8.3. With change of pump parameter G the potential is deformed in a way similar to that of a first order phase transition discussed in Section 6.7 (compare the corresponding curves there). In particular a hysteresis effect occurs.

8.8 Hierarchy of Laser Instabilities and Ultrashort Laser Pulses

We consider a laser with continuously many modes. We treat the electric field strength $\mathscr{E}$ directly, i.e., without decomposing it into cavity modes. To achieve also a "macroscopic" description of the laser material, we introduce the macroscopic polarization $\mathscr{P}(\boldsymbol{x}, t)$ and inversion density $\mathscr{D}(\boldsymbol{x}, t)$ by

$$\mathscr{P}(\boldsymbol{x}, t) = \textstyle\sum_\mu \delta(\boldsymbol{x} - \boldsymbol{x}_\mu)(\Theta_{21}\alpha_\mu + \Theta_{12}\alpha_\mu^*) \tag{8.74}$$

(Θ_{21}: atomic dipole matrix element) and

$$\mathscr{D}(\boldsymbol{x}, t) = \textstyle\sum_\mu \delta(\boldsymbol{x} - \boldsymbol{x}_\mu)\sigma_\mu. \tag{8.75}$$

The laser equations then read

$$-\Delta\mathscr{E} + \frac{1}{c^2}\partial^2\mathscr{E}/\partial t^2 + \frac{2\kappa}{c^2}\partial\mathscr{E}/\partial t = -\frac{4\pi}{c^2}\boldsymbol{\mathscr{P}}, \tag{8.76}$$

$$\partial^2\boldsymbol{\mathscr{P}}/\partial t^2 + 2\gamma\partial\boldsymbol{\mathscr{P}}/\partial t + \omega_0^2\boldsymbol{\mathscr{P}} = -\tfrac{2}{3}\omega_0(|\theta_{21}|^2/\hbar)\mathscr{E}\mathscr{D}, \tag{8.77}$$

$$\partial\mathscr{D}/\partial t = \gamma_\parallel(\mathscr{D}_0 - \mathscr{D}) + (2/\hbar\omega_0)\mathscr{E}\partial\boldsymbol{\mathscr{P}}/\partial t. \tag{8.78}$$

Eq. (8.76) follows directly from Maxwell's equation, while (8.77), (8.78) are material equations described in laser theory using quantum mechanics. To show some of the main features, we have dropped the fluctuation forces. We leave their inclusion as an exercise to the reader. κ is the "cavity" loss, γ the atomic linewidth, $\gamma_{\parallel}^{-1} = T_1$ the atomic inversion relaxation time, ω_0 the atomic transition frequency, $\hbar$ Planck's constant. The meaning of these equations is the same as those discussed on page 224. D_0 is the "unsaturated" inversion due to pump and relaxation processes. We treat only one direction of field polarization and wave propagation in x-direction. Since we anticipate that the field oscillates essentially at resonance frequency ω_0, we put

$$\mathscr{E} = e^{i\omega_0 t - ik_0 x}\hat{E}^{(-)}(x, t) + \text{c.c.}, \tag{8.79}$$

where $\omega_0 = ck_0$ and $\hat{E}^{(+)} = \hat{E}^{(-)*}$. An analogous decomposition of $\mathscr{P}$ is made. $\hat{E}^{(-)}(x, t)$ and $\hat{P}^{(-)}(x, t)$ are treated as slowly varying functions of x and t allowing us to neglect e.g., $\dot{\hat{E}}$ compared to $\omega_0\hat{E}$, etc. Thus we arrive at the following equations

$$\partial\hat{E}^{(-)}/\partial t + c\frac{\partial\hat{E}^{(-)}}{\partial x} + \kappa E^{(-)} = -2\pi i\omega_0\hat{P}^{(-)}, \tag{8.80}$$

$$\partial\hat{P}^{(-)}/\partial t + \gamma\hat{P}^{(-)} = \frac{i}{3}(|\Theta_{21}|^2/\hbar)\hat{E}^{(-)}\hat{D}, \tag{8.81}$$

$$\partial\hat{D}/\partial t = \gamma_{\parallel}(D_0 - \hat{D}) + (2i/\hbar)(\hat{E}^{(+)}\hat{P}^{(-)} - \hat{E}^{(-)}\hat{P}^{(+)}). \tag{8.82}$$

These equations are equivalent to (8.2), (8.3), (8.4) provided we assume there $\kappa_\lambda = \kappa$ and a set of discrete running modes. We now start with a small inversion, D_0, and investigate what happens to $\hat{E}$; $\hat{P}$, D, when D_0 is increased.

1) D_0 is small, $D_0 < \kappa\gamma/g^2$.

We readily find as solution

$$\hat{E} = \hat{P} = 0, \hat{D} = D_0, \tag{8.83}$$

i.e., no laser action occurs. To check the stability of the solution (8.83), we perform a linear stability analysis putting

$$\hat{E} = ae^{\lambda t - ikx}, \tag{8.84}$$

$$\hat{P} = be^{\lambda t - ikx}, \tag{8.85}$$

$$\hat{D} - D_0 = ce^{\lambda t - ikx}, \tag{8.86}$$

which yields the following characteristic values

$$\lambda^{(1)} = -\gamma_{\parallel}, \tag{8.87}$$

$$\lambda^{(2,3)} = (1/2)\{-\gamma - \kappa + ick \pm \sqrt{}\}, \tag{8.88}$$

where

$$\sqrt{} = [(\gamma + \kappa + ick)^2 + 4g^2(D_0 - \hat{D}_{\text{thr}}) + 4ic\gamma k]^{1/2} \tag{8.89}$$

At present, $\hat{D}_{\text{thr},1}$ is just an abbreviation:

$$\hat{D}_{\text{thr},1} = \kappa\gamma/g^2 \text{ where } g^2 = 2\pi\omega_0|\Theta_{12}|^2/3\hbar. \tag{8.90}$$

A closer investigation of (8.88), (8.89) reveals that $\lambda^{(2)} \geq 0$ for $D_0 \geq D_{\text{thr}}$ and an *instability* occurs, at first for $k = 0$.

2) $\hat{D}_{\text{thr},1} \leq D_0 \leq \hat{D}_{\text{thr},2}$.
It is a simple màtter to convince oneself that

$$\hat{E}(x, t) = \hat{E}_{\text{cw}},\ \hat{P}(x, t) = \hat{P}_{\text{cw}} \text{ and } \hat{D}(x, t) = \hat{D}_{\text{cw}}. \tag{8.91}$$

is now the stable solution. The rhs of (8.91) are space and time-independent constants, obeying (8.80)–(8.82). This solution corresponds to that of Sections 8.4–6 in the laser domain if fluctuations are neglected. For what follows it is convenient to normalize $\hat{E}$, $\hat{P}$, $\hat{D}$ with respect to (8.91), i.e., we put

$$E = \hat{E}/\hat{E}_{\text{c.w.}},\ P = \hat{P}/\hat{P}_{\text{c.w.}},\ D = \hat{D}/\hat{D}_{\text{thr}} \tag{8.92}$$

and introduce as effective pump parameter

$$\Lambda = (D_0 - D_{\text{thr}})/D_{\text{thr}}. \tag{8.93}$$

A detailed analysis reveals that E, P, D may be chosen real so that (8.80)–(8.82) acquire the form

$$\left(\frac{\partial}{\partial t} + \gamma\right)P = \gamma ED, \tag{8.94}$$

$$\left(\frac{\partial}{\partial t} + \gamma_{\|}\right)D = \gamma_{\|}(\Lambda + 1) - \gamma_{\|}\Lambda EP, \tag{8.95}$$

$$\left(\frac{\partial}{\partial t} + \kappa + c\frac{\partial}{\partial x}\right)E = \kappa P. \tag{8.96}$$

The solutions E, P, D are, on account of the normalization,

$$E = P = D = 1. \tag{8.97}$$

We again perform a linear stability analysis by putting

$$\begin{pmatrix} E \\ D \\ P \end{pmatrix} = \begin{pmatrix} 1 \\ 1 \\ 1 \end{pmatrix} + \begin{pmatrix} e \\ \delta \\ p \end{pmatrix}, \tag{8.98}$$

where we abbreviate

$$\boldsymbol{q}(x, t) = \begin{pmatrix} e \\ \delta \\ p \end{pmatrix}. \tag{8.99}$$

Putting $\boldsymbol{q}(x, t) = \boldsymbol{q}(0)e^{\lambda t - ikx/c}$ we obtain a characteristic equation of third order in λ which allows for an unstable solution, i.e., Re $\lambda > 0$ provided

$$\Lambda > \Lambda_c \equiv 4 + 3\varepsilon + 2\sqrt{2(1 + \varepsilon)(2 + \varepsilon)}, \tag{8.100}$$

holds, where $\varepsilon = \gamma_{\|}/\gamma$.

3) $D_0 \geq D_{\mathrm{thr.2}}$.

At $\Lambda = \Lambda_c$, the real part of the "damping" constant λ_u (u = unstable) vanishes, but λ possesses a nonvanishing imaginary part, determined by

$$\lambda_u^2 = \tfrac{1}{2}(\varepsilon - 3\Lambda_c)\gamma_{\|}\gamma. \tag{8.101}$$

Simultaneously, at $\Lambda = \Lambda_c$, also the wavenumber k is unequal zero,

$$k = k_c = -i\left(\lambda_u \frac{\kappa\gamma\gamma_{\|}}{2(\gamma + \gamma_{\|})}(\Lambda_c + \varepsilon)\right). \tag{8.102}$$

We expand $\boldsymbol{q}$ into a superposition of plane waves and eigenvectors of the linearized equations, assuming a ring laser with periodic boundary conditions with length L,

$$\boldsymbol{q} = \textstyle\sum_{k,j} \boldsymbol{O}^{(j)}\left(c\frac{\partial}{\partial x}\right)\xi_{k,j}\chi_k \tag{8.103}$$

with

$$\chi_k = \frac{1}{\sqrt{L}}e^{ikx/c}. \tag{8.104}$$

The further steps are those of Section 7.7 and 7.8. Resolving the equations for the stable modes up to fourth order in the amplitude of the unstable mode $\xi_{k_c,u}$ yields the following equation:

$$\left(\frac{\partial}{\partial t} + c\frac{\partial}{\partial x}\right)\xi_{k_c,u} = \tilde{\lambda}\xi_{k_c,u} + \tilde{a}|\xi_{k_c,u}|^2\xi_{k_c,u} - \tilde{b}|\xi_{k_c,u}|^4\xi_{k_c,u}. \tag{8.105}$$

Splitting ξ into a real amplitude and a phase factor

$$\xi_{k_c,u} = Re^{i\varphi} \tag{8.106}$$

we obtain an equation for R

$$\left(\frac{\partial}{\partial t} + c\frac{\partial}{\partial x}\right)R = \beta R + aR^3 - bR^5 \equiv -\frac{\partial V}{\partial R} \tag{8.107}$$

with

$$\begin{aligned} a &= \mathrm{Re}\,\tilde{a} > 0 \text{ or } < 0, \text{ depending on the length } L, \\ b &= \mathrm{Re}\,\tilde{b} > 0, \\ \beta &> 0 \text{ for } \Lambda > \Lambda_c, \\ &< 0 \text{ for } \Lambda < \Lambda_c. \end{aligned} \tag{8.108}$$

Introducing new variables

$$\tilde{x} = x/c - t, \tilde{t} = t \tag{8.109}$$

we transform (8.96) into a first order differential equation with respect to time, where $\tilde{x}$ plays the role of a parameter which eventually drops out from our final solution. Interpreting the rhs of (8.107) as a derivative of a potential function V and the lhs as the acceleration term of the overdamped motion of a particle, the behavior of the amplitude can immediately be discussed by means of a potential curve. We first discuss the case $\mathrm{Re}\,\tilde{a} > 0$. Because the shape of this curve is qualitatively similar to Fig. 8.3, the reader is referred to that curve. For

$$d \equiv 4\beta b/a^2 < -1 \tag{8.110}$$

there is only one minimum, $R = 0$, i.e., the continuous wave (cw) solution is stable. For $-1 < d < 0$ the cw solution is still stable; but a new minimum of the potential curve indicates that there is a second new state available by a hard excitation. If finally $d > 0$, $R = 0$ becomes unstable (or, more precisely, we have a point of marginal stability) and the new stable point lies at $R = R_2$. Thus the laser pulse amplitude jumps discontinuously from $R = 0$ to $R = R_2$. The coefficients of (8.105) are rather complicated expressions and are not exhibited here. Eq. (8.107) is solved numerically; and the resulting R (and the corresponding phase) are then inserted into (8.103). Thus the electric field strength E takes the form

$$\begin{aligned} E &= (1 + E_0) + E_1 \cos(\omega_c\tau + \varphi_1) + E_2 \cos(2\omega_c\tau + \varphi_2) \\ &\quad + E_3 \cos(3\omega_c\tau + \varphi_3), \\ \tau &= t - k_c x/(\omega_c c), \end{aligned} \tag{8.111}$$

where we have included terms up to the cubic harmonic. Fig. 8.4 exhibits the corresponding curves for the electric field strength E, the polarization P and the inversion D for $\Lambda = \Lambda_c$. These results may be compared with the numerical curves

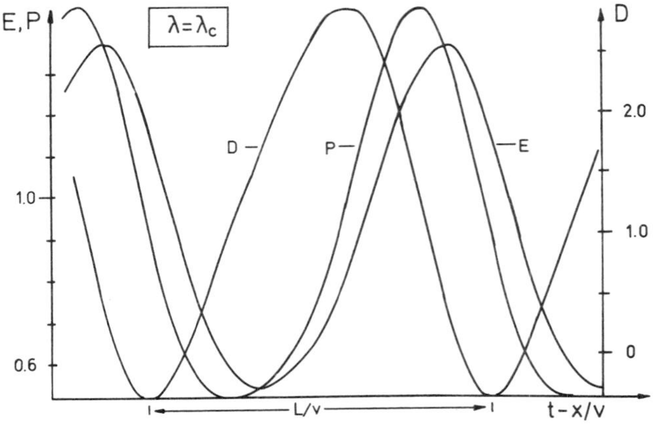

Fig. 8.4. Electric field strength E, polarization P and inversion D for $\Lambda = \Lambda_c$. (After *H. Haken, H. Ohno:* Opt. Commun. **16**, 205 (1976))

obtained by direct computer solutions of (8.94–96) for $\Lambda = 11$. Very good agreement is found, for example, with respect to the electric field strength.

We now discuss the case Re $\tilde{a} < 0$. The shapes of the corresponding potential curves for $\beta < 0$ or $\beta > 0$ are qualitatively given by the solid lines of Figs. 6.6a and 6.6b, respectively, i.e., we deal now with a second order phase transition and a soft excitation. The shapes of the corresponding field strength, polarization and inversion resemble those of Fig. 8.4.

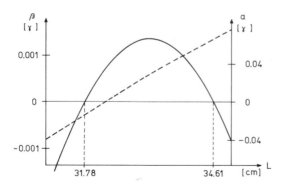

Fig. 8.4a. The coefficients β, a of (8.107) as functions of laser length L. Left ordinate refers to β, right to a. Chosen parameters: $\gamma = 2 \times 10^9\ \mathrm{s}^{-1}$; $\kappa = 0.1\gamma$; $\gamma_{\parallel} = 0.5\gamma$; $\Lambda = 11.5$. (After *H. Haken* and *H. Ohno*, unpublished)

The dependence of the coefficients a and β (8.108) on laser length L is shown in Fig. 8.4a for a given pump strength.

The method of generalized Ginzburg-Landau equations allows us also to treat the buildup of ultrashort laser pulses. Here, first the time-dependent order parameter equation (8.107) with (8.109) is solved and then the amplitudes of the slaved modes are determined. A typical result is exhibited in Fig. 8.4b.

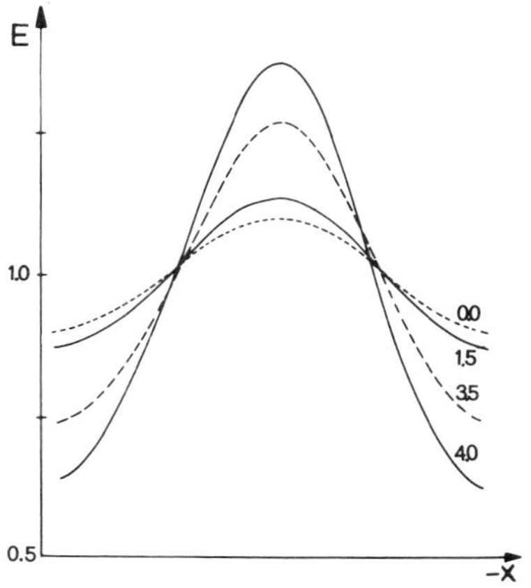

Fig. 8.4b. Temporal buildup of electric field strength of ultrashort laser pulse in the case of first order transition. (After *H. Ohno, H. Haken:* Phys. Lett. **59A**, 261(1976))

8.9 Instabilities in Fluid Dynamics: The Bénard and Taylor Problems

The following problems have fascinated physicists for at least a century because they are striking examples how thermal equilibrium states characterized by complete disorder may suddenly show pronounced order if one goes away from thermal equilibrium. There are, among others, the so-called Bénard and Taylor problems. We first describe the Bénard problem, or, using another expression for it, convection instability: Let us consider an infinitely extended horizontal fluid layer which is heated from below so that a temperature gradient is maintained. This gradient, if expressed in suitable dimensionless units, is called the Rayleigh number, R. As long as the Rayleigh number is not too large, the fluid remains quiescent and heat is transported by conduction. If the number exceeds a certain value, however, suddenly the fluid starts to convect. What is most surprising is that the convection pattern is very regular and may either show rolls or hexagons. The hexagons depicted in Fig. 1.16 are the convection cells as seen from above. One may have the fluid rising in the middle of the cell and going down at the boundaries or vice versa. An obvious task for the theoretical treatment is to explain the mechanism of this sudden disorder-order transition and to predict the form and stability of the cells. In a more refined theory one may then ask questions as to the probability of fluctuations.

A closely related problem is the so-called Taylor-problem: The flow between a long stationary outer cylinder and a concentric rotating inner cylinder takes place along circular streamlines (Couette flow) if a suitable dimensionless measure of the inner rotation speed (the Taylor number) is small enough. But Taylor vortices spaced periodically in the axial direction appear when the Taylor number exceeds a critical value.

In the following we shall explicitly treat the convection instability. The physical quantities we have to deal with are the velocity field with components u_j ($j = 1, 2, 3 \leftrightarrow x, y, z$) at space point x, y, z, the pressure p and the temperature T. Before going into the mathematical details, we shall describe its spirit. The velocity field, the pressure, and the temperature obey certain nonlinear equations of fluid dynamics which may be brought into a form depending on the Rayleigh number R which is a prescribed quantity. For small values of R, we solve by putting the velocity components equal zero. The *stability* of this solution is proven by linearizing the total equations around the stationary values of $\boldsymbol{u}$, p, T, where we obtain *damped waves*. If, however, the Rayleigh number exceeds a certain critical value R_c, the solutions become unstable. The procedure is now rather similar to the one which we have encountered in laser theory (compare Sect. 8.8). The solutions which become unstable define a set of modes. We expand the actual field ($\boldsymbol{u}$, T) into these modes with unknown amplitudes. Taking now into account the nonlinearities of the system, we obtain nonlinear equations for the mode amplitudes which quite closely resemble those of laser theory leading to certain stable-mode configurations. Including thermal fluctuations we again end up with a problem defined by nonlinear deterministic forces and fluctuating forces quite in the spirit of Chapter 7. Their interplay governs in particular the transition region, $R \approx R_c$.

8.10 The Basic Equations

We now proceed to the mathematical treatment starting from the basic equations of hydrodynamics using the so-called Boussinesq approximation. In it we neglect the temperature dependence of all coefficients except that of the density ρ. Its temperature dependence, i.e.,

$$\delta\rho = -\rho_0\alpha(T - T_0)$$

(ρ_0 average density, α thermal expansion coefficient, T absolute temperature, T_0 reference temperature for which $\rho(T) = \rho_0$) is taken into account only in a term describing the effect of the gravitational force on the fluid. To save space, we write the corresponding equations in dimensionless units.

$$\frac{\partial u_j}{\partial x_j} = 0 \tag{8.112}$$

is the *continuity equation* of an incompressible fluid (we use the convention to sum up over indices occurring doubly).

$$\frac{\partial u_i}{\partial t} = -u_j\frac{\partial u_i}{\partial x_j} - \frac{\partial \varpi}{\partial x_i} + P\Theta\lambda_i + P\Delta u_i + F_i^{(u)},$$

$$\lambda = (0, 0, 1), \tag{8.113}$$

describes the acceleration of the fluid due to external and internal forces. ϖ is defined by

$$\varpi = p/\rho_0 + gz - \tfrac{1}{2}\beta\alpha z^2. \tag{8.114}$$

g: gravitational acceleration constant. P is the Prandtl number

$$P = \nu/\kappa, \tag{8.115}$$

where ν is the kinematic viscosity, and κ the thermometric conductivity, $\Delta = \partial^2/\partial x_1^2 + \partial^2/\partial x_2^2 + \partial^2/\partial x_3^2$. $F_i^{(u)}$ is a fluctuating force. Finally

$$\frac{\partial \Theta}{\partial t} = -u_j \frac{\partial \Theta}{\partial x_j} + Ru_z + \Delta\Theta + F^{(\Theta)} \tag{8.116}$$

is the heat conduction equation, where $\Theta = T - T_0 - \beta z$ (β = temperature gradient in vertical direction). The Rayleigh number R is given by

$$R = \alpha g \beta d^4/\nu\kappa. \tag{8.117}$$

d: thickness of fluid layer. $F^{(\Theta)}$ is again a fluctuating force. The boundary conditions for $\boldsymbol{u}$, Θ are as follows: The fluid layer is horizontally infinitely extended between the planes $z = -d/2$ and $z = d/2$. For a viscous fluid we have at rigid boundaries

$$u_1 = \Theta = 0, \tag{8.118}$$

and at free boundaries

$$u_z = \frac{\partial}{\partial u_z} \varepsilon_{izk} u_k = \Theta = 0. \tag{8.119}$$

It is possible to go beyond the Boussinesq approximation by inclusion of further terms. We do not treat them here explicitly but notice that some of them are not invariant against reflection at the vertical symmetry plane of the fluid. This will have an important consequence on the structure of the final equations.

8.11 Damped and Neutral Solutions ($R \leq R_c$)

We now treat (8.112) to (8.116) as follows: We first assume that $R < R_c$. Linearising the nonlinear equations around the values $\boldsymbol{u} = \Theta = 0$, we obtain after some calculations damped solutions

$$\boldsymbol{u} \sim e^{-\gamma t}; \gamma > 0. \tag{8.120}$$

We define now the critical value R_c as the one for which γ tends to 0. $R = R_c$ thus defines the marginal (neutral) states. A lengthy but straightforward calculation shows that the neutral solutions may be defined as

$$u_i^{(1)}(\mathbf{x}) = \delta_i v^{(1)}, \tag{8.121}$$

where the operator δ_i is given by

$$\delta_i = \frac{\partial^2}{\partial x_i \partial z} - \lambda_i \Delta;\ \boldsymbol{\lambda} = (0, 0, 1), \tag{8.122}$$

and $v^{(1)}$ by

$$v^{(1)} = w(x, y)g(z). \tag{8.123}$$

The functions w and $g(z)$ are defined as follows: For free boundaries (the origin of the coordinate system is put in the middle of the layer)

$$g(z) = (\pi^2 + k^2)^2 \cos \pi z \tag{8.124}$$

and

$$w(x, y) = \sum_{\mathbf{k}} A_{\mathbf{k}} e^{i\mathbf{k}\mathbf{x}}, \tag{8.125}$$

where the coefficients $A_{\mathbf{k}}$ are still arbitrary and $\mathbf{k}$ lies in the x, y plane

$$\mathbf{k} = (k_x, k_y) \tag{8.126}$$

with the absolute value

$$k = \frac{\pi}{\sqrt{2}}. \tag{8.127}$$

Without going into all details of the neutral solution, we merely mention that the neutral solution may be described by a superposition of the formal vectors

$$\Psi_{\mathbf{k}} = \begin{pmatrix} \mathbf{u}_{\mathbf{k}}^{(1)} \\ \Theta_{\mathbf{k}}^{(1)} \end{pmatrix}. \tag{8.128}$$

For the following we define the transpose of (8.128) by

$$\overline{\Psi_{\mathbf{k}}} = (\mathbf{u}_{\mathbf{k}}^{(1)}, \Theta_{\mathbf{k}}^{(1)}). \tag{8.129}$$

The $\Theta_{\mathbf{k}}^{(1)}$'s can also explicitly be given.

8.12 Solution Near $R = R_c$ (Nonlinear Domain). Effective Langevin Equations

We may now apply the procedures of Sections 7.7 or 7.9 to eliminate the stable modes. Writing the neutral (or unstable) solutions in the form

$$\boldsymbol{u}(\boldsymbol{x}) = \sum_{\boldsymbol{k}} A_{\boldsymbol{k}} \boldsymbol{u}_{\boldsymbol{k}}^{(1)}(\boldsymbol{x}), \tag{8.130}$$

we then obtain equations for the slowly varying amplitudes $A_{\boldsymbol{k}}(\boldsymbol{x}, t)$. Neglecting all non-Boussinesq terms we obtain the following set of equations

$$\frac{\partial A_{\boldsymbol{k}}}{\partial t} = \gamma\left(\frac{\partial}{\partial x_{(\boldsymbol{k})}} - \frac{i}{\sqrt{2\pi}}\frac{\partial^2}{\partial y_{(\boldsymbol{k})}^2}\right)^2 A_{\boldsymbol{k}} + \alpha A_{\boldsymbol{k}} - \sum_{\boldsymbol{k}'} \beta_{\boldsymbol{k}\boldsymbol{k}'}|A_{\boldsymbol{k}'}|^2 A_{\boldsymbol{k}} + F_{\boldsymbol{k}}(x, y, t), \tag{8.131}$$

which has the form of Langevin equations for the $A_{\boldsymbol{k}}$'s. $\partial/\partial x_{(\boldsymbol{k})}$ is the derivative in the direction of the $\boldsymbol{k}$ vector occurring in $A_{\boldsymbol{k}}$, $\partial/\partial y_{(\boldsymbol{k})}$ the derivative perpendicular to it. We further have

$$\beta_{\boldsymbol{k}\boldsymbol{k}'} = \tilde{\beta} \quad \text{for } \boldsymbol{k} = \boldsymbol{k}', \tag{8.132}$$

and put

$$\beta_{\boldsymbol{k}\boldsymbol{k}'} = \tilde{\beta}(1 + \beta_{ij}) \quad \text{for } \boldsymbol{k} \neq \boldsymbol{k}'. \tag{8.133}$$

For what follows the explicit form of β_{ij} is not very important. We only note that all β_{ij} are positive. The coefficients α, $\tilde{\beta}$, γ in (8.131) are defined as follows

$$\gamma = \frac{4P}{P+1}, \tag{8.134}$$

$$\alpha = \frac{3P\pi^2}{2(P+1)} \cdot \frac{R - R_c}{R_c}, \tag{8.135}$$

$$\tilde{\beta} = \frac{P}{2(P+1)}. \tag{8.136}$$

The fluctuating forces $F_{\boldsymbol{k}}$ are found as projection of the original fluctuating forces on the neutral solutions, i.e., as follows:

$$F_{\boldsymbol{k}}(x, y, t) = \left\langle \left(\Psi_{\boldsymbol{k}}, \begin{pmatrix} \boldsymbol{F}^{(u)} \\ F^{(\Theta)} \end{pmatrix} \right) \right\rangle. \tag{8.137}$$

The brackets $\langle\ \rangle$ denote an average over the space coordinates over a region large compared to $1/|k|$, but small compared to the wavelength of the resulting fluctuations of $A_{\boldsymbol{k}}$. From this definition one readily derives the following correlation functions

$$\langle\langle F_{\boldsymbol{k}}(x, y, t) F_{\boldsymbol{k}'}(x', y', t') \rangle\rangle = \delta_{\boldsymbol{k}\boldsymbol{k}'}\delta(x - x')\delta(y - y')\delta(t - t')Q. \tag{8.138}$$

The double brackets $\langle\langle \cdots \rangle\rangle$ denote the statistical average over the thermal fluctuations. Using the fluctuating forces introduced by Landau and Lifshitz we obtain

$$Q = \frac{\pi^2}{4(P+1)}\left(\frac{\eta k_B T}{\rho^2\kappa^3 d} + \frac{P^2}{R_c}\frac{\alpha^2 g^2 d^3 k_B T^2}{\rho^2 c_p^2 \kappa^2 v^2}\right). \tag{8.139}$$

If non-Boussinesq terms violating the reflection symmetry are taken into account, an additional term of the form

$$-\delta \sum\nolimits_{\boldsymbol{k}_1,\boldsymbol{k}_2} A_{\boldsymbol{k}_1}^* A_{\boldsymbol{k}_2}^* \delta_{\boldsymbol{k}_1+\boldsymbol{k}_2+\boldsymbol{k},0} \tag{8.140}$$

must be added to the right hand side of (8.131). (For sake of completeness we mention that the inclusion of non-Boussinesq terms requires a "renormalization" of R_c, k and an altered function $g(z)$ in (8.123). Because this does not change the structure of our following equations, we omit a detailed discussion here).

An equation of the form (8.131) is well known to us. We already know it from the sections on laser theory and we can exploit the methods of solution which have been described there. Before doing so we write (8.131) (if necessary under inclusion of (8.140)) in the form

$$\frac{\partial A_{\boldsymbol{k}}}{\partial t} = \tilde{L} A_{\boldsymbol{k}} + N_{\boldsymbol{k}}(\{A\}) + F_{\boldsymbol{k}}, \tag{8.141}$$

where L is the linear operator occurring on the rhs of (8.131), whereas $N_{\boldsymbol{k}}$ contains all the nonlinear terms of that equation.

8.13 The Fokker-Planck Equation and Its Stationary Solution

It is now a simple matter to establish the corresponding Fokker-Planck equation which reads

$$\begin{aligned}\frac{df}{dt} = &-\iint dx\, dy \sum\nolimits_{\boldsymbol{k}} \frac{\delta}{\delta A_{\boldsymbol{k}}(x,y)}\{(\tilde{L}A_{\boldsymbol{k}}(x,y) + N_{\boldsymbol{k}}(\{A(x,y)\}))f + \text{c.c.}\}\\ &+ Q\iint dx\, dy \sum\nolimits_{\boldsymbol{k}} \frac{\delta^2}{\delta A_{\boldsymbol{k}}(x,y)\delta A_{\boldsymbol{k}}^*(x,y)} f.\end{aligned} \tag{8.142}$$

We seek its solution in the form

$$f = \mathcal{N} \exp \Phi. \tag{8.143}$$

By a slight generalization of (6.116) one derives the following explicit solution

$$\begin{aligned}\Phi = \frac{2}{Q}\iint \Big\{ &\sum\nolimits_{\boldsymbol{k}} \gamma A_{\boldsymbol{k}}^* \left(\frac{\partial}{\partial x_{(\boldsymbol{k})}} - \frac{i}{\sqrt{2\pi}}\frac{\partial^2}{\partial y_{(\boldsymbol{k})}^2}\right)^2 A_{\boldsymbol{k}}\\ &+ \sum\nolimits_{\boldsymbol{k}} \alpha |A_{\boldsymbol{k}}|^2 - \tfrac{1}{3}\sum\nolimits_{\boldsymbol{k}_1\boldsymbol{k}_2\boldsymbol{k}_3} (\delta A_{\boldsymbol{k}_1}^* A_{\boldsymbol{k}_2}^* A_{\boldsymbol{k}_3}^* \delta_{\boldsymbol{k}_1+\boldsymbol{k}_2+\boldsymbol{k}_3,0} + \text{c.c.})\\ &- \tfrac{1}{2}\sum\nolimits_{\boldsymbol{k}\boldsymbol{k}'} \beta_{\boldsymbol{k}\boldsymbol{k}'} |A_{\boldsymbol{k}}|^2 |A_{\boldsymbol{k}'}|^2 \Big\}\, dx\, dy.\end{aligned} \tag{8.144}$$

It goes, far beyond our present purpose to treat (8.144) in its whole generality. We want to demonstrate, how such expressions (8.143) and (8.144) allow us to discuss the threshold region and the stability of various mode configurations. We neglect the dependence of the slowly varying amplitudes A_k on x, y. We first put $\delta = 0$. (8.143) and (8.144) are a suitable means for the discussion of the stability of different mode configurations. Because Φ depends only on the absolute values of A_k

$$\Phi = \Phi(|A_k|^2) \tag{8.145}$$

we introduce the new variable

$$|A_k|^2 = w_k. \tag{8.146}$$

The values w_k for which Φ has an extremum are given by

$$\frac{\partial \Phi}{\partial w_k} = 0, \quad \text{or } \alpha - \sum_{k'} \beta_{kk'} w_{k'} = 0 \tag{8.147}$$

and the second derivative tells us that the extrema are all maxima

$$\frac{\partial^2 \Phi}{\partial w_k \partial w_{k'}} = -\beta_{kk'} < 0. \tag{8.148}$$

For symmetry reasons we expect

$$w_k = w/N. \tag{8.149}$$

From (8.147) we then obtain

$$\Phi(w) = \frac{1}{2}\frac{\alpha^2}{\bar{\beta}}, \tag{8.150}$$

where we use the abbreviation

$$\frac{1}{N^2} \sum_{kk'} \beta_{kk'} = \bar{\beta}.$$

We now compare the solution in which all modes participate, with the single mode solution for which

$$\alpha - \beta_{kk} w = 0 \tag{8.151}$$

holds, so that

$$\Phi(w) = \frac{1}{2}\frac{\alpha^2}{\bar{\beta}}. \tag{8.152}$$

A comparison between (8.150) and (8.152) reveals that the single mode has a greater probability than the multimode configuration. Our analysis can be generalized to different mode configurations, leading again to the result that only a single mode is stable.

Let us discuss the form of the velocity field of such a single mode configuration, using (8.121) to (8.125). Choosing $\boldsymbol{k}$ in the x direction, we immediately recognize that for example the z-component of the velocity field, u_z is independent of y, and has the form of a sine wave. Thus we obtain *rolls* as stable configurations.

We now come to the question how to explain the still more spectacular hexagons. To do this we include the cubic terms in (8.144) which stem from a spatial inhomogeneity in z-direction e.g., from non-Boussinesq terms. Admitting for the comparison only 3 modes with amplitudes A_i, A_i^*, $i = 1, 2, 3$, the potential function is given by

$$\Phi = \alpha(|A_1|^2 + |A_2|^2 + |A_3|^2) - \delta(A_1^* A_2^* A_3^* + \text{c.c.}) - \tfrac{1}{2}\textstyle\sum_{\boldsymbol{kk}'} \beta_{\boldsymbol{kk}'}|A_{\boldsymbol{k}}|^2|A_{\boldsymbol{k}'}|^2 \tag{8.153}$$

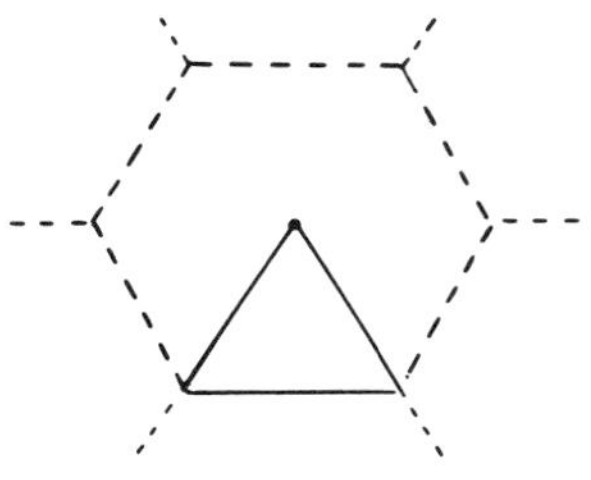

Fig. 8.5. Construction of hexagon from basic triangle (for details consult the text)

where the $\boldsymbol{k}$-sums run over the triangle of Fig. 8.5 which arises from the condition $\boldsymbol{k}_1 + \boldsymbol{k}_2 + \boldsymbol{k}_3 = 0$ and $|k_i| = \text{const.}$ (compare (8.127)). To find the extremal values of Φ we take the derivatives of (8.153) with respect to A_i, A_i^* and thus obtain six equations. Their solution is given by

$$A_i^* = A_i;\ A_1 = A_2 = A_3 = A. \tag{8.154a}$$

Using (8.154a) together with (8.121)–(8.125) we obtain for example $u_z(\boldsymbol{x})$. Concentrating our attention on its dependence on x, y, and using Fig. 8.5, we find (with $\boldsymbol{x}' = \pi/\sqrt{2}\boldsymbol{x}$)

$$u_z(\boldsymbol{x}) \propto A\{e^{ix'} + e^{-ix'} + e^{i(-x'/2 + y'\sqrt{3}/2)} + \cdots\} \tag{8.154b}$$

or in a more concise form

$$u_z(\boldsymbol{x}) \propto 2A\left\{\cos x' + \cos\left(\frac{x'}{2} + \frac{\sqrt{3}}{2}y'\right) + \cos\left(\frac{x'}{2} - \frac{\sqrt{3}}{2}y'\right)\right\}.$$

Using the solid lines in Fig. 8.5 as auxiliary pattern one easily convinces oneself, that the hexagon of that figure is the elementary cell of $u_z(\boldsymbol{x})$ (8.154b). We discuss the probability of the occurrence of this hexagon compared with that of rolls. To this end we evaluate (8.153) with aid of (8.154a) which yields (using the explicit expression for A):

$$\Phi_{\max}(A) = \hat{\beta}\{\tfrac{3}{2}\hat{\alpha}^2 + \tfrac{1}{2}\hat{\delta}^2\hat{\alpha} - A_{\max}(2\hat{\alpha}\hat{\delta} - \tfrac{1}{2}\hat{\delta}^3)\}. \tag{8.155}$$

Here we have used the abbreviations

$$\tfrac{1}{3}\hat{\beta} = \bar{\beta}, \textstyle\sum_k \beta_{k1} = \hat{\beta}, \tag{8.156}$$

$$\hat{\alpha} = \frac{\tilde{\alpha}}{\hat{\beta}}, \tag{8.157}$$

$$\hat{\delta} = \frac{\delta}{\hat{\beta}}. \tag{8.158}$$

We now discuss (8.155) for two different limiting cases.
1)

$$\hat{\delta}^2 \gg \hat{\alpha}. \tag{8.159}$$

In this case we obtain

$$\Phi_{\max}(A) \approx \tfrac{1}{2}\hat{\beta}\hat{\delta}^3 A_{\max}. \tag{8.160}$$

2)

$$\hat{\delta}^2 \ll \hat{\alpha}. \tag{8.161}$$

In this case we obtain

$$\Phi_{\max}(A) = \frac{\tilde{\alpha}^2}{2\bar{\beta}}. \tag{8.162}$$

A comparison between (8.160) or (8.162), respectively, with a single mode potential (8.152) reveals the following: for Rayleigh numbers $R > R_c$, which exceed R_c only a little, the hexagon configuration has a higher probability than the roll configuration. But a further increase of the Rayleigh number finally renders the single mode configuration (rolls) more probable.

In conclusion, let us discuss our above results using the phase transition analogy of Sections 6.7, 6.8. For $\delta = 0$ our present system exhibits all features of a second order nonequilibrium phase transition (with symmetry breaking, soft modes, critical fluctuations, etc.). In complete analogy to the laser case, the symmetry can be broken by injecting an external signal. This is achieved by superimposing a spatially, periodically changing temperature distribution on the constant temperature at the lower boundary of the fluid. By this we can prescribe the phase (i.e., the position of the rolls), and to some extent also the diameter of the rolls (↔ wavelength). For $\delta \neq 0$ we obtain a first order transition, and hysteresis is expected to occur.

8.14 A Model for the Statistical Dynamics of the Gunn Instability Near Threshold[1]

The Gunn effect, or Gunn instability, occurs in semiconductors having two conduction bands with different positions of their energy minima. We assume that donors have given their electrons to the lower conduction band. When the electrons are accelerated by application of an electric field, their velocity increases and thus the current increases with the field strength. On the other hand, the electrons are slowed down dùe to scattering by lattice vibrations (or defects). As a net effect, a non-vanishing mean velocity, v, results. For small fields E, v increases as E increases. However, at sufficiently high velocities the electrons may tunnel into the higher conduction band where they are again accelerated. If the effective mass of the higher conduction band is greater than that of the lower conduction band, the acceleration of the electrons in the higher conduction band is smaller than that of those in the lower conduction band. Of course, in both cases the electrons are slowed down again by their collisions with lattice vibrations. At any rate the effective mean velocity of the electrons is smaller in the higher conduction band than in the lower conduction band. Because with higher and higher field strength, more and more electrons get into the higher conduction band, and the mean velocity of the electrons of the whole sample becomes smaller. Thus we obtain a behavior of the mean velocity $v(E)$ as a function of the electric field strength as

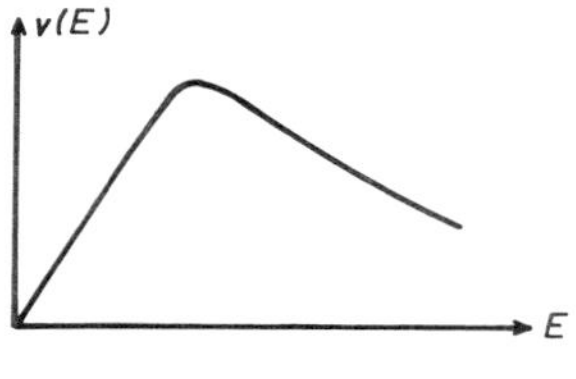

Fig. 8.6. Mean velocity of electrons as function of electric field strength

exhibited in Fig. 8.6. Multiplying v by the concentration of electrons, n_0, yields the current density. So far we have talked about the elementary effect which has nothing or little to do with a cooperative effect. We now must take into account that the electrons themselves create an electric field strength so that we are led to a feedback mechanism. It is our intention to show that the resulting equations are strongly reminiscent of the laser equations, and, in particular, that one can easily explain the occurrence of current pulses.

Let us now write down the basic equations. The first is the equation describing the conservation of the number of electrons. Denoting the density of electrons at

[1] Readers not familiar with the basic features of semiconductors may nevertheless read this chapter starting from (8.163)–(8.165). This chapter then provides an example how one may derive pulse-like phenomena from certain types of nonlinear equations.

time t and space point x by n and the electron current, divided by the electronic charge e, by J, the continuity equation reads

$$\frac{\partial n}{\partial t} + \frac{\partial J}{\partial x} = 0. \tag{8.163}$$

There are two contributions to the current J. On the one hand there is the streaming motion of the electrons, $nv(E)$, on the other hand this motion is superimposed by a diffusion with diffusion constant D. Thus we write the "current" in the form

$$J = nv(E) - D\frac{\partial^2 n}{\partial x^2}. \tag{8.164}$$

Finally the electric field is generated by charges. Denoting the concentration of ionized donors by n_0, according to electrostatics, the equation for E reads

$$\frac{dE}{dx} = e'(n - n_0), \tag{8.165}$$

where we have introduced the abbreviation

$$e' = \frac{4\pi e}{\varepsilon_0}. \tag{8.165a}$$

ε_0 is the static dielectric constant. We want to eliminate the quantities n and J from the above equations to obtain an equation for the field strength E alone. To this end we express n in (8.163) by means of (8.165) and J in (8.163) by means of (8.164) which yields

$$\frac{1}{e'}\frac{\partial^2 E}{\partial t \partial x} + \frac{d}{dx}\left(nv(E) - D\frac{d^2 n}{dx^2}\right) = 0. \tag{8.166}$$

Replacing again n in (8.166) by means of (8.165) and rearranging the resulting equation slightly we obtain

$$\frac{1}{e'}\frac{\partial}{\partial x}\left(\frac{\partial E}{\partial t} + e'n_0 v(E) + v(E)\frac{dE}{dx} - D\frac{d^2 E}{dx^2}\right) = 0. \tag{8.167}$$

This equation immediately allows for an integration over the coordinate. The integration constant denoted by

$$\frac{1}{e}I(t) \tag{8.167a}$$

can still be a function of time. After integration and a slight rearrangement we find the basic equation

$$\frac{\partial E}{\partial t} = -e'n_0 v(E) - v(E)\frac{dE}{dx} + D\frac{d^2E}{dx^2} + \frac{4\pi}{\varepsilon_0} I(t). \tag{8.168}$$

$I(t)$, which has the meaning of a current density, must be determined in such a way that the externally applied potential

$$U = \int_0^L dx\, E(x, t) \tag{8.169}$$

is constant over the whole sample. It is now our task to solve (8.168). For explicit calculations it is advantageous to use an explicit form for $v(E)$ which, as very often used in this context, has the form

$$v(E) = \frac{\mu_1 E(1 + BE/E_c)}{1 + (E/E_c)^2}. \tag{8.170}$$

μ_1 is the electron mobility of the lower band, B is the ratio between upper and lower band mobility. To solve (8.168) we expand it into a Fourier series

$$E(x, t) = E_0 + \textstyle\sum_{m \neq 0} E_m(t) e^{imk_0 x}. \tag{8.171}$$

The summation goes over all positive and negative integers, and the fundamental wave number is defined by

$$k_0 = \frac{2\pi}{L}. \tag{8.172}$$

Expanding $v(E)$ into a power series of E and comparing then the coefficients of the same exponential functions in (8.168) yields the following set of equations

$$\partial E_m/\partial t = (\alpha_m - i\omega_m)\cdot E_m - \textstyle\sum_{m_1 + \cdots + m_s = m} \frac{1}{s!} A_{s,m_1} E_{m_1} E_{m_2} \cdots E_{m_s}. \tag{8.173}$$

The different terms on the rhs of (8.173) have the following meaning

$$\alpha_m = \omega_R - Dk_0^2 \cdot m^2 \tag{8.174}$$

where

$$\omega_R = e'n_0 v_0^{(1)}. \tag{8.175}$$

$v_0^{(1)}$ is the differential mobility and ω_R is the negative dielectric relaxation frequency. ω_m is defined by

$$\omega_m = mk_0 v(E_0). \tag{8.176}$$

The coefficients A are given by

$$A_{s,m} = e'n_0 v_0^{(s)} + imsk_0 v_0^{(s-1)}, \tag{8.177}$$

where the derivatives of v are defined by

$$v_0^{(s)} = d^s v(E)/dE^s|_{E=E_0}. \tag{8.178}$$

Inserting (8.171) into the condition (8.169) fixes E_0

$$E_0 = U/L. \tag{8.179}$$

It is now our task to discuss and solve at least approximately the basic (8.173). First neglecting the nonlinear terms we perform a linear stability analysis. We observe that E_m becomes unstable if $\alpha_m > 0$. This can be the case if $v_0^{(1)}$ is negative which is a situation certainly realized as we learn by a glance at Fig. 8.6. We investigate a situation in which only one mode $m = \pm 1$ becomes unstable but the other modes are still stable. To exhibit the essential features we confine our analysis to the case of only 2 modes, with $m = \pm 1, \pm 2$, and make the hypothesis

$$E_m = c_m \mathrm{e}^{-i\omega_m t}. \tag{8.180}$$

Eqs. (8.173) reduce to

$$\partial c_1/\partial t = \alpha c_1 + V c_2 c_1^* - \frac{W}{2}|c_1|^2 c_1 - W|c_2|^2 c_1 \tag{8.181}$$

and

$$\partial c_2/\partial t = -\beta c_2 + \frac{V}{2} c_1^2 - W|c_1|^2 c_2 - \frac{W}{2}|c_2|^2 c_2. \tag{8.182}$$

Here we have used the following abbreviations

$$\alpha \equiv \alpha_1 > 0, \alpha_2 \equiv -\beta < 0, \tag{8.183}$$

$$V = 2V_k^{(3)} = 2\{-\tfrac{1}{2}e'n_0 v_0^{(2)} + imk_0 v_0^{(1)}\}, \tag{8.184}$$

$$V \approx -e'n_0 v_0^{(2)}, \tag{8.185}$$

$$W = -6V_k^{(4)} \approx e'n_0 v_0^{(3)}. \tag{8.186}$$

In realistic cases the second term in (8.184) can be neglected compared to the first so that (8.184) reduces to (8.185). Formula (8.186) implies a similar approximation. We now can apply the adiabatic elimination procedure described in Sections 7.1, 7.2. Putting

$$\partial c_2/\partial t = 0, \tag{8.187}$$

(8.182) allows for the solution

$$c_2 = \frac{V}{2} \frac{c_1^2}{\beta + W|c_1|^2}. \tag{8.188}$$

When inserting c_2 into (8.181) we obtain as final equation

$$\frac{\partial c_1}{\partial t} = \left\{ \alpha + \frac{V^2}{2} \frac{|c_1|^2}{\beta + W|c_1|^2} - \frac{W}{2}|c_1|^2 - \frac{WV^2}{4} \frac{|c_1|^4}{(\beta + W|c_1|^2)^2} \right\} c_1. \tag{8.189}$$

It is a trivial matter to represent the rhs of (8.189) as (negative) derivative of a potential Φ so that (8.189) reads

$$\frac{\partial c_1}{\partial t} = -\frac{\partial \Phi}{\partial c_1^*}.$$

With the abbreviation $I = |c_1|^2$, Φ reads

$$\Phi = \frac{W}{4} I^2 - \left(\alpha + \frac{V^2}{4W} \right) I - \frac{\beta^2 V^2/4W^2}{\beta + WI}. \tag{8.190}$$

When the parameter α is changed, we obtain a set of potential curves, $\Phi(I)$ which are similar to those of Fig. 8.3. Adding a fluctuating force to (8.189) we may take into account fluctuations. By standard procedures a Fokker-Planck equation belonging to (8.189) may be established and the stable and unstable points may be discussed. We leave it as an exercise to the reader to verify that we have again a situation of a first order phase transition implying a discontinuous jump of the equilibrium position and a hysteresis effect. Since c_1 and c_2 are connected with oscillatory terms (compare (8.171), (8.180)) the electric field strength shows undamped oscillations as observed in the Gunn effect.

8.15 Elastic Stability: Outline of Some Basic Ideas

The general considerations of Sections 5.1, 5.3 on stability and instability of systems described by a potential function find immediate applications to the nonlinear theory of elastic stability. Consider as an example a bridge. In the frame of a model we may describe a bridge as a set of elastic elements coupled by links, Fig. 8.7. We then investigate a deformation, and especially the breakdown under a load. To demonstrate how such considerations can be related to the analysis of Sections 5.1, 5.3 let us consider the simple arch (compare Fig. 8.8). It comprises two linear springs of stiffness k pinned to each other and to rigid supports. The angle between a spring and the horizontal is denoted by q. If there is no load, the corresponding angle is called α. We consider only a load in vertical direction and denote it by P. We restrict our considerations to symmetric deformations so that the system has only one degree of freedom described by the variable q. From elementary geometry

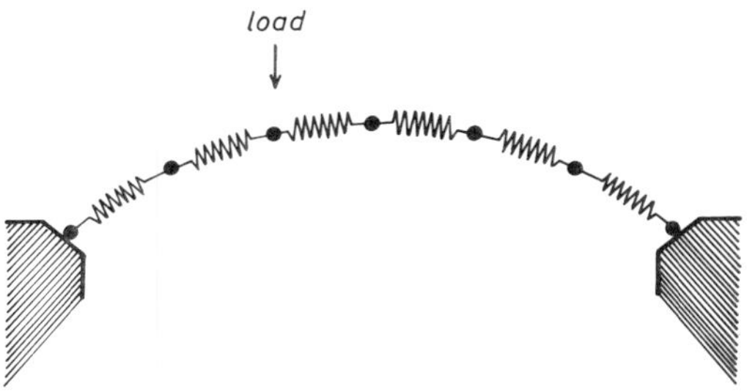

Fig. 8.7. Model of a bridge

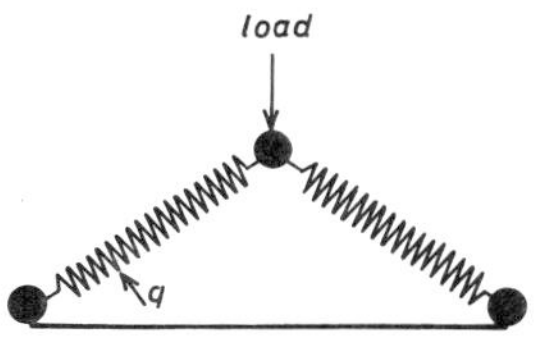

Fig. 8.8. Model of a simple arch

it follows that the strain energy of the two strings is

$$U(q) = 2\frac{1}{2}k\left(\frac{R}{\cos\alpha} - \frac{R}{\cos q}\right)^2. \tag{8.191}$$

Since we confine our analysis to small angles q we may expand the cosine functions, which yields

$$U(q) = \tfrac{1}{4}kR^2(\alpha^4 - 2\alpha^2 q^2 + q^4). \tag{8.192}$$

The deflection of the load in vertical direction reads

$$\mathcal{E}(q) = R(\tan\alpha - \tan q) \tag{8.193}$$

or after approximating $\tan x$ by x ($x = \alpha$ or q)

$$\mathcal{E}(q) = R(\alpha - q). \tag{8.194}$$

The potential energy of the load is given by $-$ load $\cdot$ deflection, i.e.,

$$-P\mathcal{E}(q). \tag{8.195}$$

The potential of the total system comprising springs and load is given by the sum of the potential energies (8.192) and (8.195)

$$V(q) = U - P\mathcal{E} \tag{8.196}$$

and thus

$$V = \tfrac{1}{4}kR^2(\alpha^4 - 2\alpha^2 q^2 + q^4) - PR(\alpha - q). \tag{8.197}$$

Evidently the behavior of the system is described by a simple potential. The system acquires that state for which the potential has a minimum

$$\frac{\partial V}{\partial q} \equiv kR^2(-\alpha^2 q + q^3) + PR = 0. \tag{8.198}$$

(8.198) allows us to determine the extremal position q as a function of P

$$q = q(P). \tag{8.199}$$

In mechanical engineering one often determines the load as a function of the deflection, again from (8.198)

$$P = P(q). \tag{8.200}$$

Taking the second derivative

$$\frac{\partial^2 V}{\partial q^2} = kR^2(3q^2 - \alpha^2) \tag{8.201}$$

tells us if the potential has a maximum or a minimum. Note that we have to insert (8.199) into (8.201). Thus when (8.201) changes sign from a positive to a negative value we reach a critical load and the system breaks down. We leave the following problem to the reader as *exercise*: Discuss the potential (8.197) as a function of the parameter P and show that at a critical P the system suddenly switches from a stable state to a different stable state. Hint: The resulting potential curves have the form of Fig. 6.7 i.e., a first order transition occurs.

Second-order transitions can also be very easily mimicked in mechanical engineering by the hinged cantilever (Fig. 8.9). We consider a rigid link of length l,

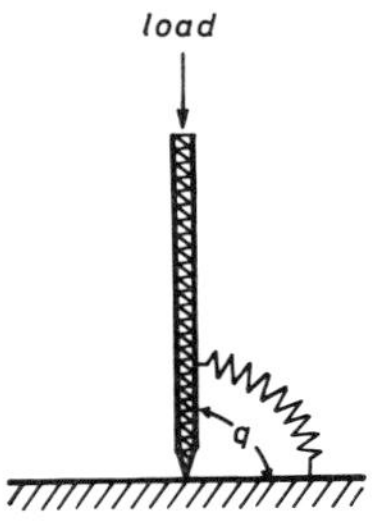

Fig. 8.9. Hinged cantilever

pinned to a rigid foundation and supported by a linear rotational spring of stiffness k. The load acts in the vertical direction. We introduce as coordinate the angle q between the deflected link and the vertical. The potential energy of the link (due to the spring) reads

$$U = \tfrac{1}{2}kq^2. \tag{8.202}$$

The deflection of the load is given by

$$\mathcal{E} = L(1 - \cos q). \tag{8.203}$$

Just as before we can easily construct the total energy which is now

$$V = U - P\mathcal{E} = \tfrac{1}{2}kq^2 - PL(1 - \cos q). \tag{8.204}$$

When we expand the cosine function for small values of q up to the fourth power we obtain potential curves of Fig. 6.6. We leave the discussion of the resulting instability which corresponds to a second order phase transition to the reader. Note that such a local instability is introduced if $PL > k$. Incidentally this example may serve as an illustration for the unfoldings introduced in Section 5.5. In practice the equilibrium position of the spring may differ slightly from that of Fig. 8.9. Denoting the equilibrium angle of the link without load by ε, the potential energy of the link is now given by

$$U(q, \varepsilon) = \tfrac{1}{2}k(q - \varepsilon)^2 \tag{8.205}$$

and thus the total potential energy under load by

$$V = \tfrac{1}{2}k(q - \varepsilon)^2 - PL(1 - \cos q). \tag{8.206}$$

Expanding the cosine function again up to 4th order we observe that the resulting potential

$$V \approx \frac{1}{2}k(q - \varepsilon)^2 - PLq^2/2 + PL\frac{1}{24}q^4 \tag{8.207}$$

is of the form (5.133) including now a linear term which we came across when discussing unfoldings. Correspondingly the symmetry inherent in (8.204) is now broken giving rise to potential curves of Fig. 5.20 depending on the sign of ε. We leave it again as an exercise to the reader to determine the equilibrium positions and the states of stability and instability as a function of load.

Remarks About the General Case

We are now in a position to define the general problem of mechanical engineering with respect to elastic stability as follows: Every static mechanical system is described by a certain set of generalized coordinates $q_1 \ldots q_n$, which may include, for instance, angles. In more advanced considerations also continuously distributed coordinates are used, for example to describe deformation of elastic shells, which are used, for instance, in cooling towers. Then the minima of the potential energy as a function of one or a set of external loads are searched. Of particular importance is the study of instability points where critical behavior shows up as we have seen in many examples before. E.g. one minimum may split into 2 minima (bifurcation) or a limit point may be reached where the system completely loses its stability. An important consequence of our general considerations of Section 5.5 should be mentioned for the characterization of the instability points. It suffices to consider only as many degrees of freedom as coefficients of the diagonal quadratic form vanish. The coordinates of the new minima may describe completely different mechanical configurations. As an example we mention a result obtained for thin shells. When the point of bifurcation is reached, the shells are deformed in such a way that a hexagonal pattern occurs. The occurrence of this pattern is a typical *post-buckling phenomenon*. Exactly the same patterns are observed for example in hydrodynamics (compare Sect. 8.13).

9. Chemical and Biochemical Systems

9.1 Chemical and Biochemical Reactions

Basically, we may distinguish between two different kinds of chemical processes:
1) Several chemical reactants are put together at a certain instant, and we are then studying the processes going on. In customary thermodynamics, one usually compares only the reactants and the final products and observes in which direction a process goes. This is not the topic we want to treat in this book. We rather consider the following situation, which may serve as a model for biochemical reactants.
2) Several reactants are continuously fed into a reactor where new chemicals are continuously produced. The products are then removed in such a way that we have steady state conditions. These processes can be maintained only under conditions far from thermal equilibrium. A number of interesting questions arise which will have a bearing on theories of formation of structures in biological systems and on theories of evolution. The questions we want to focus our attention on are especially the following:

1) Under which conditions can we get certain products in large well-controlled concentrations?
2) Can chemical reactions produce spatial or temporal or spatio-temporal patterns?

To answer these questions we investigate the following problems:

a) deterministic reaction equations without diffusion
b) deterministic reaction equations with diffusion
c) the same problems from a stochastic point of view

9.2 Deterministic Processes, Without Diffusion, One Variable

We consider a model of a chemical reaction in which a molecule of kind A reacts with a molecule of kind X so to produce an additional molecule X. Since a molecule X is produced by the same molecule X as catalyser, this process is called "auto-catalytic reaction". Allowing also for the reverse process, we thus have the scheme (cf. Fig. 9.1)

$$A + X \underset{k_1'}{\overset{k_1}{\rightleftharpoons}} 2X. \tag{9.1}$$

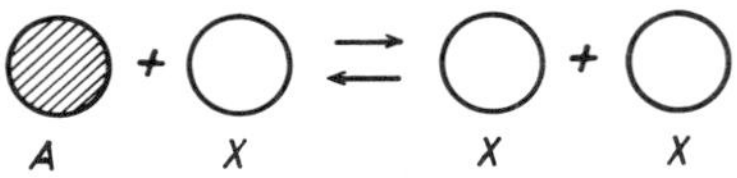

Fig. 9.1. (Compare text)

The corresponding reaction rates are denoted by k_1 and k_1', respectively. We further assume that the molecule X may be converted into a molecule C by interaction with a molecule B (Fig. 9.2)

$$B + X \underset{k_2'}{\overset{k_2}{\rightleftarrows}} C. \tag{9.2}$$

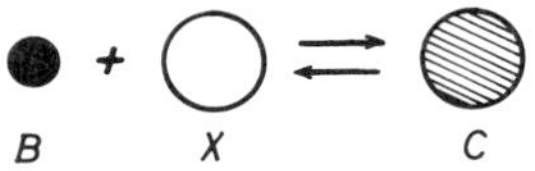

Fig. 9.2. (Compare text)

Again the inverse process is admitted. The reaction rates are denoted by k_2 and k_2'. We denote the concentrations of the different molecules A, X, B, C as follows:

$$\begin{array}{ll} A & a \\ X & n \\ B & b \\ C & c \end{array} \tag{9.3}$$

We assume that the concentrations of molecules A, B and C and the reaction rates k_j, k_j' are externally kept fixed. Therefore, what we want to study is the temporal behavior and steady state of the concentration n. To derive an appropriate equation for n, we investigate the production rate of n. We explain this by an example. The other cases can be treated similarly.

Let us consider the process (9.1) in the direction from left to right. The number of molecules X produced per second is proportional to the concentration a of molecules A, and to that of the molecules X, n. The proportionality factor is just the reaction rate, k_1. Thus the corresponding production rate is $a \cdot n \cdot k_1$. The complete list for the processes 1 and 2 in the directions indicated by the arrows reads

$$\left.\begin{array}{l} 1 \rightarrow a \cdot n \cdot k_1 \\ 1 \leftarrow -n \cdot n \cdot k_1' \end{array}\right\} r_1 = k_1 a n - k_1' n^2,$$

$$\left.\begin{array}{l} 2 \rightarrow -b \cdot n \cdot k_2 \\ 2 \leftarrow c \cdot k_2' \end{array}\right\} r_2 = -k_2 b n + k_2' c. \tag{9.4}$$

The minus signs indicate the decrease of concentration n. Taking the two processes of 1 or 2 together, we find the corresponding rates r_1 and r_2 as indicated above. The total temporal variation of n, $dn/dt \equiv \dot{n}$ is given by the sum of r_1 and r_2 so that our basic equation reads

$$\dot{n} = r_1 + r_2. \tag{9.5}$$

In view of the rather numerous constants $a, \ldots, k_1, \ldots$ appearing in (9.4) it is advantageous to introduce new variables. By an appropriate change of units of time t and concentration n we may put

$$k_1' = 1, k_1 a = 1. \tag{9.6}$$

Further introducing the abbreviations

$$k_2 b = \beta, k_2' c = \gamma \tag{9.7}$$

we may cast (9.5) in the form

$$\dot{n} = (1 - \beta)n - n^2 + \gamma \equiv \varphi(n). \tag{9.8}$$

Incidentally (9.8) is identical with a laser equation provided $\gamma = 0$ (compare (5.84)). We first study the steady state, $\dot{n} = 0$, with $\gamma = 0$. We obtain

$$n = \begin{cases} 0 & \text{for} \quad \beta > 1 \\ 1 - \beta & \text{for} \quad \beta < 1, \end{cases} \tag{9.9}$$

i.e., for $\beta > 1$ we find no molecules of the kind X whereas for $\beta < 1$ a finite concentration, n, is maintained. This transition from "no molecules" to "molecules X present" as β changes has a strong resemblance to a phase transition (cf. Section (6.7)) which we elucidate by drawing an analogy with the equation of the ferromagnet writing (9.8) with $\dot{n} = 0$ in the form

$$\gamma = n^2 - (1 - \beta)n. \tag{9.10}$$

The analogy is readily established by the following identifications

$$M \leftrightarrow n$$

$$H \leftrightarrow \gamma$$

$$T/T_c \leftrightarrow \beta$$

$$H = M^2 - \left(1 - \frac{T}{T_c}\right)M, \tag{9.11}$$

where M is the magnetization, H the magnetic field, T the absolute temperature and T_c the critical temperature.

To study the temporal behavior and the equilibrium states it is advantageous to use a potential (compare Sect. 5.1). Eq. (9.8) then acquires the form

$$\dot{n} = -\frac{\partial V}{\partial n} \tag{9.12}$$

with

$$V(n) = \frac{n^3}{3} - (1 - \beta)\frac{n^2}{2} - \gamma n. \tag{9.13}$$

We have met this type of potential at several occasions in our book and we may leave the discussion of the equilibrium positions of n to the reader. To study the temporal behavior we first investigate the case

a) $\gamma = 0$.

The equation

$$\dot{n} = -(\beta - 1)n - n^2 \tag{9.14}$$

is to be solved under the initial condition

$$t = 0;\, n = n_0. \tag{9.15}$$

In the case $\beta = 1$ the solution reads

$$n = n_0(1 + tn_0)^{-1}, \tag{9.16}$$

i.e., n tends asymptotically to 0. We now assume $\beta \neq 1$. The solution of (9.14) with (9.15) reads

$$n = \frac{(1 - \beta)}{2} - \frac{\lambda}{2} \cdot \frac{c \exp(-\lambda t) - 1}{c \exp(-\lambda t) + 1}. \tag{9.17}$$

Here we have used the abbreviations

$$\lambda = |1 - \beta|, \tag{9.18}$$

$$c = \frac{|1 - \beta| + (1 - \beta) - 2n_0}{|1 - \beta| + (1 - \beta) + 2n_0}. \tag{9.19}$$

In particular we find that the solution (9.17) tends for $t \to \infty$ to the following equilibrium values

$$n_\infty = \begin{cases} 0 & \text{for} \quad \beta > 1 \\ (1 - \beta) & \text{for} \quad \beta < 1 \end{cases}. \tag{9.20}$$

Depending on whether $\beta > 1$ or $\beta < 1$, the temporal behavior is exhibited in Figs. 9.3 and 9.4.

b) $\gamma \neq 0$.

In this case the solution reads

$$n(t) = \frac{(1 - \beta)}{2} - \frac{\lambda}{2} \cdot \frac{c \exp(-\lambda t) - 1}{c \exp(-\lambda t) + 1}, \tag{9.21}$$

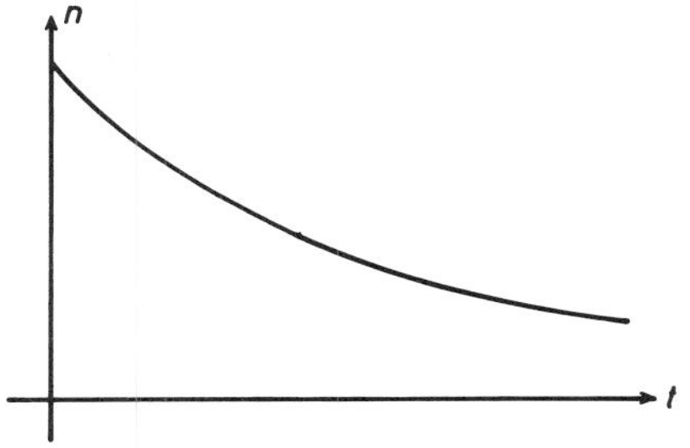

Fig. 9.3. Solution (9.17) for $\beta > 1$

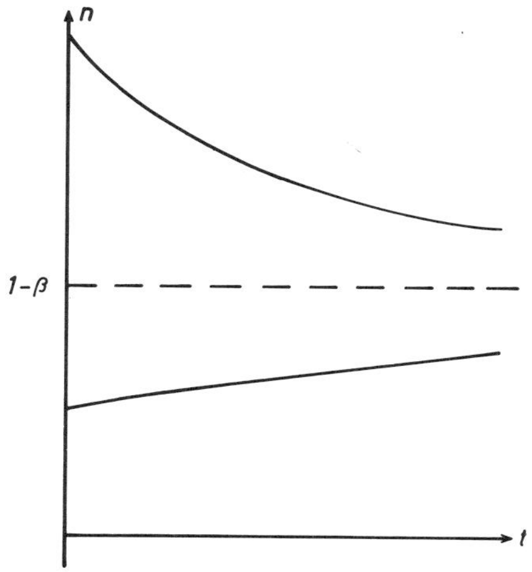

Fig. 9.4. Solution (9.17) for $\beta < 1$ and two different initial conditions

where c is the same as in (9.19) but λ is now defined by

$$\lambda = \sqrt{(1 - \beta)^2 + 4\gamma}. \tag{9.22}$$

Apparently n tends to an equilibrium solution without any oscillations.

As a second model of a similar type we treat the reaction scheme

$$A + 2X \underset{k_1'}{\overset{k_1}{\rightleftarrows}} 3X \tag{9.23}$$

$$B + X \underset{k_2'}{\overset{k_2}{\rightleftarrows}} C. \tag{9.24}$$

Eq. (9.23) implies a trimolecular process. Usually it is assumed that these are very rare and practically only bimolecular processes take place. It is, however, possible to obtain a trimolecular process from subsequent bimolecular processes, e.g., $A + X \rightarrow Y$; $Y + X \rightarrow 3X$, if the intermediate step takes place very quickly and the concentration of the intermediate product can (mathematically) be eliminated adiabatically (cf. Sect. 7.1). The rate equation of (9.23), (9.24) reads

$$\dot{n} = -n^3 + 3n^2 - \beta n + \gamma \equiv \varphi(n), \tag{9.25}$$

where we have used an appropriate scaling of time and concentration. We have met this type of equation already several times (compare Sect. 5.5) with $n \equiv q$. This equation may be best discussed using a potential. One readily finds that there may be either one stable value or two stable and one unstable values. Eq. (9.25) describes a first-order phase transition (cf. Sect. 6.7). This analogy can be still more closely exhibited by comparing the steady state equation (9.25) with $\dot{n} = 0$, with the equation of a van der Waals gas. The van der Waals equation of a real gas reads

$$p = \frac{RT}{v} - \frac{a_1}{v^2} + \frac{a_2}{v^3}. \tag{9.26}$$

The analogy is established by the translation table

$$\begin{aligned} &n \leftrightarrow 1/v,\ v\text{: volume} \\ &\gamma \leftrightarrow \text{pressure } p \\ &\beta \leftrightarrow RT \\ &(R\text{: gas constant, } T\text{: absolute temperature}) \end{aligned} \tag{9.27}$$

We leave it to readers who are familiar with van der Waals' equation, to exploit this analogy to discuss the kinds of phase transition the molecular concentration n may undergo.

9.3 Reaction and Diffusion Equations

As a first example we treat again a single concentration as variable but we now permit that this concentration may spatially vary due to diffusion. We have met a diffusion equation already in Sections 4.1 and 4.3. The temporal change of n, $\dot{n}$, is now determined by reactions which are described e.g., by the rhs of (9.8) and in addition by a diffusion term (we consider a one-dimensional model)

$$\dot{n} = \varkappa \frac{\partial^2 n}{\partial x^2} + \varphi(n). \tag{9.28}$$

$\varphi(n)$ may be derived from a potential V or from its negative Φ

$$\varphi(n) = \frac{\partial}{\partial n} \Phi(n) \quad \left(= -\frac{\partial V}{\partial n}\right). \tag{9.29}$$

We study the steady state $\dot{n} = 0$ and want to derive a criterium for the spatial coexistence of two phases, i.e., we consider a situation in which we have a change of concentration within a certain layer so that

$$\begin{aligned} n &\to n_1 \quad \text{for} \quad z \to +\infty, \\ n &\to n_2 \quad \text{for} \quad z \to -\infty. \end{aligned} \tag{9.30}$$

To study our basic equation

$$\varkappa \frac{\partial^2 n}{\partial x^2} = -\frac{\partial}{\partial n} \Phi(n) \tag{9.31}$$

we invoke an analogy with an oscillator or, more generally, with a particle in the potential field $\Phi(n)$ by means of the following correspondence:

$$\begin{aligned} x &\leftrightarrow t \text{ time} \\ \Phi &\leftrightarrow \text{potential} \\ n &\leftrightarrow q \text{ coordinate} \end{aligned} \tag{9.32}$$

Note that the spatial coordinate x is now interpreted quite formally as time while the concentration variable is now interpreted as coordinate q of a particle. The potential Φ is plotted in Fig. 9.5. We now ask under which condition is it possible

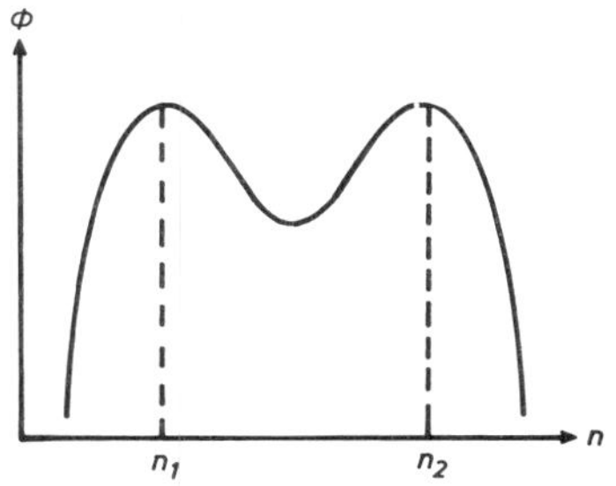

Fig. 9.5. $\Phi(n)$ in case of coexistence of two phases n_1, n_2 with plane boundary layer (After *F. Schlögl*: Z. Physik **253**, 147 (1972))

that the particle has two equilibrium positions so that when starting from one equilibrium position for $t = -\infty$ it will end at another equilibrium position for $t \to +\infty$? From mechanics it is clear that we can meet these conditions only if the potential heights at q_1 ($\equiv n_1$) and q_2 ($\equiv n_2$) are equal

$$\Phi(n_1) = \Phi(n_2). \tag{9.33}$$

Bringing (9.33) into another form and using (9.29), we find the condition

$$0 = \Phi(n_2) - \Phi(n_1) = \int_{n_1}^{n_2} \varphi(n)\, dn. \tag{9.34}$$

We now resort to our specific example (9.25) putting

$$\varphi(n) = \underbrace{-n^3 + 3n^2 - \beta n}_{-\Psi(n)} + \gamma. \tag{9.35}$$

This allows us to write (9.34) in the form

$$0 = \gamma(n_2 - n_1) - \int_{n_1}^{n_2} \Psi(n)\, dn, \tag{9.36}$$

or resolving (9.36) for γ

$$\gamma = \frac{1}{(n_2 - n_1)} \int_{n_1}^{n_2} \Psi(n)\, dn. \tag{9.37}$$

In Fig. 9.6 we have plotted γ versus n. Apparently the equilibrium condition implies

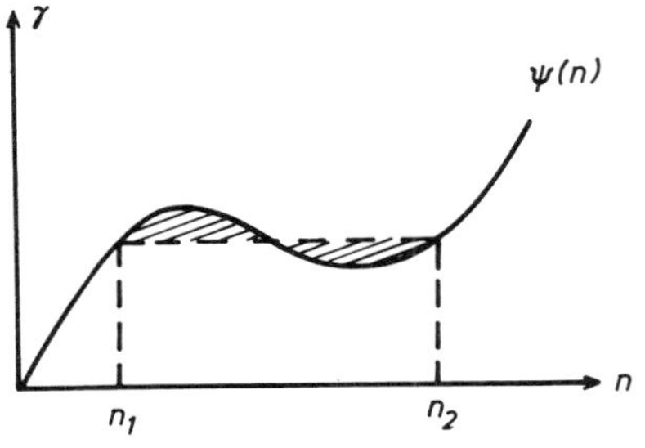

Fig. 9.6. Maxwellian construction of the coexistence value γ

that the areas in Fig. 9.6 are equal to each other. This is exactly Maxwell's construction which is clearly revealed by the comparison exhibited in (9.27)

$$\begin{aligned} &\gamma = \beta n - 3n^2 + n^3, \\ &\updownarrow \\ &p = RT\frac{1}{v} - \frac{a_1}{v^2} + \frac{a_2}{v^3}. \end{aligned} \tag{9.38}$$

This example clearly shows the fruitfulness of comparing quite different systems in and apart from thermal equilibrium.

9.4 Reaction—Diffusion Model with Two or Three Variables: The Brusselator and the Oregonator

In this section we first consider the following reaction scheme of the "Brusselator"

$$\begin{aligned} A &\rightarrow X \\ B + X &\rightarrow Y + D \\ 2X + Y &\rightarrow 3X \\ X &\rightarrow E \end{aligned} \tag{9.39}$$

between molecules of the kinds A, X, Y, B, D, E. The following concentrations enter into the equations of the chemical reaction:

$$\begin{matrix} A & a \\ B & b \\ X & n_1 \\ Y & n_2 . \end{matrix} \tag{9.40}$$

The concentrations a and b will be treated as fixed quantities, whereas n_1 and n_2 are treated as variables. Using considerations completely analogous to those of the preceding chapters the reaction-diffusion equations in one dimension, x, read

$$\frac{\partial n_1}{\partial t} = a - (b + 1)n_1 + n_1^2 n_2 + D_1 \frac{\partial^2 n_1}{\partial x^2}, \tag{9.41}$$

$$\frac{\partial n_2}{\partial t} = bn_1 - n_1^2 n_2 + D_2 \frac{\partial^2 n_2}{\partial x^2}, \tag{9.42}$$

where D_1 and D_2 are diffusion constants. We will subject the concentrations $n_1 = n_1(x, t)$ and $n_2 = n_2(x, t)$ to two kinds of boundary conditions; either to

$$n_1(0, t) = n_1(1, t) = a, \tag{9.43a}$$

$$n_2(0, t) = n_2(1, t) = \frac{b}{a}, \tag{9.43b}$$

or to

$$n_j \text{ remains finite for } x \to \pm\infty. \tag{9.44}$$

Eqs. (9.41) and (9.42) may, of course, be formulated in two or three dimensions. One easily verifies that the stationary state of (9.41), (9.42) is given by

$$n_1^0 = a, \quad n_2^0 = \frac{b}{a}. \tag{9.45}$$

To check whether new kinds of solutions occur, i.e., if new spatial or temporal structures arise, we perform a stability analysis of the (9.41) and (9.42). To this end we put

$$n_1 = n_1^0 + q_1; n_2 = n_2^0 + q_2, \tag{9.46}$$

and linearize (9.41) and (9.42) with respect to q_1, q_2. The linearized equations are

$$\frac{\partial q_1}{\partial t} = (b - 1)q_1 + a^2 q_2 + D_1 \frac{\partial^2 q_1}{\partial x^2}, \tag{9.47}$$

$$\frac{\partial q_2}{\partial t} = -bq_1 - a^2 q_2 + D_2 \frac{\partial^2 q_2}{\partial x^2}. \tag{9.48}$$

The boundary conditions (9.43a) and (9.43b) acquire the form

$$q_1(0, t) = q_1(1, t) = q_2(0, t) = q_2(1, t) = 0, \tag{9.49}$$

whereas (9.44) requires q_j finite for $x \to \pm\infty$. Putting, as everywhere in this book,

$$\boldsymbol{q} = \begin{pmatrix} q_1 \\ q_2 \end{pmatrix}, \tag{9.50}$$

(9.47) and (9.48) may be cast into the form

$$\dot{\boldsymbol{q}} = L\boldsymbol{q}, \tag{9.51}$$

where the matrix L is defined by

$$L = \begin{pmatrix} D_1 \dfrac{\partial^2}{\partial x^2} + b - 1 & a^2 \\ -b & D_2 \dfrac{\partial^2}{\partial x^2} - a^2 \end{pmatrix}. \tag{9.52}$$

To satisfy the boundary conditions (9.49) we put

$$\boldsymbol{q}(x, t) = \boldsymbol{q}_0 \exp(\lambda_l t) \sin l\pi x \tag{9.53}$$

with

$$l = 1, 2, \ldots. \tag{9.53a}$$

Inserting (9.53) into (9.51) yields a set of homogeneous linear algebraic equations for $\boldsymbol{q}_0$. They allow for nonvanishing solutions only if the determinant vanishes.

$$\begin{vmatrix} -D_1' + b - 1 - \lambda & a^2 \\ -b & -D_2' - a^2 - \lambda \end{vmatrix} = 0, \quad \lambda = \lambda_l. \tag{9.54}$$

In it we have used the abbreviation

$$D_j' = D_j l^2 \pi^2, \quad j = 1, 2. \tag{9.54a}$$

To make (9.54) vanishing, λ must obey the characteristic equation

$$\lambda^2 - \alpha\lambda + \beta = 0, \tag{9.55}$$

where we have used the abbreviations

$$\alpha = (-D_1' + b - 1 - D_2' - a^2), \tag{9.56}$$

and

$$\beta = (-D_1' + b - 1)(-D_2' - a^2) + ba^2. \tag{9.57}$$

An instability occurs if $\mathrm{Re}(\lambda) > 0$. We have in mind keeping a fixed but changing the concentration b and looking for which $b = b_c$ the solution (9.53) becomes unstable. The solution of (9.55) reads, of course,

$$\lambda = \frac{\alpha}{2} \pm \frac{1}{2}\sqrt{\alpha^2 - 4\beta}. \tag{9.58}$$

We first consider the case that λ is real. This requires

$$\alpha^2 - 4\beta > 0, \tag{9.59}$$

and $\lambda > 0$ requires

$$\alpha + \sqrt{\alpha^2 - 4\beta} > 0. \tag{9.60}$$

On the other hand, if λ is admitted to be complex, then

$$\alpha^2 - 4\beta < 0, \tag{9.61}$$

and we need for instability

$$\alpha > 0. \tag{9.62}$$

We skip the transformation of the inequalities (9.59), (9.60), (9.61) and (9.62) to the corresponding quantities a, b, D_1', D_2', and simply quote the final result: We find the following instability regions:

1) Soft-mode instability, λ real, $\lambda \geq 0$

$$(D_1' + 1)(D_2' + a^2)/D_2' < b. \tag{9.63}$$

This inequality follows from the requirement $\beta < 0$, cf. (9.59),
2) Hard-mode instability, λ complex, $\mathrm{Re}\,\lambda \geq 0$

$$D_1' + D_2' + 1 + a^2 < b < D_1' - D_2' + 1 + a^2 + 2a\sqrt{1 + D_1' - D_2'}. \tag{9.64}$$

The left inequality stems from (9.62), the right inequality from (9.61). The instabilities occur for such a wave number first for which the smallest, b, fulfils the inequalities (9.63) or (9.64) for the first time. Apparently a complex λ is associated with a hard mode excitation while λ real is associated with a soft mode. Since instability (9.63) occurs for $k \neq 0$ and real λ a static spatially inhomogeneous pattern arises. We can now apply procedures described in Sections 7.6 to 7.11. We present the final results for two different boundary conditions. For the boundary conditions (9.43) we put

$$\boldsymbol{q}(x, t) = \xi_u \boldsymbol{q}_{0,u}\sqrt{2}\sin l_c\pi x + \textstyle\sum_{j,l}' \xi_{sjl}\boldsymbol{q}_{0sjl}\sqrt{2}\sin l\pi x \tag{9.65}$$

where the index u refers to "unstable" in accordance with the notation of Section 7.7. The sum over j contains the stable modes which are eliminated adiabatically leaving us, in the *soft-mode* case, with

$$\dot{\xi}_u = c_1(b - b_c)\xi_u + c_3\xi_u^3, \tag{9.66}$$

provided l is even. The coefficients c_1 and c_3 are given by

$$c_1 = \frac{D_{2c}'^2}{a^2}(1 + D_{1c}' - D_{2c}'(a^2 + D_{2c}')/a^2)^{-1} + O[(b - b_c)] \tag{9.66a}$$

$$c_3 = -\frac{D_{2c}'^{\,2}}{[D_{2c}' - a^2(D_{1c}' + 1 - D_{2c}')]^2}\{(D_{1c}' + 1)(D_{2c}' + a^2)\}$$
$$\left\{1 + \frac{2^6 l_c^4}{\pi^2 a^2}(D_{2c}' - a^2)\sum_{l=1}^{\infty}\frac{(1 - (-1)^l)^2}{l^2(l^2 - 4l_c^2)}\frac{a^2(D_1' + 1) - D_2'(D_{1c}' + 1)}{a^2 b_c - (b_c - 1 - D_1')(D_2' + a^2)}\right\} \tag{9.66b}$$

and

$$D_{ic}' = \pi^2 l_c^2 D_i, \tag{9.66c}$$

where l_c is the critical value of l for which instability occurs first. A plot of ξ_u as a function of the parameter b is given in Fig. 5.4 (with $b \equiv k$ and $\xi_u \equiv q$). Apparently at $b = b_c$ a point of bifurcation occurs and a spatially periodic structure is established (compare Fig. 9.7). If on the other hand l is odd, the equation for ξ_u reads

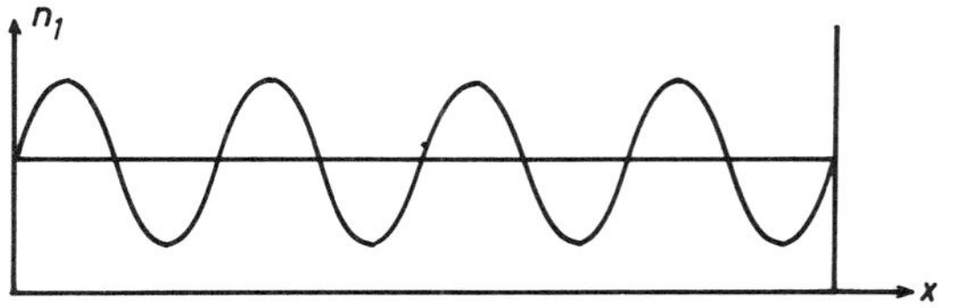

Fig. 9.7. Spatially inhomogeneous concentration beyond instability point, l_c even

$$\dot{\xi}_u = c_1(b - b_c)\xi_u + c_2\xi_u^2 + c_3\xi_u^3, \tag{9.67}$$
$$c_1, c_2, c_3 \text{ real}, \; c_1 > 0, \; c_3 < 0.$$

c_1 and c_3 are given in (9.66a), (9.66b), and c_2 by

$$c_2 = \frac{2^{5/2}}{3\pi a l_c}(1 - (-1)^{l_c})D_{2c}'^{\,3/2}(D_{2c}' + a^2)^{1/2}\frac{(D_{2c}' - a^2)(D_{1c}' + 1)}{[D_{2c}' - a^2(D_{1c}' + 1 - D_{2c}')]^{3/2}} \tag{9.67a}$$

ξ_u is plotted as function of b in Fig. 9.8. The corresponding spatial pattern is exhibited in Fig. 9.9. We leave it to the reader as an exercise to draw the potential curves corresponding to (9.66) and (9.67) and to discuss the equilibrium points in

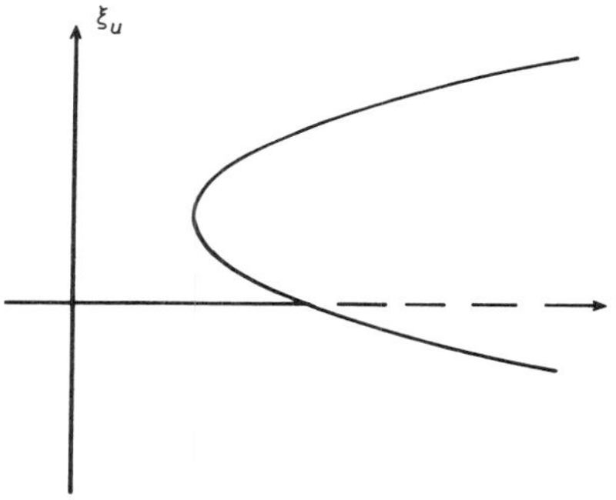

Fig. 9.8. The order parameter ξ_u as a function of the "pump" parameter b. A task for the reader: Identify for fixed b the values of ξ_u with the minima of potential curves exhibited in Section 6.3

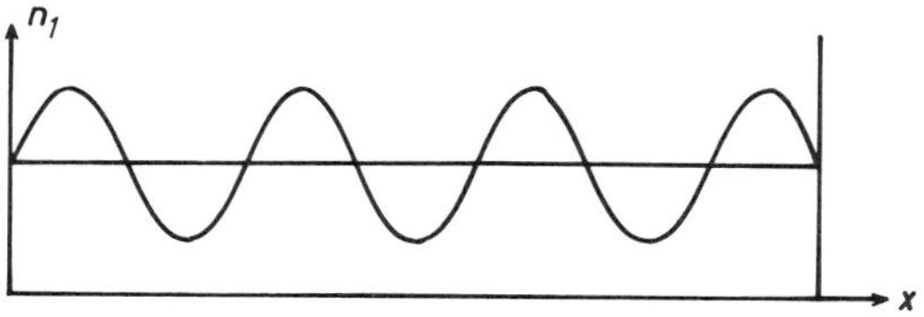

Fig. 9.9. Spatially inhomogeneous concentration beyond instability point, l_c odd

analogy to Section 5.1. So far we have considered only instabilities connected with the soft mode. If there are no finite boundaries we make the following hypothesis for q

$$\boldsymbol{q} = \xi_{u,k_c}\boldsymbol{q}_{0uk_c} \exp(ik_c x) + \sum_{j,k}' \xi_{sjk}\boldsymbol{q}_{0sjk} \exp(ikx). \tag{9.68}$$

The methods described in Section 7.8 allow us to derive the following equations for $\xi_{u,k_c} \equiv \xi$

a) Soft mode

$$\xi^{\cdot} = (\lambda_1 + \lambda_1' \nabla^2)\xi - A|\xi|^2\xi \tag{9.69}$$

where

$$\lambda_1 = (b - b_c)(1 + a^2 - \mu^2 - a\mu^3)^{-1} + O[(b - b_c)^2] \tag{9.69a}$$

$$\lambda_1' = 4a\mu((1 - \mu^2)(1 + a\mu)k_c^2)^{-1} \tag{9.69b}$$

$$A = (9(1 - \mu^2)\mu^3(1 - a\mu)^2 a)^{-1}(-8a^3\mu^3 + 5a^2\mu^2 + 20a\mu - 8) \tag{9.69c}$$

and

$$\mu = \left(\frac{D_1}{D_2}\right)^{1/2} \tag{9.69d}$$

b) Hard mode

$$\xi^{\cdot} = \frac{1 + a^2}{2}\xi + (D_1 + D_2 - ia(D_1 - D_2))\nabla^2\xi$$
$$-\frac{1}{2}\left(\frac{2 + a^2}{a^2}\right) - i\frac{4 - 7a^2 + 4a^4}{3a^3}\Big) |\xi|^2 \cdot \xi. \tag{9.70}$$

Note that A can become negative. In that case higher powers of ξ must be included. With increasing concentrations b, still more complicated temporal and spatial structures can be expected as has been revealed by computer calculations.

The above equations may serve as model for a number of biochemical reactions as well as a way to understand, at least qualitatively, the Belousov-Zhabotinski reactions where both temporal and spatial oscillations have been observed. It should be noted that these latter reactions are, however, not stationary but occur rather as a long-lasting, transitional state after the reagents have been put together. A few other solutions of equations similar to (9.41) and (9.42) have also been considered. Thus in two dimensions with polar coordinates in a configuration in which a soft and a hard mode occur simultaneously, oscillating ring patterns are found.

Let us now come to a second model, which was devised to describe some essential features of the Belousov-Zhabotinski reaction. To give an idea of the chemistry of that process we represent the following reaction scheme:

$$BrO_3^- + Br^- + 2H^+ \to HBrO_2 + HOBr \tag{C.1}$$

$$HBrO_2 + Br^- + H^+ \to 2HOBr \tag{C.2}$$

$$BrO_3^- + HBrO_2 + H^+ \to 2BRO_2 + H_2O \tag{C.3a}$$

$$Ce^{3+} + BrO_2 + H^+ \to Ce^{4+} + HBrO_2 \tag{C.3b}$$

$$2HBrO_2 \to BrO_3^- + HOBr + H^+ \tag{C.4}$$

$$nCe^{4+} + BrCH(COOH)_2 \to nCe^{3+} + Br^- + \text{oxidized products.} \tag{C.5}$$

Steps (C1) and (C4) are assumed to be bimolecular processes involving oxygen atom transfer and accompanied by rapid proton transfers; the HOBr so produced is rapidly consumed directly or indirectly with bromination of malonic acid. Step (C3a) is ratedetermining for the overall process of (C3a) + 2(C3b). The Ce^{4+} produced in step (C3b) is consumed in step (C5) by oxidation of bromomalonic acid and other organic species with production of the bromide ion. The complete chemical mechanism is considerably more complicated, but this simplified version is sufficient to explain the oscillatory behavior of the system.

Computational Model

The significant kinetic features of the chemical mechanism can be simulated by the model called the "Oregonator".

$$A + Y \to X$$

$$X + Y \to P$$

$$B + X \to 2X + Z$$

$$2X \to Q$$

$$Z \to fY.$$

This computational model can be related to the chemical mechanism by the identities $A \equiv B \equiv BrO_3^-$, $X \equiv HBrO_2$, $Y \equiv Br^-$, and $Z \equiv 2Ce^{4+}$. Here we have to deal with three variables, namely, the concentrations belonging to X, Y, Z.

Exercise

Verify, that the rate equations belonging to the above scheme are (in suitable units)

$$\dot{n}_1 = s(n_2 - n_2 n_1 + n_1 - q n_1^2),$$
$$\dot{n}_2 = s^{-1}(-n_2 - n_2 n_1 + f n_3),$$
$$\dot{n}_3 = w(n_1 - n_3).$$

9.5 Stochastic Model for a Chemical Reaction Without Diffusion. Birth and Death Processes. One Variable

In the previous sections we have treated chemical reactions from a global point of view, i.e., we were interested in the behavior of macroscopic densities. In this and the following sections we want to take into account the discrete nature of the processes. Thus we now investigate the number of molecules N (instead of the concentration n) and how this number changes by an individual reaction. Since the individual reaction between molecules is a random event, N is a random variable. We want to determine its probability distribution $P(N)$. The whole process is still treated in a rather global way. First we assume that the reaction is spatially homogeneous, or, in other words, we neglect the space dependence of N. Furthermore we do not treat details of the reaction such as the impact of local temperature or velocity distributions of molecules. We merely assume that under given conditions the reaction takes place and we want to establish an equation describing how the probability distribution $P(N)$ changes due to such events. To illustrate the whole procedure we consider the reaction scheme

$$A + X \underset{k_1'}{\overset{k_1}{\rightleftharpoons}} 2X \tag{9.71}$$

$$B + X \underset{k_2'}{\overset{k_2}{\rightleftharpoons}} C, \tag{9.72}$$

which we have encountered before. The number $N = 0, 1, 2, \ldots$ represents the number of molecules of kind X. Due to any of the reactions (9.71) or (9.72) N is changed by 1. We now want to establish the master equation for the temporal change of $P(N, t)$ in the sense of Chapter 4. To show how this can be achieved we start with the simplest of the processes, namely, (9.72) in the direction k_2'. In analogy to Section 4.1 we investigate all transitions leading to N or going away from it.

1) transition $N \to N+1$. "Birth" of a molecule X (Fig. 9.10).
The number of such transitions per second is equal to the occupation probability, $P(N, t)$, multiplied by the transition probability (per second), $w(N+1, N)$. $w(N+1, N)$ is proportional to the concentration c of molecules C and to the reaction rate k_2'. It will turn out below that the final proportionality factor must be the volume V.
2) transition $N-1 \to N$ (Fig. 9.10).
Because we start here from $N-1$, the total rate of transitions is given by $P(N-1, t)k_2'c \cdot V$. Taking into account the decrease of occupation number due to the first process $N \to N+1$ by a minus sign of the corresponding transition rate we obtain as total transition rate

$$2 \leftarrow \quad VP(N-1, t)k_2'c - P(N, t)k_2'cV. \tag{9.73}$$

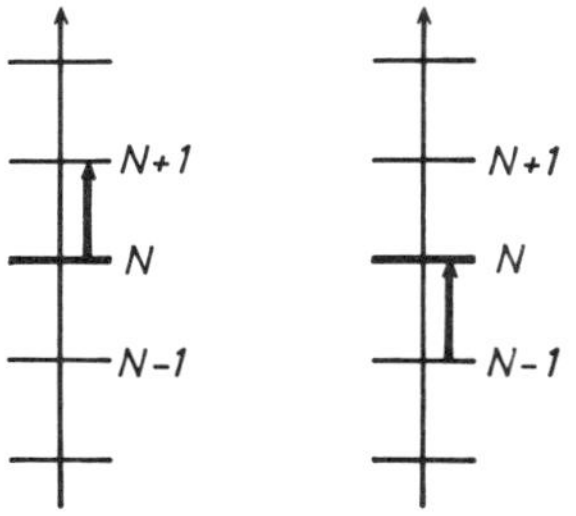

Fig. 9.10. How $P(N)$ changes due to birth of a molecule

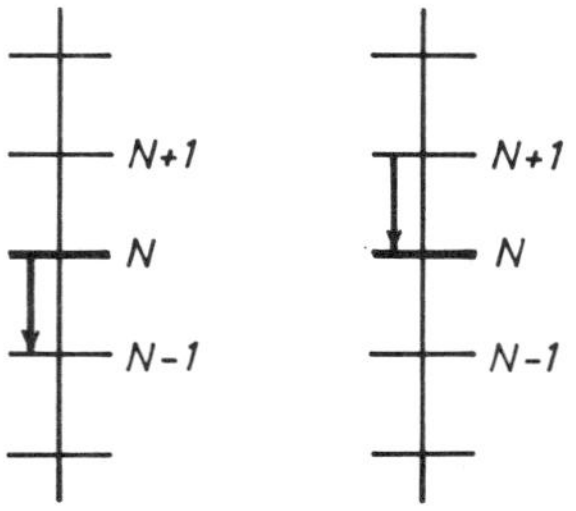

Fig. 9.11. How $P(N)$ changes due to death of a molecule

In a similar way we may discuss the first process in (9.72) with rate k_2. Here the number of molecules N is decreased by 1 ("death" of a molecule X. cf. Fig. 9.11). If we start from the level N the rate is proportional to the probability to find that state N occupied times the concentration of molecules b times the number of molecules X present times the reaction rate k_2. Again the proportionality factor must be fixed later. It is indicated in the formula written below. By the same process, however, the occupation number to level N is increased by processes starting from the level $N+1$. For this process we find as transition rate

$$2 \to \quad P(N+1, t)(N+1)bk_2 - P(N, t)\frac{N}{V}bk_2V. \tag{9.74}$$

It is now rather obvious how to derive the transition rates belonging to the processes 1. We then find the scheme

$$1 \to \quad P(N-1, t)V \cdot \frac{N-1}{V}ak_1 - P(N, t) \cdot \frac{N}{V}ak_1V \tag{9.75}$$

and

$$1 \leftarrow \quad P(N+1, t)V\frac{(N+1)N}{V^2}k_1' - P(N, t)\frac{N(N-1)}{V^2}k_1'V. \tag{9.76}$$

The rates given in (9.73)–(9.76) now occur in the master equation because they determine the total transition rates per unit time. When we write the master equation in the general form

$$\dot{P}(N, t) = w(N, N-1)P(N-1, t) + w(N, N+1)P(N+1, t) \\ - \{w(N+1, N) + w(N-1, N)\}P(N, t), \tag{9.77}$$

we find for the processes (9.71) and (9.72) the following transition probabilities per second:

$$w(N, N-1) = V\left(ak_1 \frac{(N-1)}{V} + k_2'c\right), \tag{9.78}$$

$$w(N, N+1) = V\left(k_1' \frac{(N+1)N}{V^2} + k_2 \frac{b(N+1)}{V}\right). \tag{9.79}$$

The scheme (9.78) and (9.79) has an intrinsic difficulty, namely that a stationary solution of (9.77) is just $P(0) = 1$, $P(N) = 0$ for $N \neq 0$. For a related problem compare Section 10.2. For this reason it is appropriate to include the spontaneous creation of molecules X, from molecules A with the transition rate k_1 in (9.71) and (9.72) as a third process

$$A \xrightarrow{k_1} X. \tag{9.80}$$

Thus the transition rates for the processes (9.71), (9.72) and (9.80) are

$$w(N, N-1) = V\left(ak_1 \frac{N}{V} + k_2'c\right), \tag{9.81}$$

and

$$w(N, N+1) = V\left(k_1' \frac{(N+1)N}{V^2} + k_2 b \frac{(N+1)}{V}\right). \tag{9.82}$$

The solution of the master equation (9.77) with the transition rates (9.78) and (9.79) or (9.81) and (9.82) can be easily found, at least in the stationary state, using the methods of Chapter 4. The result reads (compare (4.119))

$$P(N) = P(0) \cdot \prod_{\nu=0}^{N-1} \frac{w(\nu+1, \nu)}{w(\nu, \nu+1)}. \tag{9.83}$$

The further discussion is very simple and can be performed as in 4.6. It turns out that there is either an extremal value at $N = 0$ or at $N = N_0 \neq 0$ depending on the parameter b.

In conclusion we must fix the proportionality constants which have been left open in deriving (9.73) to (9.76). These constants may be easily found if we require that the master equation leads to the same equation of motion for the density, n, which we have introduced in Section 9.2, at least in the case of large numbers N. To achieve this we derive a mean value equation for N by multiplying (9.77) by N and summing up over N. After trivial manipulations we find

$$\frac{d}{dt}\langle N\rangle = \langle w(N+1, N)\rangle - \langle w(N-1, N)\rangle \tag{9.84}$$

where, as usual,

$$\langle N\rangle = \textstyle\sum_{N=0}^{\infty} NP(N, t)$$

and

$$\langle w(N \pm 1, N)\rangle = \textstyle\sum_{N=0}^{\infty} w(N \pm 1, N)P(N, t).$$

Using (9.78), (9.79) we obtain

$$\frac{d}{dt}\langle N\rangle = V\left\{ak_1\frac{1}{V}\langle N+1\rangle + k_2'c - k_1'\frac{1}{V^2}\langle N(N-1)\rangle - k_2b\frac{1}{V}\langle N\rangle\right\}. \tag{9.85}$$

A comparison with (9.4–5) exhibits a complete agreement provided that we put $n = (1/V)\langle N\rangle$, we neglect 1 compared to N, and we approximate $\langle N(N-1)\rangle$ by $\langle N\rangle^2$. This latter replacement would be exact if P were a Poisson distribution (compare the exercise on (2.12)). In general P is not a Poisson distribution as can be checked by studying (9.83) with the explicit forms for the w's. However, we obtain such a distribution if each of the two reactions (9.71) and (9.72) fulfils the requirement of detailed balance individually. To show this we decompose the transition probabilities (9.78) and (9.79) into a sum of probabilities referring to a process 1 or process 2

$$w(N, N-1) = w_1(N, N-1) + w_2(N, N-1), \tag{9.86}$$

$$w(N-1, N) = w_1(N-1, N) + w_2(N-1, N), \tag{9.87}$$

where we have used the abbreviations

$$w_1(N, N-1) = ak_1(N-1), \tag{9.88}$$

$$w_1(N-1, N) = k_1'N(N-1)\frac{1}{V}, \tag{9.89}$$

$$w_2(N, N-1) = Vk_2'c, \tag{9.90}$$

$$w_2(N-1, N) = k_2bN. \tag{9.91}$$

If detailed balance holds individually we must require

$$w_1(N, N-1)P(N-1) = w_1(N-1, N)P(N), \tag{9.92}$$

and

$$w_2(N, N-1)P(N-1) = w_2(N-1, N)P(N). \tag{9.93}$$

Dividing (9.92) by (9.93) and using the explicit forms (9.88)–(9.91) we find the relation

$$\frac{ak_1}{k_1' N/V} = \frac{k_2' c}{k_2 b N/V}. \tag{9.94}$$

Apparently both sides of (9.94) are of the form μ/N where μ is a certain constant. (9.94) is equivalent to the law of mass action. (According to this law,

$$\frac{\text{product of final concentrations}}{\text{product of initial concentrations}} = \text{const.}$$

In our case, the numerator is $n \cdot n \cdot c$, the denominator $a \cdot n \cdot b \cdot n$). Using that (9.94) is equal μ/N we readily verify that

$$w(N, N-1) = \frac{\mu}{N} w(N-1, N) \tag{9.95}$$

holds. Inserting this relation into (9.83) we obtain

$$P(N) = P(0)\frac{\mu^N}{N!}. \tag{9.96}$$

$P(0)$ is determined by the normalization condition and immediately found to be

$$P(0) = e^{-\mu}. \tag{9.97}$$

Thus in the present case we, in fact, find the Poisson distribution

$$P(N) = \frac{\mu^N}{N!} e^{-\mu}. \tag{9.98}$$

In general, however, we deal with a nonequilibrium situation where the individual laws of detailed balance (9.92) and (9.93) are not valid and we consequently obtain a non-Poissonian distribution. It might be shown quite generally that in *thermal equilibrium* the detailed balance principle holds so that we always get the Poisson distribution. This is, however, in other cases no more so if we are far from equilibrium.

Exercises on 9.5

1) Derive the transition rates for the master equation (9.77) for the following

processes

$$A + 2X \underset{k_1'}{\overset{k_1}{\rightleftarrows}} 3X \tag{E.1}$$

$$B + X \underset{k_2'}{\overset{k_2}{\rightleftarrows}} C. \tag{E.2}$$

2) Discuss the extrema of the probability distribution as a function of the concentration a, b.
3) Derive (9.84) explicitly.
4) Treat a set of reactions

$$A_j + l_j X \underset{k_j'}{\overset{k_j}{\rightleftarrows}} B_j + (l_j + 1)X, j = 1, \ldots, k,$$

under the requirement of detailed balance for each reaction j. Show that $P(N)$ is Poissonian.
Hint: Use $w_j(N, N - 1)$ in the form

$$w_j(N, N - 1) \propto (N - 1)(N - 2) \ldots (N - l_j)$$
$$w_j(N - 1, N) \propto N(N - 1)(N - 2) \ldots (N - l_j + 1).$$

Note that for $N \leq l_j$ a division by w_j is not possible!

9.6 Stochastic Model for a Chemical Reaction with Diffusion. One Variable

In most chemical and biochemical reactions diffusion plays an important role especially when we want to investigate the formation of spatial patterns. To obtain a proper description we divide the total volume into small cells of volume v. We distinguish the cells by an index $\boldsymbol{l}$ and denote the number of molecules in that cell by $N_{\boldsymbol{l}}$. Thus we now investigate the joint probability

$$P(\ldots, N_{\boldsymbol{l}}, N_{\boldsymbol{l}+\boldsymbol{a}}, \ldots) \tag{9.99}$$

for finding the cells $\boldsymbol{l}$ occupied by $N_{\boldsymbol{l}}$ molecules. In this chapter we consider only a single kind of molecules but the whole formalism can be readily generalized to several kinds. The number of molecules $N_{\boldsymbol{l}}$ now changes due to two causes; namely, due to chemical reactions as before, but now also on account of diffusion. We describe the diffusion again as a birth and death scheme where one molecule is annihilated in one cell and is created again in the neighboring cell. To find the total change of P due to that process, we have to sum up over all neighboring cells $\boldsymbol{l} + \boldsymbol{a}$ of the cell under consideration and we have to sum up over all cells $\boldsymbol{l}$. The temporal change of P due to diffusion thus reads

$$\dot{P}(\ldots, N_{\boldsymbol{l}}, \ldots)|_{\text{diffusion}} = \sum_{\boldsymbol{l},\boldsymbol{a}} D'\{(N_{\boldsymbol{l}+\boldsymbol{a}} + 1)P(\ldots, N_{\boldsymbol{l}} - 1, N_{\boldsymbol{l}+\boldsymbol{a}} + 1) - N_{\boldsymbol{l}}P(\ldots, N_{\boldsymbol{l}}, N_{\boldsymbol{l}+\boldsymbol{a}}, \ldots)\}. \tag{9.100}$$

The total change of P has the general form

$$\dot{P} = \dot{P}|_{\text{diffusion}} + \dot{P}|_{\text{reaction}}, \tag{9.101}$$

where we may insert for $\dot{P}|_{\text{reaction}}$ the rhs of (9.77) or any other reaction scheme. For a nonlinear reaction scheme, (9.101) cannot be solved exactly. We therefore employ another method, namely, we derive equations of motion for mean values or correlation functions. Having in mind that we go from the discrete cells to a continuum we shall replace the discrete index $\boldsymbol{l}$ by the continuous coordinate $\boldsymbol{x}$, $\boldsymbol{l} \to \boldsymbol{x}$. Correspondingly we introduce a new stochastic variable, namely, the local particle density

$$\rho(\boldsymbol{x}) = \frac{N_{\boldsymbol{l}}}{v} \tag{9.102}$$

and its average value

$$n(\boldsymbol{x}, t) = \frac{1}{v}\langle N_{\boldsymbol{l}}\rangle = \frac{1}{v}\sum_{\{N_j\}} N_{\boldsymbol{l}}P(\ldots, N_{\boldsymbol{l}}, \ldots). \tag{9.103}$$

We further introduce a correlation function for the densities at space points $\boldsymbol{x}$ and $\boldsymbol{x}'$

$$g(\boldsymbol{x}, \boldsymbol{x}', t) = \frac{1}{v^2}\langle N_{\boldsymbol{l}}N_{\boldsymbol{l}'}\rangle - \langle N_{\boldsymbol{l}}\rangle\langle N_{\boldsymbol{l}'}\rangle\frac{1}{v^2} - \delta(\boldsymbol{x} - \boldsymbol{x}')n(\boldsymbol{x}, t), \tag{9.104}$$

where we use the definition

$$\langle N_{\boldsymbol{l}}N_{\boldsymbol{l}'}\rangle = \sum_{\{N_j\}} N_{\boldsymbol{l}}N_{\boldsymbol{l}'}P(\ldots, N_{\boldsymbol{l}}, \ldots, N_{\boldsymbol{l}'}, \ldots). \tag{9.105}$$

As a concrete example we now take the reaction scheme (9.71, 72). We assume, however, that the back reaction can be neglected, i.e., $k_1' = 0$. Multiplying the corresponding equation (9.101) by (9.102) and taking the average (9.103) on both sides we obtain

$$\frac{\partial n(\boldsymbol{x}, t)}{\partial t} = D\nabla^2 n(\boldsymbol{x}, t) + (\varkappa_1 - \varkappa_2)n(\boldsymbol{x}, t) + \varkappa_1\beta, \tag{9.106}$$

where we have defined the diffusion constant D by

$$D = D'/v. \tag{9.107}$$

$$\varkappa_1 = \frac{k_1 a}{v}; \quad \varkappa_2 = \frac{k_2 b}{v} \tag{9.107a}$$

$$\beta = \frac{k_2' c}{k_1 v}. \tag{9.107b}$$

We leave the details, how to derive (9.106), to the reader as an exercise. In a similar fashion one obtains for the correlation function (9.104) the equation

$$\frac{\partial g}{\partial t} = D(\nabla_x^2 + \nabla_{x'}^2)g + 2(\varkappa_1 - \varkappa_2)g + 2\varkappa_1 n(\boldsymbol{x}, t)\delta(\boldsymbol{x} - \boldsymbol{x}'). \tag{9.108}$$

A comparison of (9.106) with (9.28), (9.5), (9.4) reveals that we have obtained exactly the same equation as in the nonstochastic treatment. Furthermore we find that putting $k_1' = 0$ amounts to neglecting the nonlinear term of (9.8). This is the deeper reason why (9.106) and also (9.108) can be solved exactly. The steady state solution of (9.108) reads

$$g(\boldsymbol{x}, \boldsymbol{x}')_{ss} = \frac{\varkappa_1 n_{ss}}{4\pi D|\boldsymbol{x} - \boldsymbol{x}'|} \exp\{-|\boldsymbol{x} - \boldsymbol{x}'|((\varkappa_2 - \varkappa_1)/D)^{1/2}\} \tag{9.109}$$

where the density of the steady state is given by

$$n_{ss} = \beta\varkappa_1/(\varkappa_2 - \varkappa_1). \tag{9.110}$$

Apparently the correlation function drops off with increasing distance between $\boldsymbol{x}$ and $\boldsymbol{x}'$. The range of correlation is given by the inverse of the factor of $|\boldsymbol{x} - \boldsymbol{x}'|$ of the exponential. Thus the correlation length is

$$l_c = (D/(\varkappa_2 - \varkappa_1))^{1/2}. \tag{9.111}$$

When we consider the effective reaction rates $\varkappa_1$ and $\varkappa_2$ (which are proportional to the concentrations of molecules a and b) we find that for $\varkappa_1 = \varkappa_2$ the coherence length becomes infinite. This is quite analogous to what happens in phase transitions of systems in thermal equilibrium. Indeed, we have already put the chemical reaction models under consideration in parallel with systems undergoing a phase transition (compare Sect. 9.2). We simply mention that one can also derive an equation for temporal correlations. It turns out that at the transition point also the correlation time becomes infinite. The whole process is very similar to the nonequilibrium phase transition of the continuous-mode laser. We now study molecule number fluctuations in small volumes and their correlation function. To this end we integrate the stochastic density (9.102) over a volume ΔV where we assume that the volume has the shape of a sphere with radius R

$$\int_{\Delta V} \rho(\mathbf{x})\, d^3x = N(\Delta V). \tag{9.112}$$

It is a simple matter to calculate the variance of the stochastic variable (9.112) which is defined as usual by

$$\sigma^2[\Delta V] \equiv \langle N(\Delta V)^2 \rangle - \langle N(\Delta V) \rangle^2. \tag{9.113}$$

Using the definition (9.104) and the abbreviation $R/l_c = r$ we obtain after elementary integrations

$$\sigma^2[\Delta V] = \langle N(\Delta V) \rangle \left\{ 1 + \frac{3\varkappa_2 l_c^2}{2Dr^3} [(1 - r^2 + \tfrac{2}{3} r^3) - e^{-2r}(1 + r)^2] \right\}. \tag{9.114}$$

We discuss the behavior of (9.114) in several interesting limiting cases. At the critical point, where $l_c \to \infty$, we obtain

$$\sigma^2[\Delta V] \to \langle N(\Delta V) \rangle \left(1 + \frac{2\varkappa_2 R^2}{5D} \right), \tag{9.115}$$

i.e., an ever increasing variance with increasing distance. Keeping l_c finite and letting $R \to \infty$ the variance becomes proportional to the square of the correlation length

$$\sigma^2[\Delta V] \to \langle N(\Delta V) \rangle \varkappa_2 l_c^2 / D. \tag{9.116}$$

For volumes with diameter small compared to the correlation length $R \ll l_c$, we obtain

$$\sigma^2[\Delta V] \approx \langle N(\Delta V) \rangle \left(1 + \frac{2\varkappa_2 R^2}{5D} \right). \tag{9.117}$$

Evidently for $R \to 0$ (9.117) becomes a Poissonian distribution which is in agreement with the postulate of local equilibrium in small volumes. On the other hand for large $R \gg l_c$ the variance reads

$$\sigma^2[\Delta V] \approx \langle N(\Delta V) \rangle \varkappa_1 |\varkappa_2 - \varkappa_1|^{-1} \left(1 - \frac{3}{2} \frac{l_c}{R} \right). \tag{9.118}$$

This result shows that for $R \to \infty$ the variance becomes independent of the distance and would obtain from a master equation neglecting diffusion. It has been proposed to measure such critical fluctuations by fluorescence spectroscopy which should be much more efficient than light-scattering measurements. The divergences, which occur at the transition point $\varkappa_1 = \varkappa_2$, are rounded off if we take the nonlinear term

$\propto n^2$, i.e., $k_1' \neq 0$, into account. We shall treat such a case in the next chapter taking a still more sophisticated model.

Exercises on 9.6

1) Derive (9.106) from (9.101) with (9.100), (9.77, 78, 79) for $k_1' = 0$.
Hint: Use exercise 3) of Section 9.5. Note that in our present dimensionless units of space

$$\tfrac{1}{2}\textstyle\sum_a \langle N_{l+a} + N_{l-a} - 2N_l\rangle = \nabla_l^2 \langle N_l\rangle .$$

2) Derive (9.108) from the same equations as in exercise 1) for $k_1' = 0$.
Hint: Multiply (9.101) by $N_l N_{l'}$ and sum over all N_j.

3) Solve (9.101) for the steady state with $\dot{P}_{\text{reaction}} = 0$.
Hint: Use the fact that detailed balance holds. Normalize the resulting probability distribution in a finite volume, i.e., for a finite number of cells.

4) Transform (9.100), one-dimensional case, into the Fokker-Planck equation:

$$f^{\cdot} = \int dx \left\{ -\frac{\delta}{\delta\rho(x)} \left(D \frac{d^2\rho(x)}{dx^2} f \right) + D \left(\frac{d}{dx} \cdot \frac{\delta}{\delta\rho(x)} \right)^2 (\rho(x) f) \right.$$

Hint:
Divide the total volume into cells which still contain a number $N_l \gg 1$. It is assumed that P changes only little for neighboring cells. Expand the rhs of (9.100) into a power series of "1" up to second order. Introduce $\rho(x)$ (9.102), and pass to the limit that l becomes a continuous variable x, hereby replacing P by $f = f\{\rho(x)\}$ and using the variational derivative $\delta/\delta\rho(x)$ instead of $\partial/\partial N_l$. (For its use see HAKEN, *Quantum Field Theory of Solids*, North-Holland, Amsterdam, 1976).

9.7* Stochastic Treatment of the Brusselator Close to Its Soft-Mode Instability

a) *Master Equation and Fokker-Planck Equation*

We consider the reaction scheme of Section 9.4

$$\begin{aligned} &A \to X \\ &B + X \to Y + D \\ &2X + Y \to 3X \\ &X \to E, \end{aligned} \tag{9.119}$$

where the concentrations of the molecules of kind A, B are externally given and kept fixed, while the numbers of molecules of kind X and Y are assumed to be variable. They are denoted by M, N respectively. Because we want to take into account diffusion, we divide the space in which the chemical reaction takes place into cells which still contain a large number of molecules (compared to unity). We distinguish the cells by an index $\boldsymbol{l}$ and denote the numbers of molecules in cell $\boldsymbol{l}$ by M_l, N_l. We again introduce dimensionless constants a, b which are proportional to the concentrations of the molecules of kind A, B. Extending the results of Sections 9.5 and 6, we obtain the following master equation for the probability distribution $P(\ldots, M_l, N_l \ldots)$ which gives us the joint probability to find $M_{l'}, N_{l'}, \ldots, M_l, N_l, \ldots$ molecules in cells $\boldsymbol{l}', \ldots, \boldsymbol{l}$

$$\begin{aligned}\dot{P}(\ldots; M_l, N_l; \ldots) = &\textstyle\sum_l v[aP(\ldots; M_l - 1, N_l; \ldots)\\ &+ b(M_l + 1)v^{-1}P(\ldots; M_l + 1, N_l - 1; \ldots) + (M_l - 2)(M_l - 1)(N_l + 1)v^{-3}\\ &\cdot P(\ldots; M_l - 1, N_l + 1; \ldots) + (M_l + 1)v^{-1}P(\ldots; M_l + 1, N_l; \ldots)\\ &- P(\ldots; M_l, N_l; \ldots)(a + (b + 1)M_l v^{-1} + M_l(M_l - 1)N_l v^{-3})]\\ &+ \textstyle\sum_{la}[D_1'\{M_{l+a} + 1)\cdot P(\ldots; M_l - 1, N_l; \ldots; M_{l+a} + 1, N_{l+a}; \ldots)\\ &- M_{l+a}P(\ldots; M_l, N_l; \ldots; M_{l+a}, N_{l+a}; \ldots)\}\\ &+ D_2'\{(N_{l+a} + 1)\cdot P(\ldots; M_l, N_l - 1; \ldots; M_{l+a}, N_{l+a} + 1, \ldots)\\ &- N_l P(\ldots; M_l, N_l; \ldots; M_{l+a}, N_{l+a}; \ldots)\}]. \end{aligned} \tag{9.120}$$

In it v is the volume of a cell, $\boldsymbol{l}$. The first sum takes into account the chemical reactions, the second sum, containing the "diffusion constants" D_1', D_2', takes into account the diffusion of the two kinds of molecules. The sum over $\boldsymbol{a}$ runs over the nearest neighboring cells of the cell $\boldsymbol{l}$. If the numbers M_l, N_l are sufficiently large compared to unity and if the function P is slowly varying with respect to its arguments we may proceed to the Fokker-Planck equation. A detailed analysis shows that this transformation is justified within a well-defined region around the soft-mode instability. This implies in particular $a \gg 1$ and $\mu \equiv (D_1/D_2)^{1/2} < 1$. To obtain the Fokker-Planck equation, we expand expressions of the type $(M_l + 1)$ $P(\ldots, M_l + 1, N_l, \ldots)$ etc. into a power series with respect to "1" keeping the first three terms (cf. Sect 4.2). Furthermore we let $\boldsymbol{l}$ become a continuous index which may be interpreted as the space coordinate $\boldsymbol{x}$. This requires that we replace the usual derivative by a variational derivative. Incidentally, we replace M_l/v, N_l/v by the densities $M(\boldsymbol{x})$, $N(\boldsymbol{x})$ which we had denoted by $\rho(\boldsymbol{x})$ in (9.102) and $P(\ldots, M_l, N_l, \ldots)$ by $f(\ldots, M(\boldsymbol{x}), N(\boldsymbol{x}), \ldots)$. Since the detailed mathematics of this procedure is rather lengthy we just quote the final result

$$\begin{aligned} f^{\cdot} = \int d^3x[&-\{(a - (b + 1)M + M^2N + D_1\cdot\nabla^2 M)f\}_{M(\boldsymbol{x})}\\ &-\{(bM - M^2N + D_2\cdot\nabla^2 N)f\}_{N(\boldsymbol{x})} + \tfrac{1}{2}\{(a + (b + 1)M + M^2N)f\}_{M(\boldsymbol{x}),M(\boldsymbol{x})}\\ &-\{(bM + M^2N)f\}_{M(\boldsymbol{x}),N(\boldsymbol{x})} + \tfrac{1}{2}\{(bM + M^2N)f\}_{N(\boldsymbol{x}),N(\boldsymbol{x})}\\ &+ D_1(\nabla(\delta/\delta M(\boldsymbol{x})))^2(Mf) + D_2(\nabla(\delta/\delta N(\boldsymbol{x})))^2(Nf)]. \end{aligned} \tag{9.121}$$

The indices $M(\boldsymbol{x})$ or $N(\boldsymbol{x})$ indicate the variational derivative with respect to $M(\boldsymbol{x})$ or $N(\boldsymbol{x})$. D_1 and D_2 are the usual diffusion constants. The Fokker-Planck equation (9.121) is still far too complicated to allow for an explicit solution. We therefore proceed in several steps: We first use the results of the stability analysis of the corresponding rate equations without fluctuations (cf. Section 9.4). According to these considerations there exist stable spatially homogeneous and time-independent solutions $M(\boldsymbol{x}) = a$, $N(\boldsymbol{x}) = b/a$ provided $b < b_c$. We therefore introduce new variables $q_j(\boldsymbol{x})$ by

$$M(\boldsymbol{x}) = a + q_1(\boldsymbol{x}),\ N(\boldsymbol{x}) = b/a + q_2(\boldsymbol{x})$$

and obtain the following Fokker-Planck equation

$$\begin{aligned} f^{\cdot} = \int dx\,[&- \{((b-1)\,q_1 + a^2 q_2 + g(q_1, q_2) + D_1\nabla^2 q_1)f\}_{q_1(\boldsymbol{x})} \\ &- \{(-bq_1 - a^2 q_2 - g(q_1, q_2) + D_2\nabla^2 q_2)f\}_{q_2(\boldsymbol{x})} \\ &+ \tfrac{1}{2}\{\hat{D}_{11}(q)f\}_{q_1(\boldsymbol{x}),q_1(\boldsymbol{x})} - \{\hat{D}_{12}(q)f\}_{q_1(\boldsymbol{x}),q_2(\boldsymbol{x})} \\ &+ \tfrac{1}{2}\{\hat{D}_{22}(q)f\}_{q_2(\boldsymbol{x}),q_2(\boldsymbol{x})} + D_1(\nabla(\delta/\delta q_1(\boldsymbol{x})))^2(a + q_1)f \\ &+ D_2(\nabla(\delta/\delta q_2(\boldsymbol{x})))^2(b/a + q_2)f]. \end{aligned} \tag{9.122}$$

f is now a functional of the variables $q_j(\boldsymbol{x})$. We have used the following abbreviations:

$$g(q_1, q_2) = 2aq_1q_2 + bq_1^2/a + q_1^2 q_2, \tag{9.123}$$

$$\hat{D}_{11} = 2a + 2ab + (3b + 1)q_1 + a^2 q_2 + 2aq_1q_2 + (b/a)q_1^2 + q_1^2 q_2, \tag{9.124}$$

$$\hat{D}_{12} = \hat{D}_{22} = 2ab + 3bq_1 + bq_1^2/a + a^2 q_2 + 2aq_1q_2 + q_1^2 q_2. \tag{9.125}$$

b) The further treatment proceeds along lines described in previous chapters. Because the details are rather space-consuming, we only indicate the individual steps:

1) We represent $\boldsymbol{q}(\boldsymbol{x}, t)$ as a superposition of the eigensolutions of the linearized equations (9.47), (9.48). Use is made of the wave-packet formulation of Section 7.7. The expansion coefficients $\xi_\mu(\boldsymbol{x}, t)$ are slowly varying functions of space and time.

2) We transform the Fokker-Planck equation to the ξ_μ's which still describe both unstable and stable modes.

3) We eliminate the ξ's of the stable modes by the method of Section 7.7. The final Fokker-Planck equation then reads:

$$f^{\cdot} = -\int d^3x \textstyle\sum_{\boldsymbol{k}} [(\delta/\delta\xi_{\boldsymbol{k}}(\boldsymbol{x}))\{\lambda_1(\nabla)\xi_{\boldsymbol{k}}(\boldsymbol{x}) + H_{\boldsymbol{k}}(\xi)\} - G_{11} \textstyle\sum_{\boldsymbol{k}} \delta^2/\delta\xi_{\boldsymbol{k}}\delta\xi_{\boldsymbol{k}}^*]f, \tag{9.126}$$

where

$$\xi_{\boldsymbol{k}}^* = \xi_{-\boldsymbol{k}}, \text{ and } \lambda_1(\nabla) = \lambda_1(b, \nabla) \approx \lambda_0 + \lambda^{(1)}\nabla^2 \tag{9.127}$$

with

$$\lambda^{(1)}\xi \approx 4a\mu((1 - \mu^2)(1 + a\mu)k_c^2)^{-1}\nabla^2\xi \tag{9.128}$$

$$\lambda_0 = (b - b_c)(1 + a^2 - \mu^2 - a\mu^3)^{-1} + O((b - b_c)^2). \tag{9.129}$$

We have further

$$G_{11} = 2(a(1 - \mu^2)^2)^{-1}\cdot\mu^2(1 + a\mu)^2$$

and

$$H_{\boldsymbol{k}}(\xi) = \sum_{\boldsymbol{k}'\boldsymbol{k}''} I_{\boldsymbol{k}'\boldsymbol{k}''\boldsymbol{k}'''}\, \bar{c}_1\xi_{\boldsymbol{k}'}(\boldsymbol{x})\xi_{\boldsymbol{k}''}(\boldsymbol{x})$$
$$- \sum_{\boldsymbol{k}'\boldsymbol{k}''\boldsymbol{k}'''} \bar{a}_1 J_{\boldsymbol{k}\boldsymbol{k}'\boldsymbol{k}''\boldsymbol{k}'''}\, \xi_{\boldsymbol{k}'}(\boldsymbol{x})\xi_{\boldsymbol{k}''}(\boldsymbol{x})\xi_{\boldsymbol{k}'''}(\boldsymbol{x}). \tag{9.130}$$

For a definition of I and J, see (7.81) and (7.83). For a discussion of the evolving spatial structures, the "selection rules" inherent in I and J are important. One readily verifies in one dimension:
$I = 0$ for boundary conditions (9.44), i.e., the χ_k's are plane waves and $I \approx 0$ for $\chi_k \propto \sin kx$ and $k \gg 1$.
Further
$J_{\boldsymbol{k}\boldsymbol{k}'\boldsymbol{k}''\boldsymbol{k}'''} = J \neq 0$ only if two pairs of k's out of $\boldsymbol{k}$, $\boldsymbol{k}'$, $\boldsymbol{k}''$, $\boldsymbol{k}'''$ satisfy:

$$k_1 = -k_2 = -k_c$$
$$k_3 = -k_4 = -k_c$$

if plane waves are used, or $k = k' = k'' = k''' = k_c$ if $\chi_k \propto \sin kx$. We have evaluated $\bar{a}_1$ explicitly for plane waves. The Fokker-Planck equation (9.126) then reduces to

$$f^{\cdot} = \left[-\int dx(\delta/\delta\xi(x))((\lambda_0 + \lambda^{(1)}\nabla^2)\xi(x) - A\xi^3(x)) + \int d^3xG_{11}\delta^2/\delta\xi^2\right]f, \tag{9.131}$$

where the coefficient A reads:

$$A = (9(1 - \mu^2)\mu^3(1 - a\mu)^2a)^{-1}\cdot(-8a^3\mu^3 + 5a^2\mu^2 + 20a\mu - 8). \tag{9.132}$$

Note that for sufficiently big $a\mu$ the coefficient A becomes negative. A closer inspection shows, that under this condition the mode with $k = 0$ approaches a marginal situation which then requires to consider the modes with $k = 0$ (and $|\boldsymbol{k}| = 2k_c$) as unstable modes. We have met eqs. like (9.131) or the corresponding Langevin equations at several instances in our book. It shows that at its soft-mode instability the chemical reaction undergoes a nonequilibrium phase transition of second order in complete analogy to the laser or the Bénard instability (Chap. 8).

9.8 Chemical Networks

In Sections 9.2–9.4 we have met several explicit examples for equations describing chemical processes. If we may assume spatial homogeneity, these equations have the form

$$\dot{n}_j = F_j(n_1, n_2, \ldots). \tag{9.133}$$

Equations of such a type occur also in quite different disciplines, where we have now in mind network theory dealing with electrical networks. Here the n's have the meaning of charges, currents, or voltages. Electronic devices, such as radios or computers, contain networks. A network is composed of single elements (e.g., resistors, tunnel diodes, transistors) each of which can perform a certain function. It can for example amplify a current or rectify it. Furthermore, certain devices can act as memory or perform logical steps, such as "and", "or", "no". In view of the formal analogy between a system of equations (9.133) of chemical reactions and those of electrical networks, the question arises whether we can devise logical elements by means of chemical reactions. In network theory and related disciplines it is shown that for a given logical process a set of equations of the type (9.133) can be constructed with well-defined functions F_j.

These rather abstract considerations can be easily explained by looking at our standard example of the overdamped anharmonic oscillator whose equation was given by

$$\dot{q} = \alpha q - \beta q^3. \tag{9.134}$$

In electronics this equation could describe e.g. the charge q of a tunnel diode of the device of Fig. 7.3. We have seen in previous chapters that (9.134) allows for two stable states $q_1 = \sqrt{\alpha/\beta}$, $q_2 = -\sqrt{\alpha/\beta}$, i.e., it describes a bistable element which can store information. Furthermore we have discussed in Section 7.3 how we can switch this element, e.g., by changing α. When we want to translate this device into a chemical reaction, we have to bear in mind that the concentration variable, n, is intrinsically nonnegative. However, we can easily pass from the variable q to a positive variable n by making the replacement

$$q = n - q_0, \quad q_0 > 0 \tag{9.135}$$

so that both stable states lie at positive values. Introducing (9.135) into (9.134) and rearranging this equation, we end up with

$$\dot{n} = \alpha_1 + \alpha_2 n + \alpha_3 n^2 - \beta n^3, \tag{9.136}$$

where we have used the abbreviations

$$\alpha_1 = \beta q_0^3 - \alpha q_0, \tag{9.137}$$

$$\alpha_2 = \alpha - 3\beta q_0^2, \tag{9.138}$$

$$\alpha_3 = 3\beta q_0. \tag{9.139}$$

Since (9.134) allowed for a bistable state for q so does (9.136) for n. The next question is whether (9.136) can be realized by chemical reactions. Indeed in the preceeding chapters we have met reaction schemes giving rise to the first three terms in (9.136). The last term can be realized by an adiabatic elimination process of a fast chemical reaction with a quickly transformed intermediate state. The steps for modeling a logical system are now rather obvious: 1) Look at the corresponding logical elements of an electrical network and their corresponding differential equations; 2) Translate them in analogy to the above example. There are two main problems. One, which can be solved after some inspection, is that the important operation points must lie at positive values of n. The second problem is, of course, one of chemistry; namely, how to find chemical processes in reality which fulfil all the requirements with respect to the directions the processes go, reaction constants etc.

Once the single elements are realized by chemical reactions, a whole network can be constructed. We simply mention a typical network which consists of the following elements: flip-flop (that is the above bistable element which can be switched), delays (which act as memory) and the logical elements "and", "or", "no".

Our above considerations referred to spatially homogeneous reactions, but by dividing space into cells and permitting diffusion, we can now construct coupled logical networks. There are, of course, a number of further extensions possible, for example, one can imagine cells separated by membranes which may be only partly permeable for some of the reactants, or whose permeability can be switched. Obviously, these problems readily lead to basic questions of biology.

10. Applications to Biology

In theoretical biology the question of cooperative effects and self-organization nowadays plays a central role. In view of the complexity of biological systems this is a vast field. We have selected some typical examples out of the following fields:

1) Ecology, population-dynamics
2) Evolution
3) Morphogenesis

We want to show what the basic ideas are, how they are cast into a mathematical form, and what main conclusions can be drawn at present. Again the vital interplay between "chance" and "necessity" will transpire, especially in evolutionary processes. Furthermore, most of the phenomena allow for an interpretation as non-equilibrium phase transitions.

10.1 Ecology, Population-Dynamics

What one wants to understand here is basically the distribution and abundance of species. To this end, a great amount of information has been gathered, for example about the populations of different birds in certain areas. Here we want to discuss some main aspects: What controls the size of a population; how many different kinds of populations can coexist?

Let us first consider a single population which may consist of bacteria, or plants of a given kind, or animals of a given kind. It is a hopeless task to describe the fate of each individual. Rather we have to look for "macroscopic" features describing the populations. The most apparent feature is the number of individuals of a population. Those numbers play the role of order parameters. A little thought will show that they indeed govern the fate of the individuals, at least "on the average". Let the number (or density) of individuals be n. Then n changes according to the growth rate, g, (births) minus the death rate, d.

$$\dot{n} = g - d. \tag{10.1}$$

The growth and death rates depend on the number of individuals present. In the simplest form we assume

$$g = \gamma n, \tag{10.2}$$

$$d = \delta n, \tag{10.3}$$

where the coefficients γ and δ are independent of n. We then speak of *density-independent* growth. The coefficients γ and δ may depend on external parameters, such as available food, temperature, climate and other environmental factors. As long as these factors are kept constant, the equation

$$\dot{n} = \alpha n \equiv (\gamma - \delta)n \tag{10.4}$$

allows for either an exponentially growing or an exponentially decaying population. (The marginal state $\gamma = \delta$ is unstable against small perturbations of γ or δ). Therefore, no steady state would exist. The essential conclusion to be drawn is that the coefficients γ or δ or both depend on the density n. An important reason among others for this lies in a limited food supply, as discussed earlier in this book by some exercises. The resulting equation is of the type

$$\dot{n} = \alpha_0 n - \beta n^2 \text{ (``Verhulst'' equation)} \tag{10.5}$$

where $-\beta n^2$ stems from a depletion of the food resources. It is assumed that new food is supplied only at a constant rate. The behavior of a system described by (10.5) was discussed in detail in Section 5.4.

We now come to several species. Several basically different cases may occur:

1) Competition and coexistence
2) Predator-prey relation
3) Symbiosis

1) *Competition and Coexistence*

If different species live on different kinds of food and do not interact with each other (e.g., by killing or using the same places for breeding, etc.) they can certainly coexist. We then have for the corresponding species equations of the type

$$\dot{n}_j = \alpha_j n_j - \beta_j n_j^2, \quad j = 1, 2, \ldots \tag{10.6}$$

Things become much more complicated if different species live or try to live on the *same* food supply, and/or depend on similar living conditions. Examples are provided by plants extracting phosphorous from soil, one plant depriving the other from sunlight by its leaves, birds using the same holes to build their nests, etc. Since the basic mathematical approach remains unaltered in these other cases, we talk explicitly only about "food". We have discussed this case previously (Sect. 5.4) and have shown that only one species survives which is defined as the fittest. Here we exclude the (unstable) case that, accidentally, all growth and decay rates coincide.

For a population to survive it is therefore vital to improve its specific rates α_j, β_j by adaption. Furthermore for a possible coexistence, additional food supply is essential. Let us consider as example two species living on two "overlapping" food supplies. This can be modelled as follows, denoting the amount of available food by N_1 or N_2:

$$\dot{n}_1 = (\alpha_{11}N_1 + \alpha_{12}N_2)n_1 - \delta_1 n_1, \tag{10.7}$$

$$\dot{n}_2 = (\alpha_{21}N_1 + \alpha_{22}N_2)n_2 - \delta_2 n_2. \tag{10.8}$$

Generalizing (Sect. 5.4) we establish equations for the food supply

$$\dot{N}_1 = \gamma_1(N_1^0 - N_1) - \mu_{11}n_1 - \mu_{12}n_2, \tag{10.9}$$

$$\dot{N}_2 = \gamma_2(N_2^0 - N_2) - \mu_{21}n_1 - \mu_{22}n_2. \tag{10.10}$$

Here $\gamma_j N_j^0$ is the rate of food production, and $-\gamma_j N_j$ is the decrease of food due to internal causes (e.g. by rotting). Adopting the adiabatic elimination hypothesis (cf. Sect. 7.1), we assume that the temporal change of the food supply may be neglected, i.e., $\dot{N}_1 = \dot{N}_2 = 0$. This allow us to express N_1 and N_2 directly by n_1 and n_2. Inserting the resulting expressions into (10.7), (10.8) leaves us with equations of the following type:

$$\dot{n}_1 = [(\alpha_{11}^0 N_1^0 + \alpha_{12}^0 N_2^0) - \delta_1 - (\eta_{11}n_1 + \eta_{12}n_2)]n_1, \tag{10.11}$$

$$\dot{n}_2 = [(\alpha_{21}^0 N_1^0 + \alpha_{22}^0 N_2^0) - \delta_2 - (\eta_{21}n_1 + \eta_{22}n_2)]n_2. \tag{10.12}$$

From $\dot{n}_1 = \dot{n}_2 = 0$ we obtain stationary states n_1^0, n_2^0. By means of a discussion of the "forces" (i.e., the rhs of (10.11/12)) in the $n_1 - n_2$ plane, one may easily discuss when coexistence is possible depending on the parameters of the system. (cf. Fig. 10.1 a–c) This example can be immediately generalized to several kinds of species and of food supply. A detailed discussion of coexistence becomes tedious, however.

From our above considerations it becomes apparent why ecological "niches"

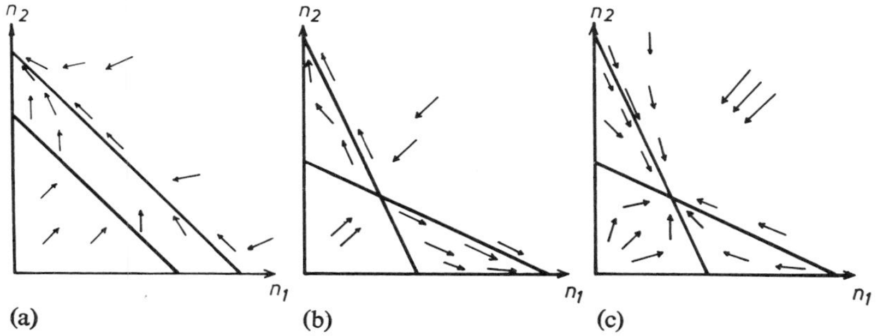

Fig. 10.1a-c. The eqs. (10.11), (10.12) for different parameters leading to different stable configurations. (a) $n_1 = 0$, $n_2 = C$ is the only stable point i.e., only one species survives. (b) $n_1 = 0$, $n_2 \neq 0$ or $n_2 = 0$, $n_1 \neq 0$ are two stable points, i.e., one species or the other one can survive. (c) $n_1 \neq 0$, $n_2 \neq 0$, the two species can coexist. If the field of arrows is made closer, one finds the trajectories discussed in Section 5.2. and the points where the arrows end are sinks in the sense of that chapter

are so important for survival and why surviving species are sometimes so highly specialized. A well-known example for coexistence and competition is the distribution of flora according to different heights in mountainous regions. There, well-defined belts of different kinds of plants are present. A detailed study of such phenomena is performed in biogeographics.

2) *Predator-Prey-Relation*

The basic phenomenon is as follows:
There are two kinds of animals: Prey animals living on plants, and predators, living on the prey. Examples are fishes in the Adriatic sea, or hares and lynxes. The latter system has been studied in detail in nature and the theoretical predictions have been substantiated. The basic Lotka-Volterra equations have been discussed in Section 5.4. They read

$$\dot{n}_1 = \alpha_1 n_1 - \alpha n_1 n_2, \tag{10.13}$$

$$\dot{n}_2 = \beta n_1 n_2 - 2\kappa_2 n_2, \tag{10.14}$$

where (10.13) refers to prey and (10.14) to predators. As was shown in Section 5.4 a periodic solution results: When predators become too numerous, the prey is eaten up too quickly. Thus the food supply of the predators decreases and consequently their population decreases. This allows for an increase of the number of prey animals so that a greater food supply becomes available for the predators whose number now increases again. When this problem is treated stochastically, a serious difficulty arises: Both populations die out (cf. Sect. 10.2).

3) *Symbiosis*

There are numerous examples in nature where the cooperation of different species facilitates their living. A well-known example is the cooperation of trees and bees. This cooperation may be modelled in this way: Since the multiplication rate of one species depends on the presence of the other, we obtain

$$\dot{n}_1 = (\alpha_1 + \alpha_1' n_2) n_1 - \delta_1 n_1, \tag{10.15}$$

$$\dot{n}_2 = (\alpha_2 + \alpha_2' n_1) n_2 - \delta_2 n_2, \tag{10.16}$$

as long as we neglect self-suppressing terms $-\beta_i n_i^2$. In the stationary case, $\dot{n}_1 = \dot{n}_2 = 0$, two types of solutions result by putting the rhs of (10.15–16) equal to zero:

a) $n_1 = n_2 = 0$, which is uninteresting,

or

b) $\alpha_1 - \delta_1 + \alpha_1' n_2 = 0,$

$\alpha_2 - \delta_2 + \alpha_2' n_1 = 0.$

It is an interesting exercise for the reader to discuss the stability properties of b). We also leave it to the reader to convince himself that for initial values of n_1 and n_2 which are large enough, an exponential explosion of the populations always occurs.

4) *Some General Remarks*

Models of the above types are now widely used in ecology. It must be mentioned that they are still on a very global level. In a next step numerous other effects must be taken into account, for example, time-lag effects, seasons, different death rates depending on age, even different reaction behavior within a single species. Even if we perform the analysis in the above-mentioned manner, in reality biological population networks are more complicated, i.e., they are organized in trophical (i.e., nourishment) levels. The first trophic level consists of green plants. They are eaten by animals, which are in turn eaten by other animals, etc. Furthermore for example, a predator may live on two or several kinds of prey. In this case the pronounced oscillations of the Lotka-Volterra model become, in general, smaller, and the system becomes more stable.

10.2 Stochastic Models for a Predator-Prey System

The analogy of our rate equations stated above to those of chemical reaction kinetics is obvious. Readers who want to treat for example (10.5) or (10.11), (10.12) stochastically are therefore referred to those chapters. Here we treat as another example the Lotka-Volterra system. Denoting the number of individuals of the two species, prey and predator, by M and N, respectively, and again using the methods of chemical reaction kinetics, we obtain as transition rates

1) Multiplication of prey

$$M \to M + 1: \quad w(M + 1, N; M, N) = \varkappa_1 M.$$

2) Death rate of predator

$$N \to N - 1: \quad w(M, N - 1; M, N) = \varkappa_2 N.$$

3) Predators eating prey

$$\left.\begin{matrix} M \to M - 1 \\ N \to N + 1 \end{matrix}\right\} \quad w(M - 1, N + 1; M, N) = \beta M N.$$

The master equation for the probability distribution $P(M, N, t)$ thus reads

$$\begin{aligned} \dot{P}(M, N; t) = {} & \varkappa_1 (M - 1) P(M - 1, N; t) \\ & + \varkappa_2 (N + 1) P(M, N + 1; t) \\ & + \beta (M + 1)(N - 1) P(M + 1, N - 1; t) \\ & - (\varkappa_1 M + \varkappa_2 N + \beta M N) P(M, N; t). \end{aligned} \tag{10.17}$$

Of course we must require $P = 0$ for $M < 0$ or $N < 0$ or both < 0. We now want to show that the only stationary solution of (10.17) is

$$P(0, 0) = 1 \text{ and all other } P\text{'s} = 0. \tag{10.18}$$

Thus both species die out, even if initially both have been present. For a proof we put $\dot{P} = 0$. Inserting (10.18) into (10.17) shows that (10.17) is indeed fulfilled. Furthermore we may convince ourselves that all points (M, N) are connected via at least one path with any other point (M', N'). Thus the solution is unique. Our rather puzzling result (10.18) has a simple explanation: From the stability analysis of the nonstochastic Lotka-Volterra equations, it is known that the trajectories have "neutral stability". Fluctuations will cause transitions from one trajectory to a neighboring one. Once, by chance, the prey has died out, there is no hope for the predators to survive, i.e., $M = N = 0$ is the only possible stationary state.

While this may indeed happen in nature, biologists have found another reason for the survival of prey: Prey animals may find a refuge so that a certain minimum number survives. For instance they wander to other regions where the predators do not follow so quickly or they can hide in certain places not accessible to predators.

10.3 A Simple Mathematical Model for Evolutionary Processes

In Section 10.1 we have learned about a few mathematical models from which we may draw several general conclusions about the development of populations. These populations may consist of highly developed plants or animals as well as of bacteria and even biological molecules "living" on certain substrates. When we try to apply these equations to evolution, an important feature is still missing. In the evolutionary process, again and again new kinds of species appear. To see how we can incorporate this fact into the equations of Section 10.1, let us briefly recollect some basic facts. We know that genes may undergo mutations, creating alleles. These mutations occur at random, though their creation frequency can be enhanced by external factors, e.g., increased temperature, irrediation with UV-light, chemical agents etc. As a consequence, a certain "mutation pressure" arises by which all the time new kinds of individuals within a species come into existence. We shall not discuss here the detailed mechanism, that is, that newly created features are first recessive and only later, after several multiplications, possibly become dominant. We rather assume simply that new kinds of individuals of a population are created at random. We denote the number of these individuals by n_j. Since these individuals may have new features, in general their growth and death factors differ. Since a new population can start only if a fluctuation occurs, we add fluctuating forces to the equations of growth:

$$\dot{n}_j = \gamma_j n_j - \delta_j n_j + F_j(t), \quad j = 1, 2, \ldots \tag{10.19}$$

The properties of $F_j(t)$ depend on both the population which was present prior to the one described by the particular equation (10.19) and on environmental factors.

The system of different "subspecies" is now exposed to a "selection pressure". To see this, we need only apply considerations and results of Section 10.1. Since the environmental conditions are the same (food supply, etc.), we have to apply equations of type (10.11/10.12). Generalizing them to N subspecies living on the *same* "food" supply, we obtain

$$\dot{n}_j = \alpha_j(g_0 - \sum g_l n_l)n_j - \kappa_j n_j + F_j(t). \tag{10.20}$$

If the mutation rate for a special mutant is small, only that mutant survives which has the highest gain factor α_j and the smallest loss factor $\varkappa_j$ and is thus the "fittest". It is possible to discuss still more complicated equations, in which the multiplication of a subspecies is replaced by a cycle $A \to B \to C \to \cdots \to A$. Such cycles have been postulated for the evolution of biomolecules. In the context of our book it is remarkable that the occurrence of a new species due to mutation ("fluctuating force") and selection ("driving force") can be put into close parallel to a second-order, nonequilibrium phase transition (e.g., that of the laser).

10.4 A Model for Morphogenesis

When reading our book the reader may have observed that each discipline has its "model" systems which are especially suited for the study of characteristic features. In the field of morphogenesis one of such "systems" is the hydra. Hydra is an animal a few mm in length, consisting of about 100,000 cells of about 15 different types. Along its length it is subdivided into different regions. At one end its "head" is located. Thus the animal has a "polar structure". A typical experiment which can be done with hydra is this: Remove part of the head region and transplant it to another part of the animal. Then, if the transplanted part is in a region close to the old head, *no* new head is formed, or, in other words, growth of a head is *inhibited*. On the other hand, if the transplant is made at a distance sufficiently far away from the old head, a new head is formed by an *activation* of cells of the hydra by the transplant. It is generally accepted that the agents causing biological processes such as morphogenesis are chemicals. Therefore, we are led to assume that there are at least two types of chemicals (or "reactants"): an *activator* and an *inhibitor*. Nowadays there is some evidence that these activator and inhibitor molecules really exist and what they possibly are. Now let us assume that both substances are produced in the head region of the hydra. Since inhibition was present still in some distance from the primary head, the inhibitor must be able to diffuse. Also the activator must be able to do so, otherwise it could not influence the neighboring cells of the transplant.

Let us try to formulate a mathematical model. We denote the concentration of the activator by a, that of the inhibitor by h. The basic features can be seen in the frame of a one-dimensional model. We thus let a and h depend on the coordinate x and time t. Consider the rate of change of a, $\partial a/\partial t$. This change is due to

1) generation by a source (head):

$$\text{production rate: } \rho, \tag{10.21}$$

2) decay: $-\mu a$, (10.22)

where μ is the decay constant

3) diffusion: $D_a \frac{\partial^2 a}{\partial x^2}$, (10.23)

D_a diffusion constant.

Furthermore it is known from other biological systems (e.g., slime mold, compare Section 1.1) that autocatalytic processes ("stimulated emission") can take place. They can be described—depending on the process—by the production rate

$$k_1 a, \tag{10.24}$$

or

$$k_2 a^2, \text{ etc.} \tag{10.25}$$

Finally, the effect of inhibition has to be modelled. The most direct way the inhibitor can inhibit the action of the activator is by lowering the concentration of a. A possible "ansatz" for the inhibition rate could be

$$-ah. \tag{10.26}$$

Another way is to let h hinder the autocatalytic rates (10.24) or (10.25). The higher h, the lower the production rates (10.21) or (10.25). This leads us in the case (10.25) to

$$k\frac{a^2}{h}. \tag{10.27}$$

Apparently there is some arbitrariness in deriving the basic equations and a final decision can only be made by detailed chemical analysis. However, selecting typical terms, such as (10.21), (10.22), (10.23), (10.27), we obtain for the total rate of change of a

$$\frac{\partial a}{\partial t} = \rho + k\frac{a^2}{h} - \mu a + D_a \frac{\partial^2 a}{\partial x^2}. \tag{10.28}$$

Let us now turn to derive an equation for the inhibitor h. It certainly has a decay time, i.e., a loss rate

$$-\nu h, \tag{10.29}$$

and it can diffuse:

$$D_h \frac{\partial^2 h}{\partial x^2}. \tag{10.30}$$

Again we may think of various generation processes. Gierer and Meinhard, whose equations we present here, suggested (among other equations)

$$\text{production rate: } ca^2, \tag{10.31}$$

i.e., a generation by means of the activator. We then obtain

$$\frac{\partial h}{\partial t} = ca^2 - vh + D_h \frac{\partial^2 h}{\partial x^2}. \tag{10.32}$$

Before we represent our analytical results using the order parameter concept in Section 10.5, we exhibit some computer solutions whose results are not restricted to the hydra, but may be applied also to other phenomena of morphogenesis. We simply exhibit two typical results: In Fig. 10.2 the interplay between activator and inhibitor leads to a growing periodic structure. Fig. 10.3 shows a resulting two-dimensional pattern of activator concentration. Obviously, in both cases the inhibitor suppressed a second center (second head of hydra!) close to a first center (primary head of hydra!). To derive such patterns it is essential that h diffuses more easily than a, i.e , $D_h > D_a$. With somewhat further developed activator-inhibitor models, the structures of leaves, for example, can be mimicked.

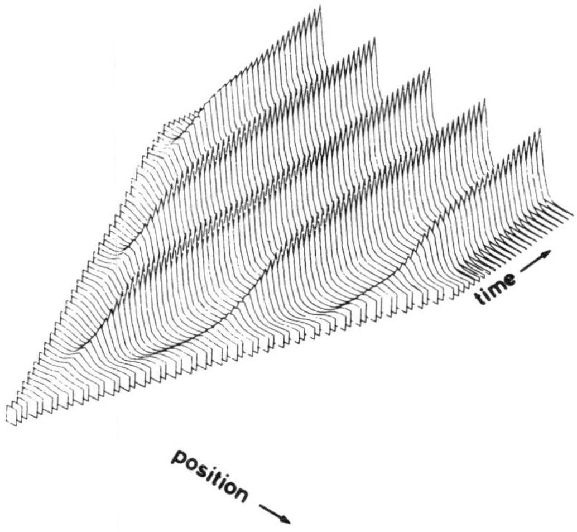

Fig. 10.2. Developing activator concentration as a function of space and time (computer solution). (After *H. Meinhardt, A. Gierer:* J. Cell Sci. **15**, 321 (1974))

In conclusion we mention an analogy which presumably is not accidental but reveals a general principle used by nature: The action of neural networks (e.g., the cerebral cortex) is again governed by the interplay between short-range activation and long-range inhibition but this time the activators and inhibitors are neurons.

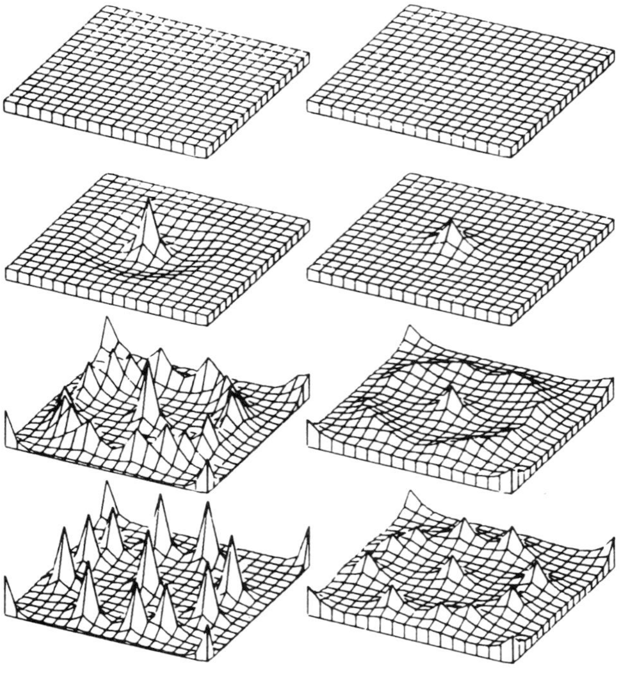

Fig. 10.3. Results of the morphogenetic model. Left column: activator concentration plottet over two dimensions. Right column: same for inhibitor. Rows refer to different times growing from above to below (computer solution). (After *H. Meinhardt*, *A. Gierer*: J. Cell Sci. **15**, 321 (1974))

10.5 Order Parameters and Morphogenesis

In this chapter we apply the analytical methods developed in Sections 7.6–7.8 to determine the evolving patterns described by (10.28) and (10.32). Since we want to treat the two-dimensional case, we replace $\partial^2 a/\partial x^2$ and $\partial^2 h/\partial x^2$ by

$$\Delta a \equiv \frac{\partial^2 a}{\partial x^2} + \frac{\partial^2 a}{\partial y^2} \quad \text{and} \quad \Delta h \equiv \frac{\partial^2 h}{\partial x^2} + \frac{\partial^2 h}{\partial y^2},$$

respectively. We assume that ρ is the control parameter which can be changed arbitrarily whereas all the other constants are prescribed quantities. It is convenient to go over to new variables so to reduce the number of parameters by the transformations

$$\boldsymbol{x}' = \sqrt{\frac{\nu}{D_a}}\,\boldsymbol{x}, \qquad t' = \nu t, \qquad a' = \frac{k}{c}\,a, \qquad h' = \frac{\nu c}{k^2}\,h. \tag{10.33}$$

Then we have

$$\dot{a}' = \rho' + \frac{a'^2}{h'} - \mu' a' + \Delta' a', \tag{10.34}$$

$$\dot{h}' = a'^2 - h' + D' \Delta' h', \tag{10.35}$$

where we have used the abbreviations

$$\rho' = \frac{\rho c}{v k}, \quad \mu' = \frac{\mu}{v}, \tag{10.36}$$

$$D' = \frac{D_h}{D_a}. \tag{10.37}$$

From now on we shall drop the primes. The stationary homogeneous solution of (10.34) and (10.35) reads

$$a_0 = \frac{1}{\mu}(\rho + 1), \tag{10.38}$$

$$h_0 = a_0^2. \tag{10.39}$$

To perform the stability analysis we introduce an expansion around the stationary solution

$$q_1 = a - a_0, \quad q_2 = h - h_0. \tag{10.40}$$

Eqs. (10.34) and (10.35) then can be cast in the form (compare (7.62))

$$\dot{\boldsymbol{q}} = K(\Delta)\boldsymbol{q} + \boldsymbol{g}(\boldsymbol{q}) \tag{10.41}$$

where K is given by

$$K(\Delta) = \begin{pmatrix} \mu\left(\frac{2}{\rho+1} - 1\right) + \Delta & -\frac{\mu^2}{(\rho+1)^2} \\ \frac{2}{\mu}(\rho+1) & -1 + D\Delta \end{pmatrix} \tag{10.42}$$

and $\boldsymbol{g}(\boldsymbol{q})$ contains the nonlinearities. For the linear stability analysis we drop the nonlinear term $\boldsymbol{g}(\boldsymbol{q})$ and make the hypothesis

$$\boldsymbol{q} = \boldsymbol{O}\, e^{i\boldsymbol{k}\boldsymbol{x} + \lambda t}. \tag{10.43}$$

The resulting eigenvalue equation yields

$$\lambda^{\pm}(k) = \frac{\alpha(k)}{2} \pm \sqrt{\frac{\alpha^2(k)}{4} - \beta(k)} \tag{10.44}$$

where

$$\alpha(k) = -(D+1)k^2 + \frac{2\mu}{\rho+1} - \mu - 1\,, \tag{10.45}$$

$$\beta(k) = (k^2+\mu)(1+Dk^2) - \frac{2\mu D k^2}{\rho+1}\,. \tag{10.46}$$

The condition for the first occurrence of a soft-mode instability is

$$\mathrm{Re}\{\lambda^+(k)\} \geq 0\,, \qquad \mathrm{Im}\{\lambda^j(k)\} = 0\,. \tag{10.47}$$

A brief analysis of (10.44) reveals that (10.47) is fulfilled if

$$(1) \quad \alpha < 0 \quad \text{and} \quad (2) \quad \beta \leq 0\,. \tag{10.48}$$

From condition (1) we obtain

$$\rho > \frac{2\mu}{(\mu+1)+(D+1)k^2} - 1\,, \tag{10.49}$$

whereas condition (2) yields

$$\rho \leq \frac{2\mu D k^2}{(1+Dk^2)(\mu+k^2)} - 1\,. \tag{10.50}$$

The dependence of the critical ρ on the wave vector $\boldsymbol{k}$ (10.49), is exhibited in Fig. 10.4. When $\rho > \rho_c$, the instability condition (10.47) cannot be fulfilled.

For a critical ρ_c the instability condition $\beta = 0$ can first be met for two critical values of k, namely $k = +k_c$ and $k = -k_c$. For $\rho = \rho_{\max}$ the condition $\alpha = 0$, which indicates the onset of a hard-mode instability, can be met. In our following

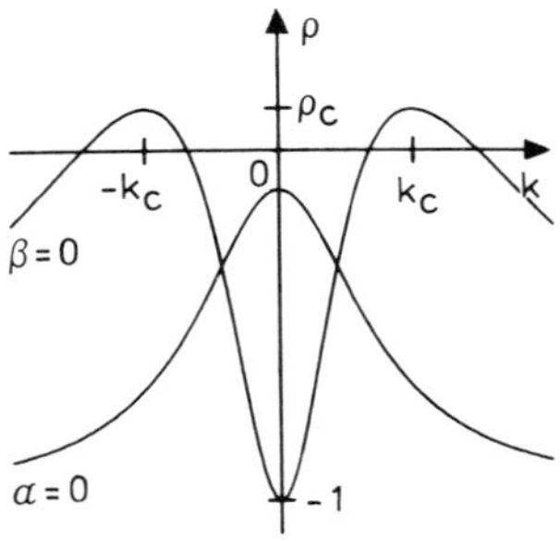

Fig. 10.4. This figure shows the curves defined by (10.45) = 0, and (10.46) = 0 in the k, ρ plane. The parameters D and μ are kept fixed. The area above the curve $\beta = 0$ defines the stable region. The area below it defines the unstable region. Since the condition $\alpha < 0$ is met for all k the onset of the instability occurs when $\rho = \rho_c$

analysis we will focus our attention on the soft mode case. k_c and ρ_c and $\rho_{\max}$ are given by

$$k_c = \sqrt[4]{\frac{\mu}{D}}, \tag{10.51}$$

$$\rho_c = \frac{2\sqrt{\mu D}}{2+\sqrt{\mu D}+\dfrac{1}{\sqrt{\mu D}}} - 1, \tag{10.52}$$

$$\rho_{\max} = \frac{\mu-1}{\mu+1}. \tag{10.53}$$

In order that the soft-mode instability occurs first we have to require $\rho_c > \rho_{\max}$, from which it follows

$$D > 2\mu + 1 + 2\sqrt{\mu+1}. \tag{10.54}$$

Using (10.37) and (10.36) we derive from (10.54) that the diffusion constant of the inhibitor must be bigger than that of the activator by a certain amount. In other words, "long range inhibition" and "short range activation" are required for a nonoscillating pattern.

We assume a *two-dimensional* layer with side lengths L_1 and L_2 and first adopt periodic boundary conditions. The detailed method of solution has been described in Sections 7.6–7.8 and we repeat here only the main steps of the whole procedure. We assume ρ close to ρ_c.

We make the hypothesis (cf. (7.72))

$$\boldsymbol{q} = \sum_j \boldsymbol{O}^j(\Delta) \sum_{\boldsymbol{k}} \xi_{\boldsymbol{k}}^j(t) e^{i\boldsymbol{k}\boldsymbol{x}}, \quad j = \pm. \tag{10.55}$$

To exhibit the essential features, we neglect "small band excitations" and assume $\xi_{\boldsymbol{k}}^j$ independent of $\boldsymbol{x}$. The coefficients $\boldsymbol{O}^j$ obey the equation

$$K(\Delta)\boldsymbol{O}^j(\Delta) = \lambda^j(\Delta)\boldsymbol{O}^j(\Delta), \tag{10.56}$$

and the wave vector $\boldsymbol{k}$ is assumed in the form

$$\boldsymbol{k} = 2\pi \begin{pmatrix} \dfrac{n}{L_1} \\ \dfrac{m}{L_2} \end{pmatrix}, \quad n, m = 0, \pm 1, \pm 2, \ldots. \tag{10.57}$$

Since the solution must be real we have to require

$$\xi_{\boldsymbol{k}}^j = \xi_{-\boldsymbol{k}}^{j*}, \quad \xi_0^+ = \xi_0^{-*}. \tag{10.58}$$

Inserting (10.55) into (10.41) and multiplying the resulting expressions from the left with the conjugate complex of e^{ikx} and the adjoint of $\boldsymbol{O}^j$, we obtain after some analysis the equations

$$\left(\frac{\partial}{\partial t} - \lambda^j(\boldsymbol{k})\right)\xi^j_{\boldsymbol{k}} = (\text{N.L.T})^j_{\boldsymbol{k}}\,. \tag{10.59}$$

The nonlinear term on the right hand side has the form

$$\begin{aligned}(\text{N.L.T})^j_{\boldsymbol{k}} &= \sum_{j',j''}\sum_{\boldsymbol{k}',\boldsymbol{k}''} a^{jj'j''}_{\boldsymbol{k}\boldsymbol{k}'\boldsymbol{k}''}\,\xi^{j'}_{\boldsymbol{k}'}\xi^{j''}_{\boldsymbol{k}''}\, I_{\boldsymbol{k},\boldsymbol{k}',\boldsymbol{k}''}\\ &+ \sum_{j',j'',j'''}\sum_{\boldsymbol{k}',\boldsymbol{k}'',\boldsymbol{k}'''} b^{jj'j''j'''}_{\boldsymbol{k}\boldsymbol{k}'\boldsymbol{k}''\boldsymbol{k}'''}\,\xi^{j'}_{\boldsymbol{k}'}\xi^{j''}_{\boldsymbol{k}''}\xi^{j'''}_{\boldsymbol{k}'''}\, J_{\boldsymbol{k},\boldsymbol{k}',\boldsymbol{k}'',\boldsymbol{k}'''}\end{aligned} \tag{10.60}$$

where we have kept only the important terms including up to third order.

The integrals I, J are given by

$$I_{\boldsymbol{k},\boldsymbol{k}',\boldsymbol{k}''} = \frac{1}{L_1 L_2}\int_F d^2\boldsymbol{x}\, e^{i(\boldsymbol{k}'+\boldsymbol{k}''-\boldsymbol{k})\boldsymbol{x}} = \delta_{\boldsymbol{k},\boldsymbol{k}'+\boldsymbol{k}''}\,, \tag{10.61}$$

$$J_{\boldsymbol{k},\boldsymbol{k}',\boldsymbol{k}'',\boldsymbol{k}'''} = \frac{1}{L_1 L_2}\int_F d^2\boldsymbol{x}\, e^{i(\boldsymbol{k}'+\boldsymbol{k}''+\boldsymbol{k}'''-\boldsymbol{k})\boldsymbol{x}} = \delta_{\boldsymbol{k},\boldsymbol{k}'+\boldsymbol{k}''+\boldsymbol{k}'''}\,. \tag{10.62}$$

We eliminate the stable modes as in Section 7.7. The great advantage of the "slaving principle" consists in an enormous reduction of the degrees of freedom because we keep now only the unstable modes with index $\boldsymbol{k}=\boldsymbol{k}_c$. These modes serve as everywhere in this book as *order parameters*. Their cooperation or competition determines the resulting patterns as we will show now. We introduce a new notation by which we replace the vector $\boldsymbol{k}_c$ by its modulus and the angle φ which this vector forms with a fixed axis: $\xi_{\boldsymbol{k}_c}\to\xi_{\boldsymbol{k}_c,\varphi}$. We let φ run from 0 to π. The resulting order parameter equations read

$$\dot{\xi}^+_{k_c,\varphi} = \lambda\xi^+_{k_c,\varphi} + c\,\xi^+_{k_c,\varphi+\frac{\pi}{3}}\xi^+_{k_c,\varphi-\frac{\pi}{3}} + \xi^+_{k_c,\varphi}\sum_{\varphi'} d(|\varphi-\varphi'|)\,|\xi^+_{k_c,\varphi'}|^2\,. \tag{10.63}$$

λ is proportional to $(\rho-\rho_c)$, whereas C can be considered as independent of ρ. The constants $d(|\varphi-\varphi'|)$ have been evaluated by computer calculations and are exhibited in Fig. 10.5. Eq. (10.63) represents a set of coupled equations for the time dependent functions $\xi^+_{k_c,\varphi}$. These equations can be written in the form of potential equations

$$\dot{\xi}_\varphi = -\frac{\partial V}{\partial\xi^*_\varphi}\,,\qquad \xi^+_{k_c,\varphi}\equiv\xi_\varphi \tag{10.64}$$

where the potential function V is given by

$$\begin{aligned}V = -\sum_{\varphi=0}^{\pi}\Bigg[\lambda\,|\xi_\varphi|^2 &+ \frac{c}{3}\left(\xi_\varphi\xi_{\varphi+\frac{\pi}{3}}\xi_{\varphi-\frac{\pi}{3}} + \text{c.c.}\right)\\ &+ \tfrac{1}{2}\,|\xi_\varphi|^2\sum_{\varphi'=0}^{\pi} d(|\varphi-\varphi'|)\,|\xi_{\varphi'}|^2\Bigg].\end{aligned} \tag{10.65}$$

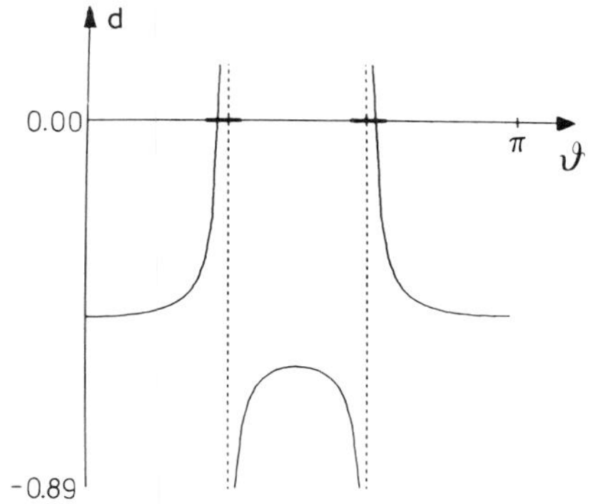

Fig. 10.5. $d(\vartheta)$ is plotted as the function of ϑ. In the practical calculations the region of divergence is cut out as indicated by the bars. This procedure can be justified by the buildup of wave packets

We may assume as elsewhere in this book (cf. especially Sect. 8.13) that the eventually resulting pattern is determined by such a configuration of ξ's where the potential V acquires a (local) minimum. Thus we have to seek such ξ's for which

$$\frac{\partial V}{\partial \xi_\varphi^*} = 0 \tag{10.66}$$

and

$$\sum_{\varphi,\varphi'} \frac{\partial V}{\partial \xi_\varphi \partial \xi_{\varphi'}^*} \delta \xi_\varphi \delta \xi_{\varphi'}^* > 0 \, . \tag{10.67}$$

The system is globally stable if

$$d(|\varphi - \varphi'|) < 0 \tag{10.68}$$

or if the matrix

$$-d(|\varphi - \varphi'|)$$

has only positive eigenvalues.

We discuss three typical examples, which are strongly reminiscent of pattern formation in hydrodynamics (cf. Sect. 8.13). 1) The entirely homogeneous state for which all ξ_φ are equal to 0 is stable for $\lambda < 0$.
2) We obtain a "roll" pattern if

$$\xi_{\varphi_1} = x_1 \, , \tag{10.69}$$

$$\xi_\varphi = 0 \quad \text{for} \quad \varphi \neq \varphi_1 \tag{10.70}$$

where

$$x_1^2 = -\frac{\lambda}{d(\pi)} \, . \tag{10.71}$$

The angle φ_1 between the roll axis and a fixed axis is arbitrary, i.e., symmetry breaking with respect to φ occurs. This configuration is locally stable for

$$\lambda>0, \quad d(\vartheta+\pi)<d(\pi)<0\,. \tag{10.72}$$

The resulting spatial pattern can be obtained by inserting (10.69, 70) into (10.55). In our present treatment and in the corresponding figures we neglect the impact of the slaved modes. They would give rise to a slight sharpening of the individual peaks. The corresponding pattern is exhibited in Fig. 10.6.

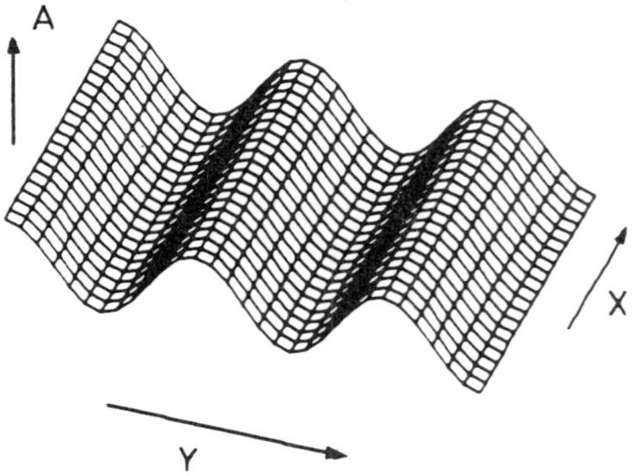

Fig. 10.6. A roll-type pattern

3) Another pattern, in which we find hexagons, is realized when the minimum of V is attained for

$$\xi_{\varphi_1}=\xi_{\varphi_1+(\pi/3)}=\xi_{\varphi_1-(\pi/3)}=x_1\,, \tag{10.73}$$

$$\xi_\varphi=0 \quad \text{otherwise.} \tag{10.74}$$

x_1 is given by

$$x_1=\frac{1}{2F}(-c\pm\sqrt{c^2-4F\lambda}), \quad F=d(\pi)+2d\left(\frac{\pi}{3}\right). \tag{10.75}$$

This configuration is locally stable for

$$\begin{aligned} &F_\varphi<F<0\,, \quad cx_1>\frac{c^2}{-F}\,, \\ &F_\varphi=d(|\varphi-\varphi_1|)+d\left(\left|\varphi-\varphi_1+\frac{\pi}{3}\right|\right)+d\left(\left|\varphi-\varphi_1-\frac{\pi}{3}\right|\right). \end{aligned} \tag{10.76}$$

The bifurcation diagram of the solution (10.75) is shown in Fig. 10.7. The solid line indicates the stable configuration, the dashed line the unstable configuration. The order parameter equations allow us to determine not only the stationary

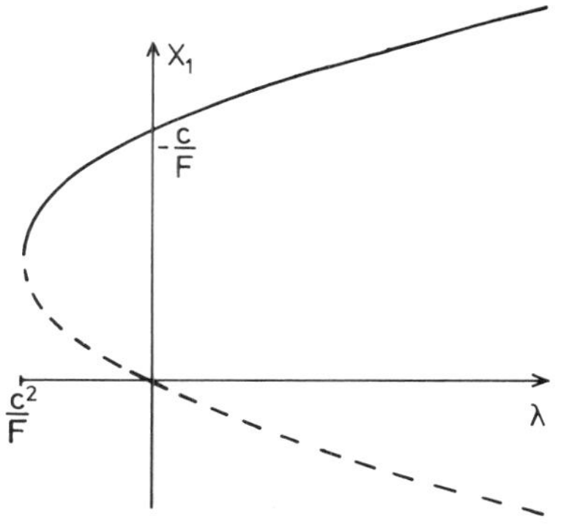

Fig. 10.7. The amplitude x_1 as a function of λ. The solid line indicates a stable solution, the dotted line an unstable solution

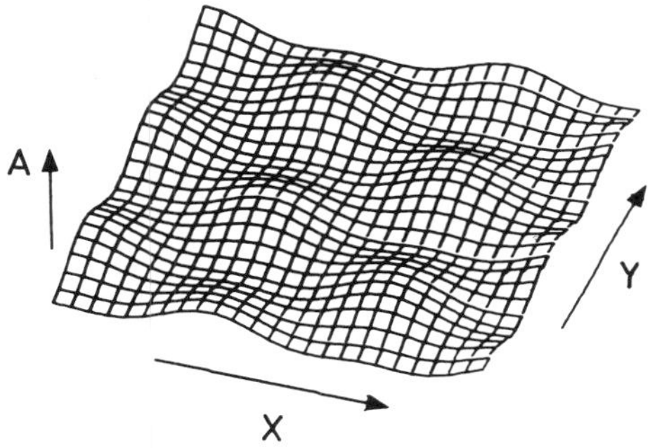

Fig. 10.8

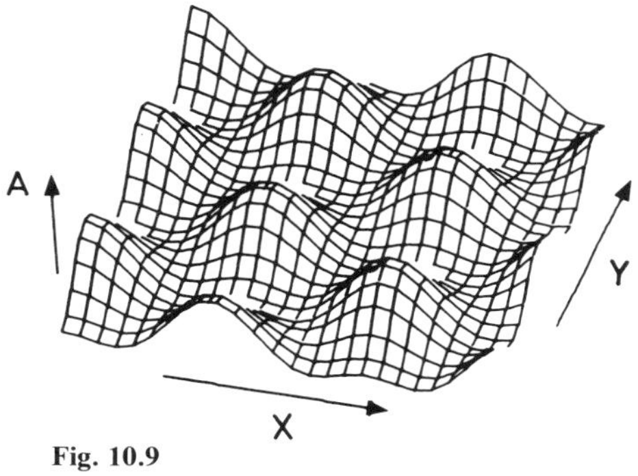

Fig. 10.9

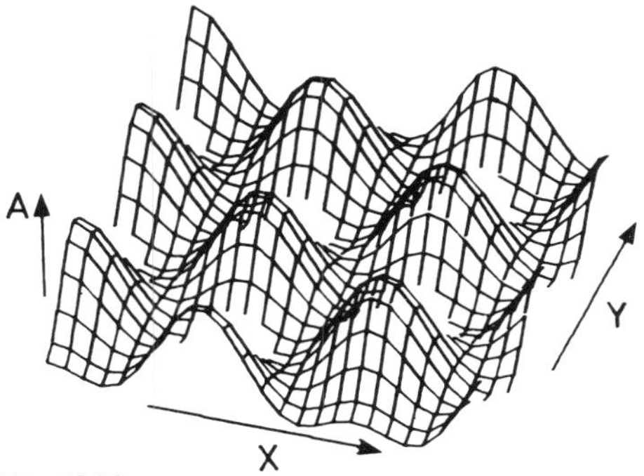

Fig. 10.10

Figs. 10.8. to 10.10. Buildup of a hexagonal pattern of activator concentration for three subsequent times

solution but also the transient. We start from a homogeneous solution on which a small inhomogeneity of the form (10.73, 74) is superimposed and solve the time-dependent equations (10.63). The resulting solutions of the spatial pattern are exhibited in Figs. 10.8, 10.9, and 10.10. This shows again the usefulness of order parameter equations.

In our next example we present the solution of the nonlinear equations within a rectangular domain with *nonflux boundary conditions* (close to the instability point). We now expand the wanted solution into a complete orthogonal system with nonflux boundary conditions, i.e., with respect to functions of the form

$$\cos k_x x \cdot \cos k_y y, \quad \text{where} \quad \begin{pmatrix} k_x \\ k_y \end{pmatrix} = \pi \begin{pmatrix} \frac{n}{L_1} \\ \frac{m}{L_2} \end{pmatrix}. \tag{10.77}$$

The procedure goes through in complete analogy to the above. k_x and k_y must be chosen so that $k_x^2 + k_y^2$ comes close to k_c^2. When $L_1 \approx L_2$, different modes may simultaneously become unstable ("degeneracy"). For simplicity, we treat here the case of a *single* unstable mode (i.e., $L_1 \neq L_2$). Its amplitude obeys the equation

$$\dot{\xi} = \lambda \xi + d' \xi^3 . \tag{10.78}$$

The solution of this time-dependent equation describes the growth of the spatial pattern. Examples of the resulting patterns are given by Figs. 10.11 to 10.13.

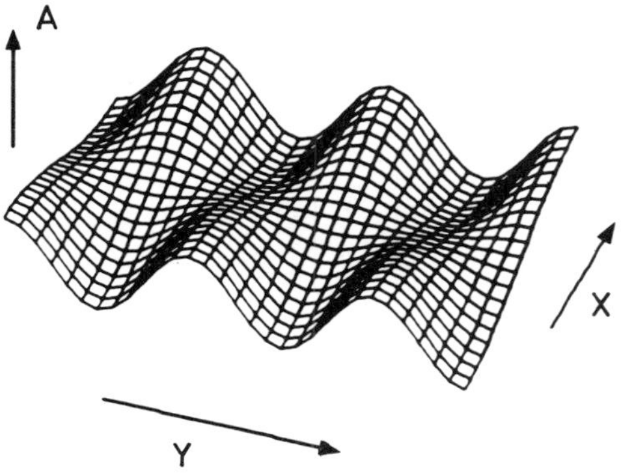

Fig. 10.11. The activator concentration belonging to the mode (10.77) with $k_x = \pi/L_1$ and $k_y = 5\pi/L_2$

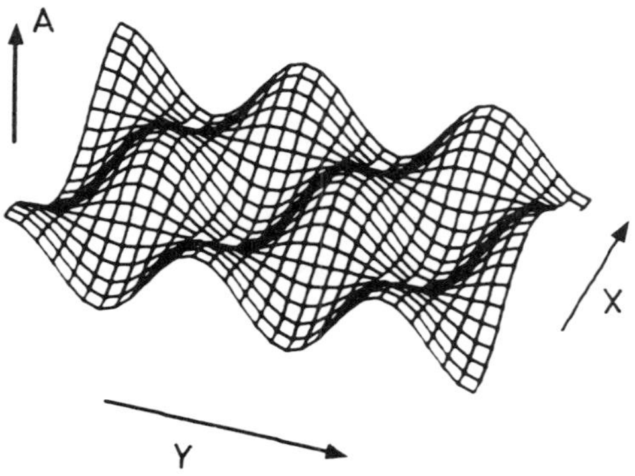

Fig. 10.12. The activator concentration belonging to the mode (10.77) with $k_x = 2\pi/L_1$ and $k_y = 5\pi/L_2$

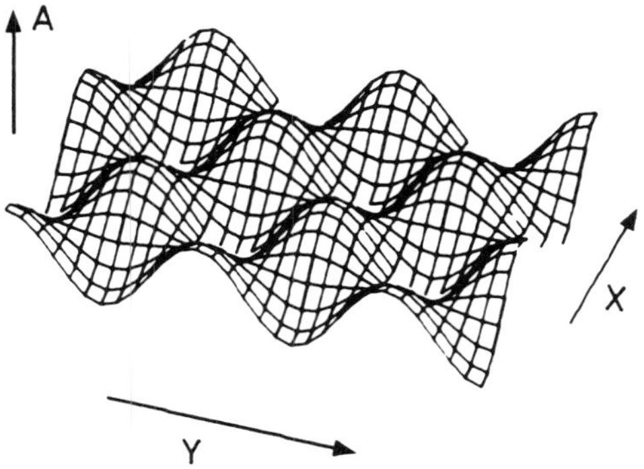

Fig. 10.13. The activator concentration belonging to the mode (10.77) with $k_x = 3\pi/L_1$ and $k_y = 5\pi/L_2$

Our last example refers to cylindrical nonflux boundary conditions. In this case we introduce polar coordinates r and φ and replace the formerly used plane waves of the expansion (10.55) by cylinder functions of the form $e^{im\varphi} J_m(kr)$, where $J_m(kr)$ is the Bessel function. The nonflux boundary condition requires

$$\left. \frac{\partial}{\partial r} J_m(kr) \right|_{r=R} = 0 . \tag{10.79}$$

This equation fixes a series of k-values for which (10.79) is fulfilled. The expansion of $\boldsymbol{q}$ now reads

$$\boldsymbol{q} = \textstyle\sum_j \sum_k \boldsymbol{O}^j(k) \sum_m \xi^j_{k,m}(t) e^{im\varphi} J_m(kr) . \tag{10.80}$$

Inserting (10.80) into the original equation (10.41) leads eventually to equations for the ξ's. The slaving principle allows us to do away the stable modes and we thus obtain equations for the order parameter alone, since we have a discrete sequence of k-values and for symmetry reasons we may assume that only one mode becomes unstable first. The resulting order parameter equation has again the form (10.78).

A marked difference occurs depending on whether the unstable mode (order parameter) has $m \equiv m_c = 0$ or $\neq 0$. If $m_c = 0$ the right hand side of equation (10.78) must be supplemented by a term quadratic in ξ which is absent if $m_c \neq 0$. As is well known from phase transition theory (cf. Sect. 6.7), in the former case ($m_c = 0$) we obtain a first-order phase transition connected with an abrupt change of the homogeneous state into the inhomogeneous state connected with hysteresis effects. In the latter case ($m_c \neq 0$) we obtain a second-order phase transition, and the pattern grows continuously out of the homogeneous state when ρ passes through ρ_c. Some typical patterns are exhibited in Figs. 10.14, to 10.16.

In conclusion we mention that we can also take into account small band excitations and fluctuations in complete analogy to the Bénard instability (cf. Sects. 8.13 and 8.14) using the methods developed in this book. A comparison between the figures of this section and those of Section 10.3 shows a qualitative resemblance but no exact agreement. The reason for this lies in the fact that the analytical

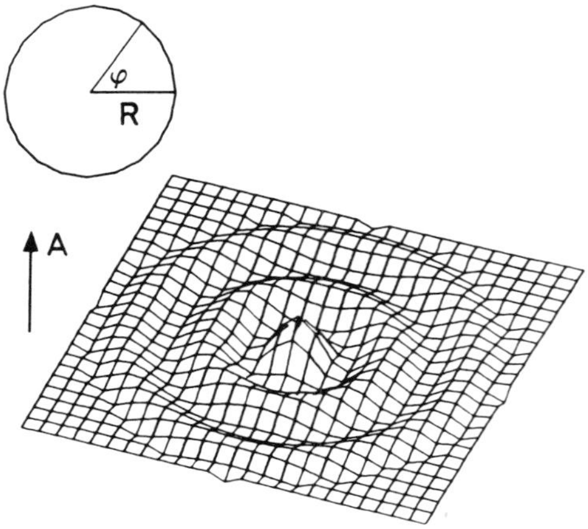

Fig. 10.14. The activator concentration with cylindrical nonflux boundary conditions described by a rotation symmetric Bessel function with $m=0$

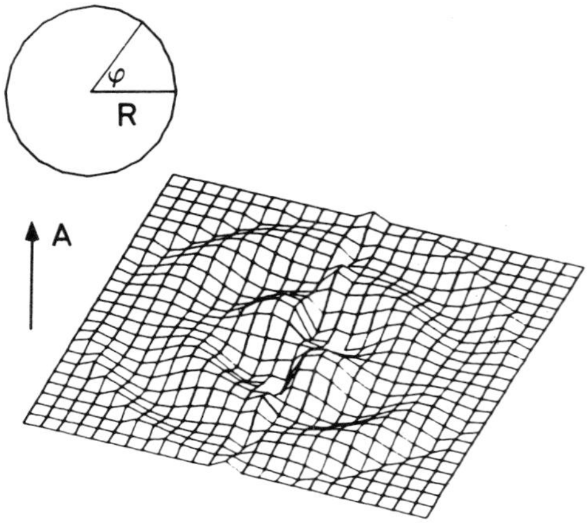

Fig. 10.15. The activator concentration with cylindrical nonflux boundary conditions described by a rotation symmetric Bessel function with $m=1$

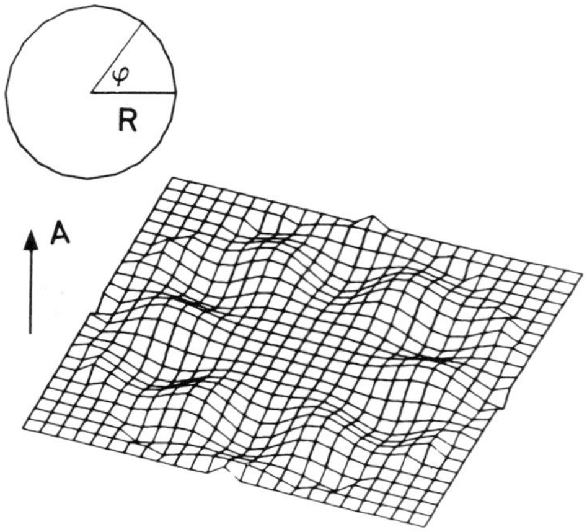

Fig. 10.16. The activator concentration with cylindrical nonflux boundary conditions described by a rotation symmetric Bessel function with $m=3$

approach yields "pure cases", whereas the computer solution makes use of (artificially) introduced random fluctuations. As we know, in that latter case the analytical approach would yield a probability distribution (of patterns) of a whole ensemble, whereas a computer solution will be equivalent to a "simple event", i.e., a specific realization.

10.6 Some Comments on Models of Morphogenesis

Present-day modelling of morphogenesis is based on the idea that by diffusion and reaction of certain chemicals a prepattern (or "morphogenetic field") is formed. This prepattern switches genes on to cause cell differentiation. Such models are partly substantiated by direct observation of certain chemicals, for instance the neural growth factor. On the other hand it might be necessary to consider also other mechanisms of cell communications, for instance cell-cell contacts using recognition sites.

Aside from this comment, looking at the general outline of the present book we can draw the following important conclusion. On the one hand we have seen that a single model, for instance in hydrodynamics or now in morphogenesis, can produce quite different patterns depending on the individual parameters, on boundary conditions, and on fluctuations. On the other hand, quite different systems may produce the same pattern, for instance hexagons. As a consequence we must conclude that different models of morphogenetic processes may lead to the same resulting pattern. In each case there exists a total *class of models* (differential equations) giving rise to the *same* pattern. For this reason it seems very important to develop other criteria in morphogenesis to decide which kind of model is adequate from the theoretical point of view. This can be done, for instance by invoking general principles about fundamental processes, for example the principle of "long range inhibition and short range activation". Possibly other mechanisms or principles will also have to be discussed in future developments. Furthermore the importance of experiments to decide between different mechanisms is quite evident.

In view of the general spirit of the present book the following phase transition analogy seems to be interesting:

physical system	**biological system**
full symmetry	totipotent cells
symmetry breaking	cell differentiation
first-order transition	irreversible change

This analogy suggests that cell differentiation might occur spontaneously in very much the same way as a ferromagnet acquires its spontaneous magnetization. Unfortunately, lack of space does not allow us to elaborate further on this interesting analogy, which, of course, is purely formal.

Next steps to morphogenetic models might include the morphogenesis of neural nets taking into account irreversible storage of information, i.e., the formation of the long time memory or, more generally, the process of learning and its connection with the formation of, for instance, chemical patterns in the brain.

Let us conclude with the following remark. In this book we have stressed the profound analogies between quite different systems, and one is tempted to treat biological systems in complete analogy to physical or chemical systems far from thermal equilibrium. One important difference should be stressed, however. While the physical and chemical systems under consideration lose their structure when the flux of energy and matter is switched off, much of the structure of biological systems is still preserved for an appreciable time. Thus biological systems seem rather to combine nondissipative and dissipative structures. Furthermore, biological systems serve certain purposes or tasks and it will be more appropriate to consider them as *functional structures*. Future research will have to develop adequate methods to cope with such functional structures. It can be hoped, however, that the ideas and methods outlined in this book may serve as a first step in that direction.

11. Sociology: A Stochastic Model for the Formation of Public Opinion

Intuitively it is rather obvious that formation of public opinion, actions of social groups, etc., are of a cooperative nature. On the other hand it appears extremely difficult if not impossible to put such phenomena on a rigorous basis because the actions of individuals are determined by quite a number of very often unknown causes. On the other hand, within the spirit of this book, we have seen that in systems with many subsystems there exist at least two levels of description: One analysing the individual system and its interaction with its surrounding, and the other one describing the statistical behavior using macroscopic variables. It is on this level that a quantitative description of interacting social groups becomes possible.

As a first step we have to seek the macroscopic variables describing a society. First we must look for the relevant, characteristic features, for example of an opinion. Of course "the opinion" is a very weak concept. However, one can measure public opinion, for example by polls, by votes, etc. In order to be as clear as possible, we want to treat the simplest case, that of only two kinds of opinions denoted by plus and minus. An obvious order parameter is the number of individuals n_+, n_- with the corresponding opinions $+$ and $-$, respectively. The basic concept now to be introduced is that the formation of the opinion, i.e., the change of the numbers n_+, n_- is a cooperative effect: The formation of an individual's opinion is influenced by the presence of groups of people with the same or the opposite opinion. We thus assume that there exists a probability per unit time, for the change of the opinion of an individual from plus to minus or vice versa. We denote these transition probabilities by

$$p_{+-}(n_+, n_-) \text{ and } p_{-+}(n_+, n_-). \tag{11.1}$$

We are interested in the probability distribution function $f(n_+, n_-, t)$. One may easily derive the following master equation

$$\begin{aligned} \frac{df[n_+, n; t]}{dt} &= (n_+ + 1)p_{+-}[n_+ + 1, n_- - 1]f[n_+ + 1, n_- - 1; t] \\ &+ (n_- + 1)p_{-+}[n_+ - 1, n_- + 1]f[n_+ - 1, n_- + 1; t] \\ &- \{n_+p_{+-}[n_+, n_-] + n_-p_{-+}[n_+, n_-]\}f[n_+, n_-; t]. \end{aligned} \tag{11.2}$$

The crux of the present problem is, of course, not so much the solution of this equation which can be done by standard methods but the determination of the transition probability. Similar to problems in physics, where not too much is known about the individual interaction, one may now introduce plausibility arguments to derive p. One possibility is the following: Assume that the rate of change of the opinion of an individual is enhanced by the group of individuals with an opposite opinion and diminished by people of his own opinion. Assume furthermore that there is some sort of social overall climate which facilitates the change of opinion or makes it more difficult to form. Finally one can think of external influences on each individual, for example, informations from abroad etc. It is not too difficult to cast these assumptions into a mathematical form, if we think of the Ising model of the ferromagnet. Identifying the spin direction with the opinion $+$, $-$, we are led to put in analogy to the Ising model

$$\begin{aligned} p_{+-}[n_+, n_-] \equiv p_{+-}(q) &= \nu \exp\left\{\frac{-(Iq + H)}{\Theta}\right\} \\ &= \nu \exp\{-(kq + h)\}, \\ p_{-+}[n_+, n_-] \equiv p_{-+}(q) &= \nu \exp\left\{\frac{+(Iq + H)}{\Theta}\right\} \\ &= \nu \exp\{+(kq + h)\}, \end{aligned} \tag{11.3}$$

where I is a measure of the strength of adaptation to neighbours. H is a preference parameter ($H > 0$ means that opinion $+$ is preferred to $-$), Θ is a collective climate parameter corresponding to $k_B T$ in physics (k_B is the Boltzmann constant and T the temperature), ν is the frequency of the "flipping" processes. Finally

$$q = (n_+ - n_-)/2n, \quad n = n_+ + n_-. \tag{11.4}$$

For a quantitative treatment of (11.2) we assume the social groups big enough so that q may be treated as a continuous parameter. Transforming (11.2) to this continuous variable and putting

$$\begin{aligned} w_{+-}(q) &\equiv n_+ p_{+-}[n_+, n_-] = n(\tfrac{1}{2} + q)p_{+-}(q), \\ w_{-+}(q) &\equiv n_- p_{-+}[n_+, n_-] = n(\tfrac{1}{2} - q)p_{-+}(q), \end{aligned} \tag{11.5}$$

we transform (11.2) into a partial differential equation (see, e.g., Section 4.2.) Its solution may be found by quadratures in the form

$$f_{st}(q) = cK_2^{-1}(q) \exp\left\{2 \int_{-\frac{1}{2}}^{q} \frac{K_1(y)}{K_2(y)} dy\right\} \tag{11.6}$$

with

$$\begin{aligned} K_1(q) &= \nu\{\sinh(kq + h) - 2q \cosh(kq + h)\} \\ K_2(q) &= (\nu/n)\{\cosh(kq + h) - 2q \sinh(kq + h)\}. \end{aligned} \tag{11.7}$$

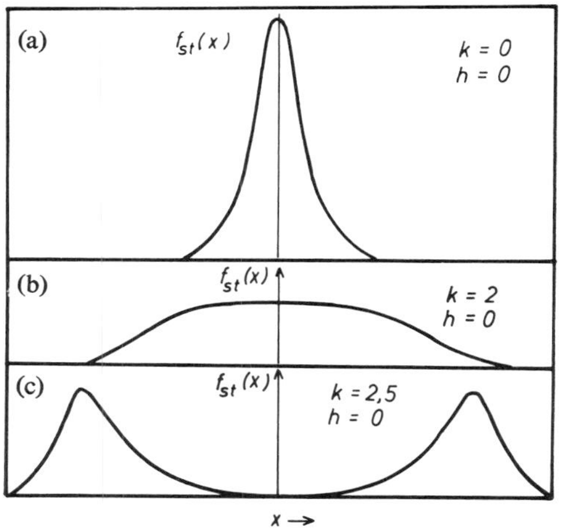

Fig. 11.1. (a) Centered distribution in the case of rather frequent changes of opinion (independent decision), (b) Distribution at the transition between independent and strongly adaptive decision, (c) "Polarization phenomenon" in the case of strong neighbor-neighbor interaction. (After *W. Weidlich*: Collective Phenomena **1**, 51 (1972))

Fig. 11.1 shows a plot of the result when there is no external parameter. As one may expect from a direct knowledge of the Ising model, there are typically two results. The one corresponds to the high-temperature limit: on account of rather frequent changes of opinion we find a centered distribution of opinions. If the social climate factor Θ is lowered or if the coupling strength between individuals is increased, two pronounced groups of opinions occur which clearly describe the by now well-known "polarization phenomenon" of society. It should be noted that the present model allows us to explain, at least in a qualitative manner, further processes, for example unstable situations where the social climate parameter is changed to a critical value. Here suddenly large groups of a certain opinion are formed which are dissolved only slowly and it remains uncertain which group (+ or −) finally wins. Using the considerations of Section 6.7 it is obvious again that here concepts of phase transition theory become important, like critical slowing down (Remember the duration of the 1968 French student revolution?), critical fluctuations, etc. Such statistical descriptions certainly do not allow unique predictions due to the stochastic nature of the process described. Nevertheless, such models are certainly most valuable for understanding general features of cooperative behavior, even that of human beings, though the behavior of an individual may be extremely complicated and not accessible to a mathematical description. Quite obviously the present model allows for a series of generalizations.

12. Chaos

12.1 What is Chaos?

Sometimes scientists like to use dramatic words of ordinary language in their science and to attribute to them a technical meaning. We already saw an example in Thom's theory of "catastrophes". In this chapter we become acquainted with the term "chaos". The word in its technical sense refers to irregular motion. In previous chapters we encountered numerous examples for *regular* motions, for instance an entirely periodic oscillation, or the regular occurrence of spikes with well-defined time intervals. On the other hand, in the chapters about Brownian motion and random processes we treated examples where an irregular motion occurs due to random, i.e., in principle unpredictable, causes. Surprisingly the *irregular* motion represented in Fig. 12.1 stems from completely *deterministic equations*. To

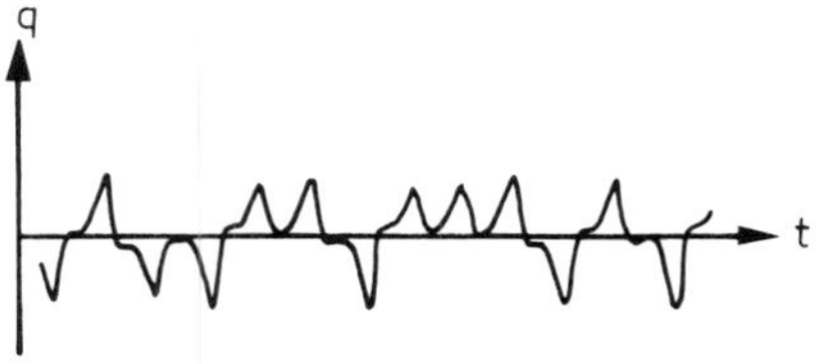

Fig. 12.1. Example of chaotic motion of a variable q (versus time)

characterize this new phenomenon, we define chaos as *irregular motion stemming from deterministic equations*. The reader should be warned that somewhat different definitions of chaos and the criteria to check its occurrence are available in the literature. The difficulty rests mainly in the problem of how to define "irregular motion" properly. For instance, the superposition of motions with different frequencies could mimic to some extent an irregular behavior and one wants to preclude such a case from representing chaos. We shall come back to this question in Section 12.5, where we shall discuss the typical behavior of the correlation function of chaotic processes. A good deal of present-day analysis of "chaos" rests on computer calculations.

In the following we shall present one of the most famous examples, namely the so-called Lorenz model of turbulence, which reveals some of the most interesting features of chaos.

12.2 The Lorenz Model. Motivation and Realization

In this section we describe how the Lorenz model can be motivated. It turns out that it is an instructive but not quite realistic model for turbulence in fluids. It can probably be realized in lasers and spin systems. The reader who is more interested in the mathematics than in the physics of this model is advised to go straight to Section 12.3.
A long-standing and still unsolved problem is the explanation of turbulence in fluids. The original purpose of the Lorenz equations is to provide a model for turbulence. To this end we recall briefly some of the results of Section 8.12 on the Bénard instability. There we have seen that out of the quiescent state first a single mode, namely a certain component of the velocity field in vertical direction, becomes unstable. (Here we neglect mode degeneracies with respect to the horizontal direction). This "mode" serves as order parameter. As one may show in more detail, this mode slaves in particular two further modes which are connected with temperature deviations. To obtain equations for the amplitudes of these three variables in a systematic manner we first decompose the components of the velocity field into a Fourier series

$$u_j(x,y,z)=i\sum_{l,m,n=-\infty}^{\infty} u_j(l,m,n)\exp\{i(k_1 lx+k_2 my+n\pi z)\} \tag{12.1}$$

and similarly the temperature deviation field. We then insert these expressions into the Navier-Stokes equations (in the Boussinesq approximation) and keep only the three terms

$$u_1(1,0,1)\equiv X, \quad \Theta(1,0,1)\equiv Y, \quad \Theta(0,0,2)\equiv Z\,. \tag{12.2}$$

After some analysis and using properly scaled variables we obtain the Lorenz equations

$$\dot{X}=\sigma Y-\sigma X\,, \tag{12.3}$$

$$\dot{Y}=-XZ+rX-Y, \tag{12.4}$$

$$\dot{Z}=X\cdot Y-bZ\,. \tag{12.5}$$

$\sigma=\nu/\varkappa'$ is the Prandtl number (where ν is the kinematic viscosity, $\varkappa'$ the thermometric conductivity), $r=R/R_c$ (where R is the Rayleigh number, R_c the critical Rayleigh number), $b=4\pi^2/(\pi^2+k_1^2)$. When we put

$$X=\xi, \quad Y=\eta, \quad Z=r-\zeta\,,$$

(12.3)–(12.5) acquire the form

$$\dot{\xi}=\sigma\eta-\sigma\xi, \quad \dot{\eta}=\xi\zeta-\eta, \quad \dot{\zeta}=b(r-\zeta)-\xi\eta\,. \tag{12.6}$$

Astonishingly equations entirely equivalent to the set of equations (12.3–5) occur in laser physics. We start from the laser equations (8.94–96) for the field strength E, the polarization P, and the inversion D (in properly chosen units). We assume single mode operation by putting $\partial E/\partial x=0$. In the following we assume that E and P are real quantities, i.e., that their phases can be kept constant, which can be proved by computer calculations. The thus resulting equations read

$$\dot{E}=\varkappa P-\varkappa E\,, \tag{12.7}$$

$$\dot{P}=\gamma ED-\gamma P\,, \tag{12.8}$$

$$\dot{D}=\gamma_{\parallel}(\Lambda+1)-\gamma_{\parallel}D-\gamma_{\parallel}\Lambda EP\,. \tag{12.9}$$

These equations are identical with those of the Lorenz model in the form (12.6) which can be realized by the following identifications

$$t\to t'\sigma/\varkappa,\quad E\to\alpha\xi \quad \text{where} \quad \alpha=\{b(r-1)\}^{-1/2},\quad r>1$$

$$P\to\alpha\eta,\quad D\to\zeta,\quad \gamma_{\parallel}=\varkappa b/\sigma,\quad \gamma=\varkappa/\sigma,\quad \Lambda=r-1.$$

In particular the following correspondence holds:

Table 12.1

Bénard problem	**Laser**
σ: Prandtl number	$\sigma=\varkappa/\gamma$
$r=\frac{R}{R_c}$ (R Rayleigh number)	$r=\Lambda+1$
$b=\frac{4\pi^2}{\pi^2+k_1^2}$, $k_1^2=k_c^2=\frac{\pi^2}{2}$	$b=\gamma_{\parallel}/\gamma$

Eqs. (12.6) describe at least two instabilities which have been found independently in fluid dynamics and in lasers. For $\Lambda<0$ ($r<1$) there is no laser action (the fluid is at rest), for $\Lambda\geq 0$ ($r\geq 1$) laser action (convective motion starts) with stable, time-independent solutions ξ,η,ζ occurs. As we will see in Section 12.3, besides this well known instability a new one occurs provided

$$\text{laser:}\quad \varkappa>\gamma+\gamma_{\parallel} \qquad \text{fluid:}\quad \sigma>b+1 \tag{12.10}$$

$$\text{and} \qquad\qquad \text{and}$$

$$\Lambda>(\gamma+\gamma_{\parallel}+\varkappa)(\gamma+\varkappa)/\gamma(\varkappa-\gamma-\gamma_{\parallel}) \qquad r>\sigma(\sigma+b+3)/(\sigma-1-b). \tag{12.11}$$

This instability leads to the irregular motion, an example of which we have shown in Fig. 12.1. When numerical values are used in the conditions (12.10) and (12.11) it turns out that the Prandtl number must be so high that it cannot be realized by realistic fluids. On the other hand, in lasers and masers it is likely that conditions (12.10) and (12.11) can be met. Furthermore it is well known that the two-level atoms used in lasers are mathematically equivalent to spins. Therefore, such phenomena may also be obtainable in spin systems coupled to an electromagnetic field.

12.3 How Chaos Occurs

Since we can scale the variables in different ways, the Lorenz equations occur in different shapes. In this section we shall adopt the following form

$$\dot{q}_1 = -\alpha q_1 + q_2 , \tag{12.12}$$

$$\dot{q}_2 = -\beta q_2 + q_1 q_3' , \tag{12.13}$$

$$\dot{q}_3' = d_0' - q_3' - q_1 q_2 . \tag{12.14}$$

These equations result from (12.3–5) by the scaling

$$X = b q_1 ; \quad Y = \frac{b^2}{\sigma} q_2 ; \quad Z = r - \frac{b^2}{\sigma} q_3' ; \quad t = \frac{1}{b} t' ;$$

$$\alpha = \frac{\sigma}{b} ; \quad d_0' = r \frac{\sigma}{b^2} ; \quad \beta = \frac{1}{b} . \tag{12.15}$$

The stationary state of (12.12–14) with $\dot{q}_1, \dot{q}_2, \dot{q}_3' = 0$ is given by

$$q_1^0 = \pm \sqrt{\frac{1}{\alpha}(d_0' - \alpha\beta)} , \quad q_2^0 = \pm \sqrt{\alpha(d_0' - \alpha\beta)} , \quad q_3'^0 = \alpha\beta . \tag{12.16}$$

A linear stability analysis reveals that the stationary solution becomes unstable for

$$d_0' = \alpha^2 \beta \cdot \frac{\alpha + 3\beta + 1}{\alpha - \beta - 1} . \tag{12.17}$$

In this domain, (12.12–14) were solved by computer calculations. An example of one variable was shown in Fig. 12.1. Since X, Y, Z span a three-dimensional space, a direct plot of the trajectory is not possible. However, Fig. 12.2 shows projections of the trajectory on two planes. Apparently the representative point (X, Y, Z), or (q_1, q_2, q_3'), circles first in one part of space and then suddenly jumps to another part where it starts a circling motion again. The origin of that kind of behavior, which is largely responsible for the irregular motion, can be visualized as follows. From laser physics we know that the terms on the right hand side of (12.12–14) have a different origin. The last terms

$$q_2 ,$$

$$q_1 \cdot q_3' ,$$

$$-q_1 \cdot q_2$$

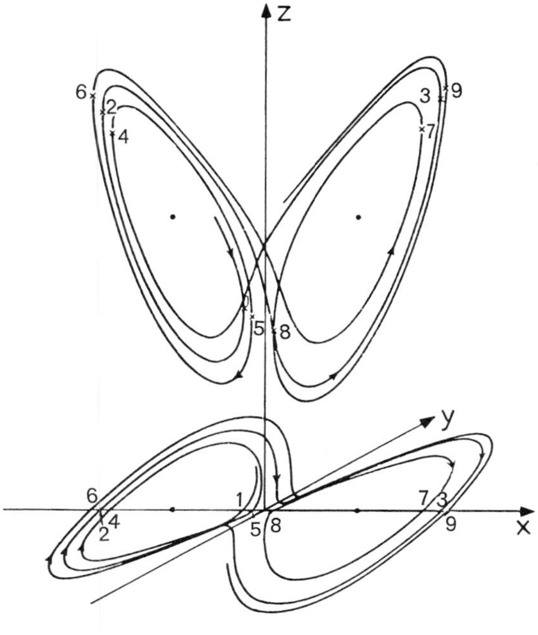

Fig. 12.2. Upper half: Trajectories projected on the $X-Z$ plane.
Lower half: Trajectories projected on the $X-Y$ plane.
The points represent the steady state solution (after *M. Lücke*)

stem from the coherent interaction between atoms and field. As is known in laser physics coherent interaction allows for two conservation laws, namely energy conservation and conservation of the total length of the so-called pseudo spin. What is of relevance for us in the following is this. The conservation laws are equivalent to two constants of motion

$$R^2 = q_1^2 + q_2^2 + q_3^2, \tag{12.18}$$

$$\rho^2 = q_2^2 + (q_3 - 1)^2, \tag{12.19}$$

where

$$q_3' + 1 = q_3, \qquad d_0' + 1 = d_0. \tag{12.20}$$

On the other hand the first terms in (12.12–14)

$$-\alpha q_1,$$

$$-\beta q_2,$$

$$d_0 - q_3$$

stem from the coupling of the laser system to reservoirs and describe damping and pumping terms, i.e., nonconservative interactions. When we first ignore these terms in (12.12–14) the point (q_1, q_2, q_3) must move in such a way that the conservation laws (12.18) and (12.19) are obeyed. Since (12.18) describes a sphere in q_1, q_2, q_3 space and (12.19) a cylinder, the representative point must move on the cross section of sphere and cylinder. There are two possibilities depending on the relative size of the diameters of sphere and cylinder. In Fig. 12.3 we have two well-separated trajectories, whereas in Fig. 12.4 the representative point can

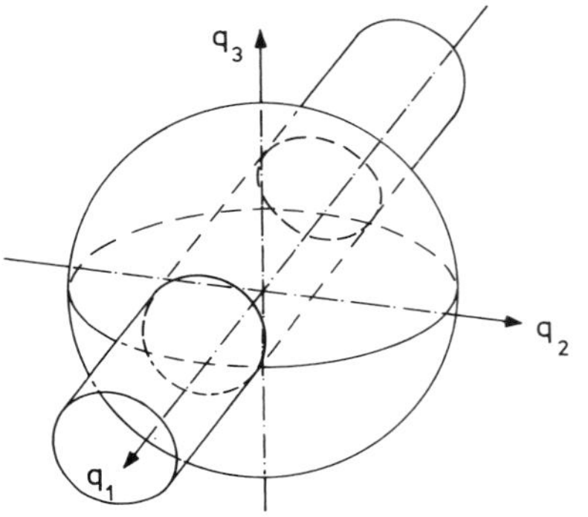

Fig. 12.3. The cross section of the sphere (12.18) and the cylinder (12.19) is drawn for $R>1+\rho$. One obtains two separated trajectories

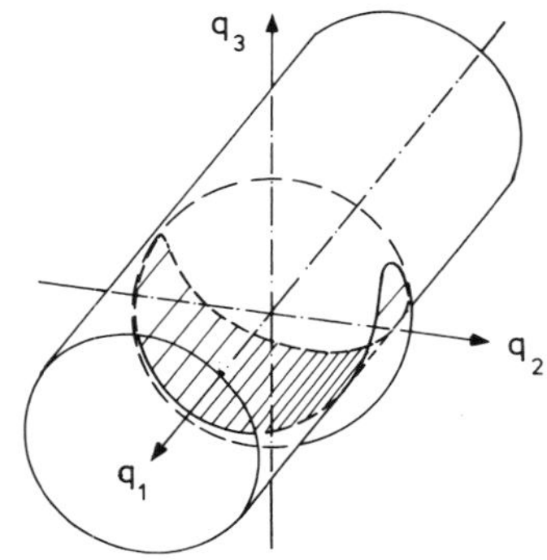

Fig. 12.4. The case $R<1+\rho$ yields a single closed trajectory

move from one region of space continuously to the other region. When we include damping and pump terms the conservation laws (12.18) and (12.19) are no longer valid. The radii of sphere and cylinder start a "breathing" motion. When this motion takes place, apparently both situations of Fig. 12.3 and 12.4 are accessible. In the situation of Fig. 12.3 the representative point circles in one region of space, while it may jump to the other region when the situation of Fig. 12.4 becomes realized. The jump of the representative point depends very sensitively on where it is when the jump condition is fulfilled. This explains at least intuitively the origin of the seemingly random jumps and thus of the random motion. This interpretation is fully substantiated by movies produced in my institute. As we will show below, the radii of cylinder and sphere cannot grow infinitely but are bounded. This implies that the trajectories must lie in a finite region of space. The shape of such a region has been determined by a computer calculation and is shown and explained in Fig. 12.5.

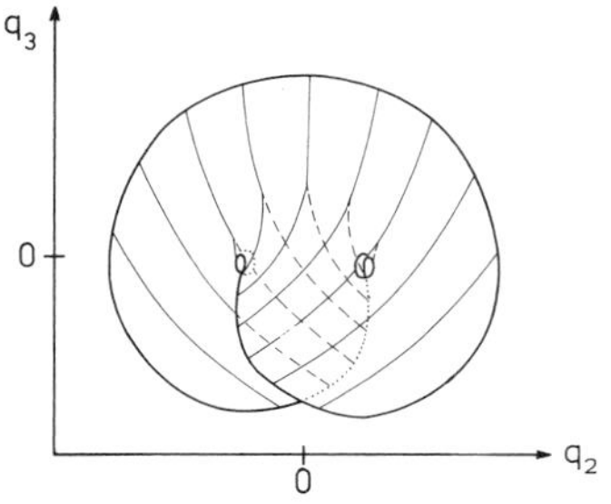

Fig. 12.5. Projection of the Lorenz surface on the q_2, q_3 plane. The heavy solid curve and extensions as dotted curves indicate natural boundaries. The isopleths of q_1 (i.e., lines of constant q_1) as function of q_2 and q_3 are drawn as solid or dashed thin curves (redrawn after *E.N. Lorenz*)

When we let start the representative point with an initial value outside of this region, after a while it enters this region and will never leave it. In other words, the representative point is attracted to that region. Therefore the region itself is called attractor. We have encountered other examples of attracting regions in earlier sections. Fig. 5.11a shows a stable focus to which all trajectories converge. Similarly we have seen that trajectories can converge to a limit cycle. The Lorenz attractor has a very strange property. When we pick a trajectory and follow the representative point on its further way, it is as if we stick a needle through a ball of yarn. It finds its path without hitting (or asymptotically approaching) this trajectory. Because of this property the Lorenz attractor is called a strange attractor.

There are other examples of strange attractors available in the mathematical literature. We just mention for curiosity that some of these strange attractors can be described by means of the so-called Cantor set. This set can be obtained as follows (compare Fig. 12.6). Take a strip and cut out the middle third of it.

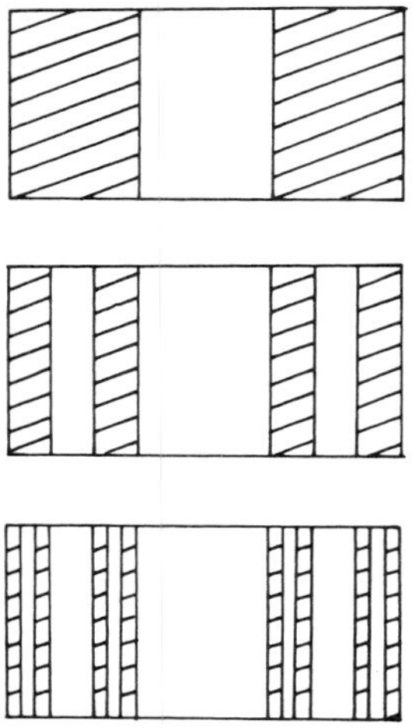

Fig. 12.6. Representation of a Cantor set. The whole area is first divided into three equal parts. The right and left intervals are closed, whereas the interval in the middle is open. We remove the open interval and succeed in dividing the two residual intervals again into three parts, and again remove the open intervals in the middle. We proceed by again dividing the remaining closed intervals into three parts, removing the open interval in the center, etc. If the length of the original area is one, the Lebesgues measure of the set of all open intervals is given by $\sum_{n=1}^{\infty} 2^{n-1}/3^n = 1$. The remaining set of closed intervals (i.e., the Cantor set) therefore has the measure zero

Then cut out the middle third of each of the resulting parts and continue in this way ad infinitum. Unfortunately, it is far beyond the scope of this book to present more details here. We rather return to an analytical estimate of the size of the Lorenz attractor.

The above considerations using the conservation laws (12.18) and (12.19) suggest introduction of the new variables R, ρ, and q_3. The original equations (12.12–14) transform into

$$\tfrac{1}{2}(R^2)^{\cdot} = -\alpha+\beta-\alpha R^2+(\alpha-\beta)\rho^2+(2(\alpha-\beta)+d_0)q_3-(1-\beta)q_3^2\,, \tag{12.21}$$

$$\tfrac{1}{2}(\rho^2)^{\cdot} = -d_0+\beta-\beta\rho^2+(d_0+1-2\beta)q_3-(1-\beta)q_3^2\,, \tag{12.22}$$

$$\dot{q}_3 = d_0-q_3-(\pm)(R^2-\rho^2+1-2q_3)^{1/2}((1+\rho-q_3)(q_3-1+\rho))^{1/2}\,. \tag{12.23}$$

The equation for q_3 fixes allowed regions because the expressions under the roots in (12.23) must be positive. This yields

$$1-\rho<q_3<1+\rho; \qquad R>1+\rho \tag{12.24}$$

(two separate trajectories of a limit cycle form), and

$$1-\rho<q_3<\tfrac{1}{2}(R^2-\rho^2+1); \qquad R<1+\rho \tag{12.25}$$

(a single closed trajectory). Better estimates for R and ρ can be found as follows. We first solve (12.22) formally

$$\rho^2(t)=\rho^2(0)\cdot e^{-2\beta t}-2\frac{d_0-\beta}{2\beta}(1-e^{-2\beta t})+2\int_0^t e^{-2\beta(t-\tau)}g(q_3(\tau))\,d\tau \tag{12.26}$$

with

$$g(q_3(\tau))=Aq_3-Bq_3^2\,, \qquad A=d_0+1-2\beta, \qquad B=1-\beta\,. \tag{12.27}$$

We find an upper bound for the right hand side of (12.26) by replacing g by its maximal value

$$g_{\max}(q_3)=\frac{A^2}{4B}\,. \tag{12.28}$$

This yields

$$\rho^2(t)\le\rho_M^2(t)=\rho^2(0)\cdot e^{-2\beta t}+\frac{(d_0-1)^2}{4\beta(1-\beta)}(1-e^{-2\beta t}). \tag{12.29}$$

For $t\to\infty$, (12.29) reduces to

$$\rho_M^2(\infty)=\frac{(d_0-1)}{4\beta(1-\beta)}\,. \tag{12.30}$$

In an analogous way we may treat (12.21) which after some algebra yields the estimate

$$R_M^2(\infty) = \frac{1}{4\alpha\beta(1-\beta)} \{\alpha d_0^2 + (\alpha-\beta)[2(2\beta-1)d_0 + 1 + 4\beta(\alpha-1)]\} . \qquad (12.31)$$

These estimates are extremely good, as we can substantiate by numerical examples. For instance we have found (using the parameters of Lorenz' original paper)

a) our estimate

$$\rho_M(\infty) = 41.03$$

$$R_M(\infty) = 41.89$$

b) direct integration of Lorenz equations over some time

$$\rho = 32.84$$

$$R = 33.63$$

c) estimate using Lorenz' surface (compare Fig. 12.7)

$$\rho_L = 39.73$$

$$R_L = 40.73 .$$

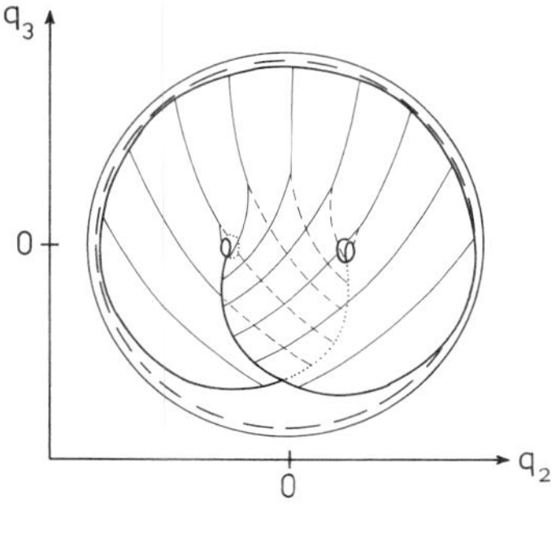

Fig. 12.7. Graphical representation of estimates (12.30), (12.31). Central figure: Lorenz attractor (compare Fig. 12.5). Outer solid circle: estimated maximal radius of cylinder. ρ_M (12.30). Dashed circle: Projection of cylinder (12.29) with radius ρ_L constructed from the condition that at a certain time $q_1 = q_2 = 0$ and $q_3 = d_0$.

12.4 Chaos and the Failure of the Slaving Principle

A good deal of the analysis of the present book is based on the adiabatic elimination of fast relaxing variables, technique which we also call the slaving principle. This slaving principle had allowed us to reduce the number of degrees of freedom considerably. In this sense one can show that the Lorenz equations result from that principle. We may, however, ask if we can again apply that principle to the

Lorenz equations themselves. Indeed, we have observed above that at a certain threshold value (12.11) the steady-state solution becomes unstable. At such an instability point we can distinguish between stable and unstable modes. It turns out that two modes become unstable while one mode remains stable. Since the analysis is somewhat lengthy we describe only an important new feature. It turns out that the equation for the stable mode has the following structure:

$$\dot{\xi}_s = (-|\lambda_s| + \xi_u)\xi_s + \text{nonlinear terms}, \tag{12.32}$$

ξ_u: amplitude of an unstable mode.

As we remarked on page 200, the adiabatic elimination principle remains valid only if the order parameter remains small enough so that

$$|\xi_u| \ll |\lambda_s| . \tag{12.33}$$

Now let us compare numerical results obtained by a direct computer solution of the Lorenz equation and those obtained by means of the slaving principle (Figs. 12.8, 12.9). We observe that for a certain time interval there is rather good agreement, but suddenly a discrepancy occurs which persists for all later times. A detailed analysis shows that this discrepancy occurs when (12.33) is

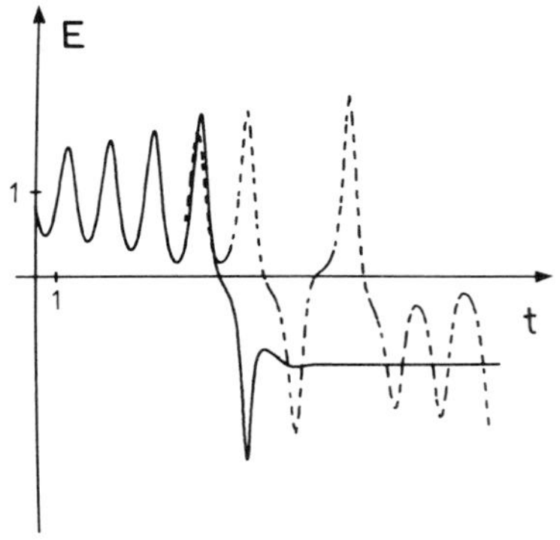

Fig. 12.8. The field strength $E(\propto q_1)$ of a single mode laser under the condition of chaos.
Solid line: solution by means of "slaving principle"; dashed line: direct integration of Lorenz equations. Initially, solid and dashed lines coincide, but then "slaving principle" fails

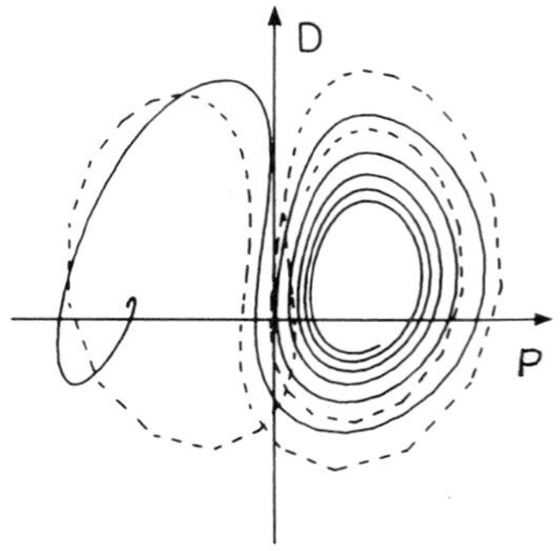

Fig. 12.9. Same physical problem as in Fig. 12.2. Trajectories projected on P (polarization) $-$ D (inversion) plane.
Solid line: solution by means of slaving principle; dashed line: Direct integration of Lorenz equations. Initially, solid and dashed lines coincide

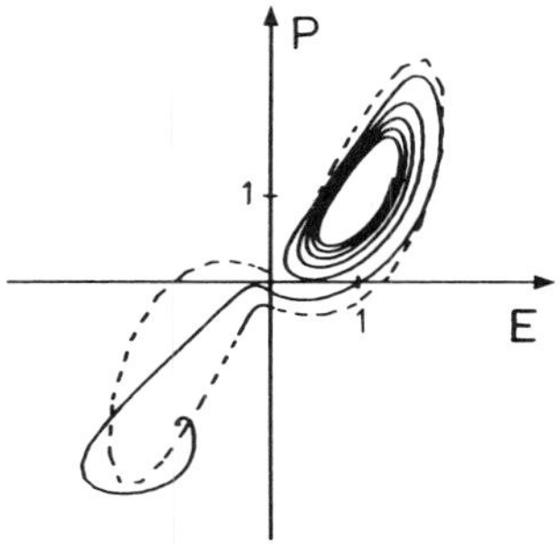

Fig. 12.10. Same as in Fig. 12.8, but trajectories projected on $E-P$ plane

violated. Furthermore, it turns out that at that time the representative point just jumps from one region to the other one in the sense discussed in Section 12.3. Thus we see that chaotic motion occurs when the slaving principle fails and the formerly stable mode can no longer be slaved but is destabilized.

12.5 Correlation Function and Frequency Distribution

The foregoing sections have given us at least an intuitive insight into what chaotic motion looks like. We now want to find a more rigorous description of its properties. To this end we use the correlation function between the variable $q(t)$ at a time t and at a later time $t+t'$. We have encountered correlation functions already in the sections on probability. There they were denoted by

$$\langle q(t)q(t+t')\rangle . \tag{12.34}$$

In the present case we do not have a random process and therefore the averaging process indicated by the brackets seems to be meaningless. However, we may replace (12.34) by a time average in the form

$$\lim_{T\to\infty} \frac{1}{2T} \int_{-T}^{T} q(t)q(t+t')dt , \tag{12.35}$$

where we first integrate over t and then let the time interval $2T$ become very large or, more strictly speaking, let T go to infinity. Taking purely periodic motion as a first example, i.e., for instance

$$q(t)=\sin\omega_1 t , \tag{12.36}$$

we readily obtain

$$(12.35)=\tfrac{1}{2}\cos\omega_1 t'. \tag{12.37}$$

That means we obtain again a periodic function (compare Fig. 12.11). One may simply convince oneself that a motion containing several frequencies such as

$$q(t)=\sin\omega_1 t \sin\omega_2 t \tag{12.38}$$

also gives rise to an oscillating nondecaying behavior of (12.34).

On the other hand, when we think of a purely diffusive process based on random events which we encountered in the section on probability theory, we should expect that (12.34) goes to 0 for $t'\to\infty$ (compare Fig. 12.12). Because

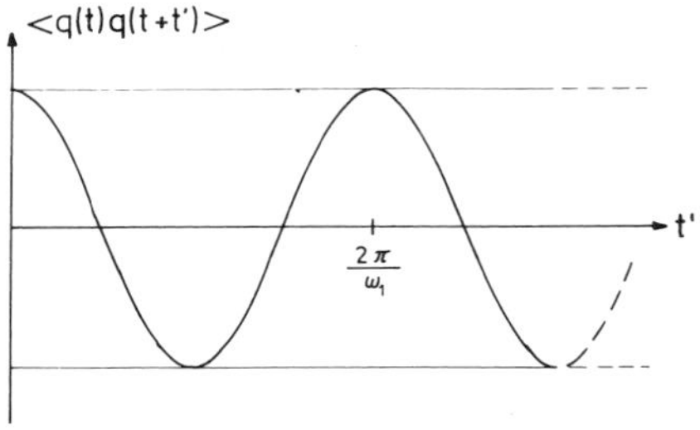

Fig. 12.11. The correlation function as function of time t' for the periodic function $q(t)$, (12.36). Note that there is no decay of the amplitude even if $t'\to\infty$

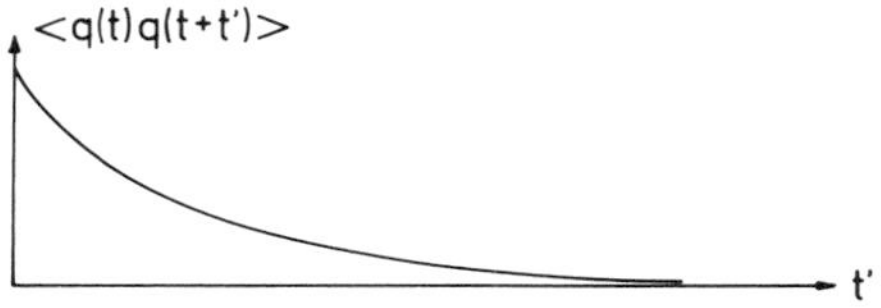

Fig. 12.12. The correlation function for t' for a chaotic motion. Note that an oscillating behavior need not be excluded but that the correlation function must vanish as $t'\to\infty$

we wish to characterize chaotic motion as seemingly random (though caused by deterministic forces), we may adopt as a criterion for chaotic motion a behavior as shown in Fig. 12.12. A further criterion results if we decompose $q(t)$ into a Fourier integral

$$q(t)=\frac{1}{2\pi}\int_{-\infty}^{+\infty} c(\omega)\mathrm{e}^{i\omega t}d\omega\,. \tag{12.39}$$

Inserting it into (12.36) we find two infinitely high peaks at $\omega=\pm\omega_1$ (compare Fig. 12.13). Similarly (12.38) would lead to peaks at $\omega=\pm\omega_1\pm\omega_2$. On the

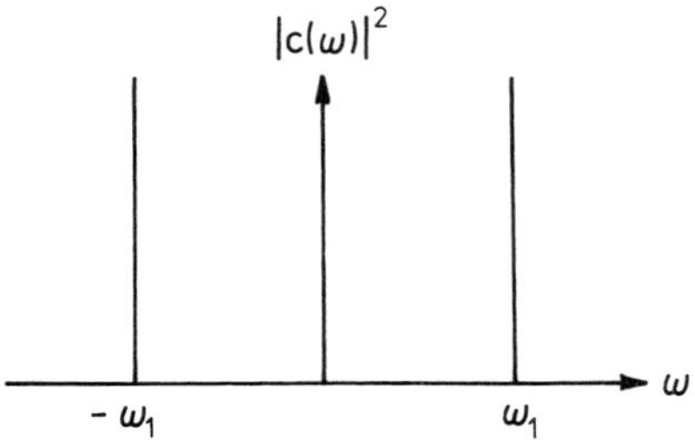

Fig. 12.13. The power spectrum $|c(\omega)|^2$ for a purely periodic variable $q(t)$ (compare (12.36)). It shows only two peaks at $-\omega_1$ and ω_1. For multiperiodic variables a set of peaks would appear

other hand, chaotic motion should contain a continuous broad band of frequencies. Fig. 12.14 shows an example of the intensity distribution of frequencies of the Lorenz model. Both criteria using (12.34) and (12.39) in the way described above are nowadays widely used, especially when the analysis is based on computer solutions of the original equations of motion. For the sake of completeness we mention a third method, namely that based on the so-called Poincaré map. A description of this method is, however, beyond the scope of this book.

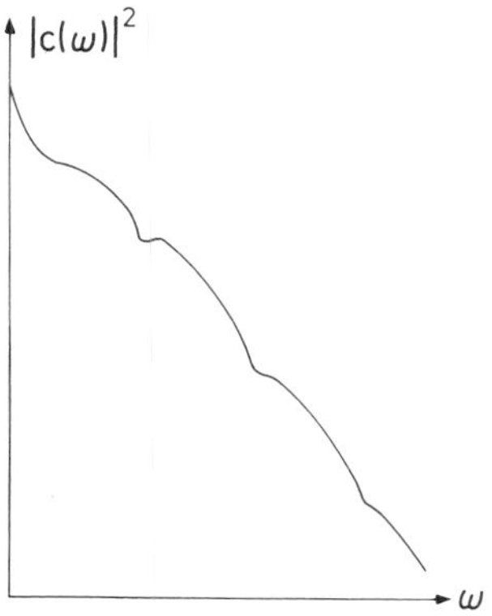

Fig. 12.14. The power spectrum $|c(\omega)|^2$ for the Lorenz attractor (redrawn after *Y. Aizawa* and *I. Shimada*). Their original figure contains a set of very closely spaced points. Here the studied variable is $q_2(t)$ of the Lorenz model as described in the text

12.6 Further Examples of Chaotic Motion

Chaotic motion in the sense discussed above occurs in quite different disciplines. In the last century Poincaré discovered irregular motion in the three-body problem. Chaos is also observed in electronic devices and is known to electrical engineers. It occurs in the Gunn oscillator, whose regular spikes we discussed in Section 8.14. More recently numerous models of chemical reactions have been developed showing chaos. It may occur in models both with and without diffusion. Chaos occurs also in chemical reactions when they are externally modulated, for instance by photochemical effects. Another example are equations describing the flipping of the earth magnetic field which again shows a chaotic motion, i.e., a random flipping. Certain models of population dynamics show entirely irregular changes of populations. It seems that such models can account for certain fluctuation phenomena of insect populations.

Some of these models are particularly simple. The best known refers to a single variable q which is taken not as a continuous function of time but at discrete times labelled by an index n. The equation for this variable then reads

$$q_{n+1} = a_n q_n (1 - q_n) .$$

It can be expected that models of chaotic variations may also find applications in economics and even sociology though, amusingly, so far no theories of this type seem to have been published. In view of the widespread phenomena of chaos, which becomes more and more evident, one may ask why biological systems are apparently capable of avoiding chaos which, in this author's opinion, is a still unsolved problem.

13. Some Historical Remarks and Outlook

The reader who has followed us through our book has been most probably amazed by the profound analogies between completely different systems when they pass through an instability. This instability is caused by a change of external parameters and leads eventually to a new macroscopic spatio-temporal pattern of the system. In many cases the detailed mechanism can be described as follows: close to the instability point we may distinguish between stable and unstable collective motions (modes). The stable modes are slaved by the unstable modes and can be eliminated. In general, this leads to an enormous reduction of the degrees of freedom. The remaining unstable modes serve as order parameters determining the macroscopic behavior of the system. The resulting equations for the order parameters can be grouped into a few universality classes which describe the dynamics of the order parameters. Some of these equations are strongly reminiscent of those governing first and second order phase transitions of physical systems in thermal equilibrium. However, new kinds of classes also occur, for instance describing pulsations or oscillations. The interplay between stochastic and deterministic "forces" ("chance and necessity") drives the systems from their old states into new configurations and determines which new configuration is realized.

The General Scheme

Old structure ⟶ Instability ⟶ New structure
Change of external parameters

Old structure → { Unstable modes / slave / → Stable modes } →

⟶ { Unstable modes / Order parameters } : { Same type of equations for different systems } →

⟶ Universality classes : New structure

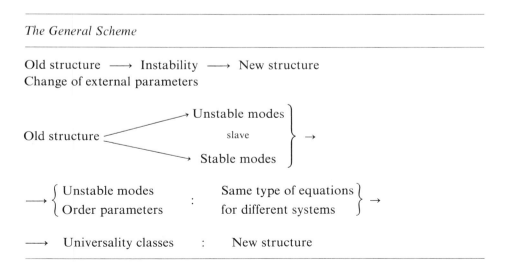

A first detailed and explicit account of the phase-transition analogy between a system far from thermal equilibrium (the laser) and systems in thermal equilibrium (superconductors, ferromagnets) was given in independent papers by Graham and Haken (1968, 1970), and by DeGiorgio and Scully (1970).[1] When we now browse through the literature, knowing of these analogies, we are reminded of Kohelet's words: There is nothing new under this sun. Indeed we now discover that such analogies have been inherent, more or less visibly, in many phenomena (and theoretical treatments).

In the realm of general systems theory (with emphasis on biology), its founder von Bertalanffi observed certain analogies between closed and open systems. In particular, he coined the concept of "flux equilibrium" (Fließgleichgewicht). In other fields, e.g., computers, such analogies have been exploited in devices. Corresponding mathematical results were obtained in papers by Landauer on tunnel diodes (1961, 1962) and by myself on lasers (1964). While the former case referred to a certain kind of switching, for instance analogous to a Bloch wall motion in ferromagnets, the latter paved the way to compare the laser threshold with a second order phase transition.

Knowing of the laser phase transition analogy, a number of authors established similar analogies in other fields, in particular non-equilibrium chemical reactions (Schlögl, Nicolis, Nitzan, Ortoleva, Ross, Gardiner, Walls and others).

The study of models of chemical reactions producing spatial or temporal structures had been initiated in the fundamental work by Turing (1952) and it was carried further in particular by Prigogine and his coworkers. In these latter works the concept of excess entropy production which allows to find instabilities has played a central role. The present approach of synergetics goes beyond these concepts in several respects. In particular it investigates what happens at the instability point and it determines the new structure beyond it. Some of these problems can be dealt with by the mathematical theory of bifurcation, or, more generally, by a mathematical discipline called dynamic systems theory. In many cases presented in this book we had to treat still more complex problems, however. For instance, we had to take into account fluctuations, small band excitations and other features. Thus, synergetics has established links between dynamic systems theory and statistical physics. Undoubtedly, the marriage between these two disciplines has started.

When it occurred to me that the cooperation of many subsystems of a system is governed by the same principles irrespective of the nature of the subsystems I felt that the time had come to search for and explore these analogies within the frame of an interdisciplinary field of research which I called synergetics. While I was starting from physics and was led into questions of chemistry and biology, quite recently colleagues of some other disciplines have drawn my attention to the fact that a conception, called synergy, has long been discussed in fields such as sociology and economics. Here for instance the working together of different parts of a company, to improve the performance of the company, is studied. It thus appears that we are presently from two different sides digging a tunnel under a big

[1] The detailed references to this chapter are listed on page 349.

mountain which has so far separated different disciplines, in particular the "soft" from the "hard" sciences.

It can be hoped that synergetics will contribute to the mutual understanding and further development of seemingly completely different sciences. How synergetics might proceed shall be illustrated by the following example taken from philology. Using the terminology of synergetics, languages are the order parameters slaving the subsystems which are the human beings. A language changes only little over the duration of the life of an individual. After his birth an individual learns a language, i.e., he is slaved by it, and for his lifetime contributes to the survival of the language. A number of facts about languages such as competition, fluctuations (change of meaning of words, etc.) can now be investigated in the frame established by synergetics.

Synergetics is a very young discipline and many surprising results are still ahead of us. I do hope that my introduction to this field will stimulate and enable the reader to make his own discoveries of the features of self-organizing systems.

References, Further Reading, and Comments

Since the field of Synergetics has ties to many disciplines, an attempt to provide a more or less complete list of references seems hopeless. Indeed, they would fill a whole volume. We therefore confine the references to those papers which we used in the preparation of this book. In addition we quote a number of papers, articles or books which the reader might find useful for further study. We list the references and further reading according to the individual chapters.

1. Goal

H. Haken, R. Graham: Synergetik-Die Lehre vom Zusammenwirken. Umschau **6**, 191 (1971)
H. Haken (ed.): *Synergetics* (Proceedings of a Symposium on Synergetics, Elmau 1972) (B. G. Teubner, Stuttgart 1973)
H. Haken (ed.): *Cooperative Effects, Progress in Synergetics* (North Holland, Amsterdam 1974)
H. Haken: Cooperative effects in systems far from thermal equilibrium and in nonphysical systems. Rev. Mod. Phys. **47**, 67 (1975)
H. Haken (ed.): *Synergetics, A Workshop*, Proceedings of a workshop on Synergetics, Elmau 1977 (Springer, Berlin-Heidelberg-New York 1977)

An approach, entirely different from ours, to treat the formation of structures in physical, chemical an biochemical systems is due to Prigogine and his school, see

P. Glansdorff, I. Prigogine: *Thermodynamic Theory of Structure, Stability and Fluctuations* (Wiley, New York 1971)

Prigogine has coined the word "dissipative structures". Glansdorff and Prigogine base their work on entropy production principles and use the excess entropy production as means to search for the onset of an instability. The validity of such criteria has been critically investigated by R. Landauer: Phys. Rev. A **12**, 636 (1975). The Glansdorff-Prigogine approach does not give an answer to what happens at the instability point and how to determine or classify the new evolving structures. An important line of research by the Brussels school, namely chemical reaction models, comes closer to the spirit of Synergetics (compare Chapter 9).

1.1 *Order and Disorder. Some Typical Phenomena*

For literature on thermodynamics see Section 3.4. For literature on phase transitions see Section 6.7. For detailed references on lasers, fluid dynamics, chemistry and biology consult the references of the corresponding chapters of our book. Since the case of slime-mold is not treated any further here, we give a few references:

J. T. Bonner, D. S. Barkley, E. M. Hall, T. M. Konijn, J. W. Mason, G. O'Keefe, P. B. Wolfe: Develop. Biol. **20**, 72 (1969)
T. M. Konijn: Advanc. Cycl. Nucl. Res, **1**, 17 (1972)
A. Robertson, D. J. Drage, M. H. Cohen: Science **175**, 333 (1972)
G. Gerisch, B. Hess: Proc. nat. Acad. Sci. (Wash.) **71**, 2118 (1974)
G. Gerisch: Naturwissenschaften **58**, 430 (1971)

2. Probability

There are numerous good textbooks on probability. Here are some of them:

Kai Lai Chung: *Elementary Probability Theory with Stochastic Processes* (Springer, Berlin-Heidelberg-New York 1974)

W. Feller: *An Introduction to Probability Theory and Its Applications*, Vol. 1 (Wiley, New York 1968), Vol. 2 (Wiley, New York 1971)

R.C. Dubes: *The Theory of Applied Probability* (Prentice Hall, Englewood Cliffs, N.J. 1968)

Yu. V. Prokhorov, Yu. A. Rozanov: *Probability Theory*. In *Grundlehren der mathematischen Wissenschaften in Einzeldarstellungen*, Bd. 157 (Springer, Berlin-Heidelberg-New York 1968)

J. L. Doob: *Stochastic Processes* (Wiley, New York-London 1953)

M. Loève: *Probability Theory* (D. van Nostrand, Princeton, N.J.-Toronto-New York-London 1963)

R. von Mises: *Mathematical Theory of Probability and Statistics* (Academic Press, New York-London 1964)

3. Information

3.1. *Some Basic Ideas*

Monographs on this subject are:

L. Brillouin: *Science and Information Theory* (Academic Press, New York-London 1962)

L. Brillouin: *Scientific Uncertainty and Information* (Academic Press, New York-London 1964)

Information theory was founded by

C. E. Shannon: A mathematical theory of communication. Bell System Techn. J. **27**, 370–423, 623–656 (1948)

C. E. Shannon: Bell System Techn. J. **30**, 50 (1951)

C. E. Shannon, W. Weaver: *The Mathematical Theory of Communication* (Univ. of Illin. Press, Urbana 1949)

Some conceptions, related to information and information gain (H-theorem!) were introduced by

L. Boltzmann: *Vorlesungen über Gastheorie*, 2 Vols. (Leipzig 1896, 1898)

3.2 *Information Gain: An Illustrative Derivation*

For a detailed treatment and definition see

S. Kullback: Ann. Math. Statist. **22**, 79 (1951)

S. Kullback: *Information Theory and Statistics* (Wiley, New York 1951)

Here we follow our lecture notes.

3.3 *Information Entropy and Constraints*

We follow in this chapter essentially

E. T. Jaynes: Phys. Rev. **106**, 4, 620 (1957); Phys. Rev. **108**, 171 (1957)

E. T. Jaynes: In *Delaware Seminar in the Foundations of Physics* (Springer, Berlin-Heidelberg-New York 1967)

Early ideas on this subject are presented in

W. Elsasser: Phys. Rev. **52**, 987 (1937); Z. Phys. **171**, 66 (1968)

3.4 *An Example from Physics: Thermodynamics*

The approach of this chapter is conceptually based on Jaynes' papers, l.c. Section 3.3. For textbooks giving other approaches to thermodynamics see

Landau-Lifshitz: In *Course of Theoretical Physics*, Vol. 5: Statistical Physics (Pergamon Press, London-Paris 1952)

R. Becker: *Theory of Heat* (Springer, Berlin-Heidelberg-New York 1967)

A. Münster: *Statistical Thermodynamics*, Vol. 1 (Springer, Berlin-Heidelberg-New York 1969)

H. B. Callen: *Thermodynamics* (Wiley, New York 1960)

P. T. Landsberg: *Thermodynamics* (Wiley, New York 1961)

R. Kubo: *Thermodynamics* (North Holland, Amsterdam 1968)

W. Brenig: *Statistische Theorie der Wärme* (Springer, Berlin-Heidelberg-New York 1975)
W. Weidlich: *Thermodynamik und statistische Mechanik* (Akademische Verlagsgesellschaft, Wiesbaden 1976)

3.5 *An Approach to Irreversible Thermodynamics*

An interesting and promising link between irreversible thermodynamics and network theory has been established by

A. Katchalsky, P. F. Curran: *Nonequilibrium Thermodynamics in Biophysics* (Harvard University Press, Cambridge Mass. 1967)

For a recent representation including also more current results see

J. Schnakenberg: *Thermodynamic Network Analysis of Biological Systems*, Universitext (Springer, Berlin-Heidelberg-New York 1977)

For detailed texts on irreversible thermodynamics see

I. Prigogine: *Introduction to Thermodynamics of Irreversible Processes* (Thomas, New York 1955)
I. Prigogine: *Non-equilibrium Statistical Mechanics* (Interscience, New York 1962)
S. R. De Groot, P. Mazur: *Non-equilibrium Thermodynamics* (North Holland, Amsterdam 1962)
R. Haase: *Thermodynamics of Irreversible Processes* (Addison-Wesley, Reading, Mass. 1969)
D. N. Zubarev: *Non-equilibrium Statistical Thermodynamics* (Consultants Bureau, New York-London 1974)

Here, we present a hitherto unpublished treatment by the present author.

3.6 *Entropy—Curse of Statistical Mechanics?*

For the problem subjectivistic-objectivistic see for example

E. T. Jaynes: Information Theory. In *Statistical Physics*, Brandeis Lectures, Vol. 3 (W. A. Benjamin, New York 1962)

Coarse graining is discussed by

A. Münster: In *Encyclopedia of Physics*, ed. by S. Flügge, Vol. III/2: Principles of Thermodynamics and Statistics (Springer, Berlin-Göttingen-Heidelberg 1959)

The concept of entropy is discussed in all textbooks on thermodynamics, cf. references to Section 3.4.

4. Chance

4.1 *A Model of Brownian Motion*

For detailed treatments of Brownian motion see for example

N. Wax, ed.: *Selected Papers on Noise and Statistical Processes* (Dover Publ. Inc., New York 1954) with articles by S. Chandrasekhar, G. E. Uhlenbeck and L. S. Ornstein, Ming Chen Wang and G. E. Uhlenbeck, M. Kac
T. T. Soong: *Random Differential Equations in Science and Engineering* (Academic Press, New York 1973)

4.2 *The Random Walk Model and Its Master Equation*

See for instance

M. Kac: Am. Math. Month. **54**, 295 (1946)
M. S. Bartlett: *Stochastic Processes* (Univ. Press, Cambridge 1960)

4.3 *Joint Probability and Paths. Markov Processes. The Chapman-Kolmogorov Equation. Path Integrals*

See references on stochastic processes, Chapter 2. Furthermore

R. L. Stratonovich: *Topics in the Theory of Random Noise* (Gordon Breach, New York-London, Vol. I 1963, Vol. II 1967)
M. Lax: Rev. Mod. Phys. **32**, 25 (1960); **38**, 358 (1965); **38**, 541 (1966)

Path integrals will be treated later in our book (Section 6.6), where the corresponding references may be found.

4.4 *How to Use Joint Probabilities. Moments. Characteristic Function. Gaussian Processes*

Same references as on Section 4.3.

4.5 *The Master Equation*

The master equation does not only play an important role in (classical) stochastic processes, but also in quantum statistics. Here are some references with respect to quantum statistics:

H. Pauli: *Probleme der Modernen Physik*. Festschrift zum 60. Geburtstage A. Sommerfelds, ed. by P. Debye (Hirzel, Leipzig 1928)
L. van Hove: Physica **23**, 441 (1957)
S. Nakajiama: Progr. Theor. Phys. **20**, 948 (1958)
R. Zwanzig: J. Chem. Phys. **33**, 1338 (1960)
E. W. Montroll: *Fundamental Problems in Statistical Mechanics*, compiled by E. D. G. Cohen (North Holland, Amsterdam 1962)
P. N. Argyres, P. L. Kelley: Phys. Rev. **134**, A98 (1964)

For a recent review see

F. Haake: In *Springer Tracts in Modern Physics*, Vol. 66 (Springer, Berlin-Heidelberg-New York 1973) p. 98.

4.6 *Exact Stationary Solution of the Master Equation for Systems in Detailed Balance*

For many variables see

H. Haken: Phys. Lett. **46**A, 443 (1974); Rev. Mod. Phys. **47**, 67 (1975),

where further discussions are given.

For one variable see

R. Landauer: J. Appl. Phys. **33**, 2209 (1962)

4.8 *Kirchhoff's Method of Solution of the Master Equation*

G. Kirchhoff: Ann. Phys. Chem., Bd. LXXII 1847, Bd. 12, S. 32
G. Kirchhoff: Poggendorffs Ann. Phys. **72**, 495 (1844)
R. Bott, J. P. Mayberry: *Matrices and Trees, Economic Activity Analysis* (Wiley, New York 1954)
E. L. King, C. Altmann: J. Phys. Chem. **60**, 1375 (1956)
T. L. Hill: J. Theor. Biol. **10**, 442 (1966)

A very elegant derivation of Kirchhoff's solution was recently given by

W. Weidlich; Stuttgart (unpublished)

4.9 *Theorems About Solutions of the Master Equation*

I. Schnakenberg: Rev. Mod. Phys. **48**, 571 (1976)
J. Keizer: *On the Solutions and the Steady States of a Master Equation* (Plenum Press, New York 1972)

4.10 *The Meaning of Random Processes. Stationary State, Fluctuations, Recurrence Time*

For Ehrenfest's urn model see

P. and T. Ehrenfest: Phys. Z. **8**, 311 (1907)

and also

A. Münster: In *Encyclopedia of Physics*, ed. by S. Flügge, Vol. III/2; Principles of Thermodynamics and Statistics (Springer, Berlin-Göttingen-Heidelberg 1959)

5. Necessity

Monographs on dynamical systems and related topics are

N. N. Bogoliubov, Y. A. Mitropolsky: *Asymptotic Methods in the Theory of Nonlinear Oscillations* (Hindustan Publ. Corp., Delhi 1961)
N. Minorski: *Nonlinear Oscillations* (Van Nostrand, Toronto 1962)
A. Andronov, A. Vitt, S. E. Khaikin: *Theory of Oscillators* (Pergamon Press, London-Paris 1966)
D. H. Sattinger In *Lecture Notes in Mathematics*, Vol. 309: Topics in Stability and Bifurcation Theory, ed. by A. Dold, B. Eckmann (Springer, Berlin-Heidelberg-New York 1973)

M. W. Hirsch, S. Smale: *Differential Equations, Dynamical Systems, and Linear Algebra* (Academic Press, New York-London 1974)
V. V. Nemytskii, V. V. Stepanov: *Qualitative Theory of Differential Equations* (Princeton Univ. Press, Princeton, N.J. 1960)

Many of the basic ideas are due to

H. Poincaré: *Oeuvres*, Vol. 1 (Gauthiers-Villars, Paris 1928)
H. Poincaré: Sur l'equilibre d'une masse fluide animée d'un mouvement de rotation. Acta Math. **7** (1885)
H. Poincaré: Figures d'equilibre d'une masse fluide (Paris 1903)
H. Poincaré: Sur le problème de trois corps et les équations de la dynamique. Acta Math. **13** (1890)
H. Poincaré: *Les méthodes nouvelles de la méchanique céleste* (Gauthier-Villars, Paris 1892–1899)

5.3 *Stability*

J. La Salle, S. Lefshetz: *Stability by Ljapunov's Direct Method with Applications* (Academic Press, New York-London 1961)
W. Hahn: Stability of Motion. In *Die Grundlehren der mathematischen Wissenschaften in Einzeldarstellungen*, Bd. 138 (Springer, Berlin-Heidelberg-New York 1967)
Exercises 5.3: F. Schlögl: Z. Phys. **243**, 303 (1973)

5.4 *Examples and Exercises on Bifurcation and Stability*

A. Lotka: Proc. Nat. Acad. Sci. (Wash.) **6**, 410 (1920)
V. Volterra: *Leçons sur la théorie mathematiques de la lutte pour la vie* (Paris 1931)
N. S. Goel, S. C. Maitra, E. W. Montroll: Rev. Mod. Phys. **43**, 231 (1971)
B. van der Pol: Phil. Mag. **43**, 6, 700 (1922); **2**, 7, 978 (1926); **3**, 7, 65 (1927)
H. T. Davis: *Introduction to Nonlinear Differential and Integral Equations* (Dover Publ. Inc., New York 1962)

5.5 *Classification of Static Instabilities, or an Elementary Approach to Thom's Theory of Catastrophes*

R. Thom: *Structural Stability and Morphogenesis* (W. A. Benjamin, Reading, Mass. 1975)

Thom's book requires a good deal of mathematical background. Our "pedestrian's" approach provides a simple access to Thom's classification of catastrophes. Our interpretation of how to apply these results to natural sciences, for instance biology is, however, entirely different from Thom's.

6. Chance and Necessity

6.1 *Langevin Equations: An Example*

For general approaches see

R. L. Stratonovich: *Topics in the Theory of Random Noise*, Vol. 1 (Gordon & Breach, New York-London 1963)
M. Lax: Rev. Mod. Phys. **32**, 25 (1960); **38**, 358, 541 (1966); Phys. Rev. **145**, 110 (1966)
H. Haken: Rev. Mod. Phys. **47**, 67 (1975)

with further references

6.2 *Reservoirs and Random Forces*

Here we present a simple example. For general approaches see

R. Zwanzig: J. Stat. Phys. **9**, 3, 215 (1973)
H. Haken: Rev. Mod. Phys. **47**, 67 (1975)

6.3 *The Fokker-Planck Equation*

Same references as for Section 6.1

6.4 *Some Properties and Stationary Solution of the Fokker-Planck Equation*

The "potential case" is treated by

R. L. Stratonovich: *Topics in the Theory of Random Noise*, Vol. 1 (Gordon & Breach, New York-London 1963)

The more general case for systems in detailed balance is treated by
R. Graham, H. Haken: Z. Phys. **248**, 289 (1971)
H. Risken: Z. Phys. **251**, 231 (1972);
see also
H. Haken: Rev. Mod. Phys. **47**, 67 (1975)

6.5 *Time-Dependent Solutions of the Fokker-Planck Equation*

The solution of the n-dimensional Fokker-Planck equation with linear drift and constant diffusion coefficients was given by
M. C. Wang, G. E. Uhlenbeck: Rev. Mod. Phys. **17**, 2 and 3 (1945)
For a short representation of the results see
H. Haken: Rev. Mod. Phys. **47**, 67 (1975)

6.6 *Solution of the Fokker-Planck Equation by Path Integrals*

L. Onsager, S. Machlup: Phys. Rev. **91**, 1505, 1512 (1953)
I. M. Gelfand, A. M. Yaglome: J. Math. Phys. **1**, 48 (1960)
R. P. Feynman, A. R. Hibbs: *Quantum Mechanics and Path Integrals* (McGraw-Hill, New York 1965)
F. W. Wiegel: *Path Integral Methods in Statistical Mechanics*, Physics Reports 16C, No. 2 (North Holland, Amsterdam 1975)
R. Graham: In *Springer Tracts in Modern Physics*, Vol. 66 (Springer, Berlin-Heidelberg-New York 1973) p. 1
A critical discussion of that paper gives W. Horsthemke, A. Bach: Z. Phys. **B22**, 189 (1975)
We follow essentially H. Haken: Z. Phys. **B24**, 321 (1976) where also classes of solutions of Fokker-Planck equations are discussed.

6.7 *Phase Transition Analogy*

The theory of phase transitions of systems *in thermal equilibrium* is presented, for example, in the following books and articles
L. D. Landau, I. M. Lifshitz: In *Course of Theoretical Physics*, Vol. 5: Statistical Physics (Pergamon Press, London-Paris 1959)
R. Brout: *Phase Transitions* (Benjamin, New York 1965)
L. P. Kadanoff, W. Götze, D. Hamblen, R. Hecht, E. A. S. Lewis, V. V. Palcanskas, M. Rayl, J. Swift, D. Aspnes, J. Kane: Rev. Mod. Phys. **39**, 395 (1967)
M. E. Fischer: Repts. Progr. Phys. **30**, 731 (1967)
H. E. Stanley: *Introduction to Phase Transitions and Critical Phenomena.* Internat. Series of Monographs in Physics (Oxford University, New York 1971)
A. Münster: *Statistical Thermodynamics*, Vol. 2 (Springer, Berlin-Heidelberg-New York and Academic Press, New York-London 1974)
C. Domb, M. S. Green, eds.: *Phase Transitions and Critical Phenomena*, Vols. 1–5 (Academic Press, London 1972–76)
The modern and powerful renormalization group technique of Wilson is reviewed by
K. G. Wilson, J. Kogut: Phys. Rep. **12**C, 75 (1974)
The profound and detailed analogies between a second order phase transition of a system in thermal equilibrium (for instance a superconductor) and transitions of a non-equilibrium system were first derived in the laser-case in independent papers by
R. Graham, H. Haken: Z. Phys. **213**, 420 (1968) and in particular Z. Phys. **237**, 31 (1970),
who treated the continuum mode laser, and by
V. DeGiorgio, M.O. Scully: Phys. Rev. **A2**, 1170 (1970),
who treated the single mode laser.
For further references elucidating the historical development see Section 13.

6.8 *Phase Transition Analogy in Continuous Media: Space Dependent Order Parameter*

a) References to Systems in Thermal Equilibrium

The Ginzburg-Landau theory is presented, for instance, by

N. R. Werthamer: In *Superconductivity*, Vol. 1, ed. by R. D. Parks (Marcel Dekker Inc., New York 1969) p. 321

with further references

The exact evaluation of correlation functions is due to

D. J. Scalapino, M. Sears, R. A. Ferrell: Phys. Rev. **B6**, 3409 (1972)

Further papers on this evaluation are:

L. W. Gruenberg, L. Gunther: Phys. Lett. **38A**, 463 (1972)
M. Nauenberg, F. Kuttner, M. Fusman: Phys. Rev. **A 13**, 1185 (1976)

b) References to Systems Far from Thermal Equilibrium (and Nonphysical Systems)

R. Graham, H. Haken: Z. Phys. **237**, 31 (1970)

Furthermore the Chapters 8 and 9

7. Self-Organization

7.1 *Organization*

H. Haken: unpublished material

7.2 *Self-Organization*

A different approach to the problem of self-organization has been developed by

J. v. Neuman: *Theory of Self-reproducing Automata*, ed. and completed by Arthur W. Burks (University of Illinois Press, 1966)

7.3 *The Role of Fluctuations: Reliability or Adaptability? Switching*

For a detailed discussion of reliability as well as switching, especially of computer elements, see

R. Landauer: IBM Journal **183** (July 1961)
R. Landauer: J. Appl. Phys. **33**, 2209 (1962)
R. Landauer, J. W. F. Woo: In *Synergetics*, ed. by H. Haken (Teubner, Stuttgart 1973)

7.4 *Adiabatic Elimination of Fast Relaxing Variables from the Fokker-Planck Equation*

H. Haken: Z. Phys. **B 20**, 413 (1975)

7.5 *Adiabatic Elimination of Fast Relaxing Variables from the Master Equation*

H. Haken: unpublished

7.6 *Self-Organization in Continuously Extended Media. An Outline of the Mathematical Approach*

7.7 *Generalized Ginzburg-Landau Equations for Nonequilibrium Phase Transitions*

H. Haken: Z. Phys. **B 21**, 105 (1975)

7.8 *Higher-Order Contributions to Generalized Ginzburg-Landau Equations*

H. Haken: Z. Phys. **B 22**, 69 (1975); **B 23**, 388 (1975)

7.9 *Scaling Theory of Continuously Extended Nonequilibrium Systems*

We follow essentially

A. Wunderlin, H. Haken: Z. Phys. **B 21**, 393 (1975)

For related work see

E. Hopf: Berichte der Math.-Phys. Klasse der Sächsischen Akademie der Wissenschaften, Leipzig XCIV, 1 (1942)
A. Schlüter, D. Lortz, F. Busse: J. Fluid Mech. **23**, 129 (1965)
A. C. Newell, J. A. Whitehead: J. Fluid Mech. **38**, 279 (1969)
R. C. Diprima, W. Eckhaus, L. A. Segel: J. Fluid Mech. **49**, 705 (1971)

8. Physical Systems

For related topics see
H. Haken: Rev. Mod. Phys. **47**, 67 (1975)
and the articles by various authors in
H. Haken, ed.: *Synergetics* (Teubner, Stuttgart 1973)
H. Haken, M. Wagner, eds.: *Cooperative Phenomena* (Springer, Berlin-Heidelberg-New York 1973)
H. Haken, ed.: *Cooperative Effects* (North Holland, Amsterdam 1974)

8.1 *Cooperative Effects in the Laser: Self-Organization and Phase Transition*

The dramatic change of the statistical properties of laser light at laser threshold was first derived and predicted by
H. Haken: Z. Phys. **181**, 96 (1964)

8.2 *The Laser Equations in the Mode Picture*

For a detailed review on laser theory see
H. Haken: In *Encyclopedia of Physics*, Vol. XXV/2c: Laser Theory (Springer, Berlin-Heidelberg-New York 1970)

8.3 *The Order Parameter Concept*

Compare especially
H. Haken: Rev. Mod. Phys. **47**, 67 (1975)

8.4 *The Single Mode Laser*

Same references as of Sections 8.1–8.3.

The laser distribution function was derived by
H. Risken: Z. Phys. **186**, 85 (1965) and
R. D. Hempstead, M. Lax: Phys. Rev. **161**, 350 (1967)

For a fully quantum mechanical distribution function cf.
W. Weidlich, H. Risken, H. Haken: Z. Phys. **201**, 396 (1967)
M. Scully, W. E. Lamb: Phys. Rev. **159**, 208 (1967); **166**, 246 (1968)

8.5 *The Multimode Laser*

H. Haken: Z. Phys. **219**, 246 (1969)

8.6 *Laser with Continuously Many Modes. Analogy with Superconductivity*

For a somewhat different treatment see
R. Graham, H. Haken: Z. Phys. **237**, 31 (1970)

8.7 *First-Order Phase Transitions of the Single Mode Laser*

J. F. Scott, M. Sargent III, C. D. Cantrell: Opt. Commun. **15**, 13 (1975)
W. W. Chow, M. O. Scully, E. W. van Stryland: Opt. Commun. **15**, 6 (1975)

8.8 *Hierarchy of Laser Instabilities and Ultrashort Laser Pulses*

We follow essentially
H. Haken, H. Ohno: Opt. Commun. **16**, 205 (1976)
H. Ohno, H. Haken: Phys. Lett. **59A**, 261 (1976), and unpublished work

For a machine calculation see
H. Risken, K. Nummedal: Phys. Lett. **26A**, 275 (1968); J. appl. Phys. **39**, 4662 (1968)

For a discussion of that instability see also
R. Graham, H. Haken: Z. Phys. **213**, 420 (1968)

For temporal oscillations of a single mode laser cf.
K. Tomita, T. Todani, H. Kidachi: Phys. Lett. **51A**, 483 (1975)

8.9 *Instabilities in Fluid Dynamics: The Bénard and Taylor Problems*

8.10 *The Basic Equations*

8.11 *Damped and Neutral Solutions* ($R \leq R_c$)

Some monographs in hydrodynamics:

L. D. Landau, E. M. Lifshitz: In *Course of Theoretical Physics*, Vol. 6: Fluid Mechanics (Pergamon Press, London-New York-Paris-Los Angeles 1959)
Chia-Shun-Yih: *Fluid Mechanics* (McGraw Hill, New York 1969)
G.K. Batchelor: *An Introduction to Fluid Dynamics* (University Press, Cambridge 1970)
S. Chandrasekhar: *Hydrodynamic and Hydromagnetic Stability* (Clarendon Press, Oxford 1961)

Stability problems are treated particularly by Chandrasekhar l.c. and by

C. C. Lin: *Hydrodynamic Stability* (University Press, Cambridge 1967)

8.12 *Solution Near* $R = R_c$ *(Nonlinear Domain). Effective Langevin Equations*

8.13 *The Fokker-Planck Equation and Its Stationary Solution*

We follow essentially

H. Haken: Phys. Lett. **46A**, 193 (1973) and in particular Rev. Mod. Phys. **47**, 67 (1976)

For related work see

R. Graham: Phys. Rev. Lett. **31**, 1479 (1973); Phys. Rev. **10**, 1762 (1974)
A. Wunderlin: Thesis, Stuttgart University (1975)

For the analysis of mode-configurations, but without fluctuations, cf.

A. Schlüter, D. Lortz, F. Busse: J. Fluid Mech. **23**, 129 (1965)
F. H. Busse: J. Fluid Mech. **30**, 625 (1967)
A. C. Newell, J. A. Whitehead: J. Fluid Mech. **38**, 279 (1969)
R. C. Diprima, H. Eckhaus, L. A. Segel: J. Fluid Mech. **49**, 705 (1971)

Higher instabilities are discussed by

F. H. Busse: J. Fluid Mech. **52**, 1, 97 (1972)
D. Ruelle, F. Takens: Comm. Math. Phys. **20**, 167 (1971)
J. B. McLaughlin, P. C. Martin: Phys. Rev. **A 12**, 186 (1975)

where further references may be found.

A review on the present status of experiments and theory gives the book

Fluctuations, Instabilities and Phase Transitions, ed. by T. Riste (Plenum Press, New York 1975)

8.14 *A Model for the Statistical Dynamics of the Gunn Instability Near Threshold*

J. B. Gunn: Solid State Commun. **1**, 88 (1963)
J. B. Gunn: IBM J. Res. Develop. **8**, 141 (1964)

For a theoretical discussion of this and related effects see for instance

H. Thomas: In *Synergetics*, ed. by H. Haken (Teubner, Stuttgart 1973)

Here, we follow essentially

K. Nakamura: J. Phys. Soc. Jap. **38**, 46 (1975)

8.15 *Elastic Stability: Outline of Some Basic Ideas*

Introductions to this field give

J. M. T. Thompson, G. W. Hunt: *A General Theory of Elastic Stability* (Wiley, London 1973)
K. Huseyin: *Nonlinear Theory of Elastic Stability* (Nordhoff, Leyden 1975)

9. Chemical and Biochemical Systems

In this chapter we particularly consider the occurrence of spatial or temporal structures in chemical reactions.

Concentration oscillation were reported as early as 1921 by

C. H. Bray: J. Am. Chem. Soc. **43**, 1262 (1921)

A different reaction showing oscillations was studied by

B. P. Belousov: Sb. ref. radats. med. Moscow (1959)

This work was extended by Zhabotinsky and his coworkers in a series of papers

V. A. Vavilin, A. M. Zhabotinsky, L. S. Yaguzhinsky: *Oscillatory Processes in Biological and Chemical Systems* (Moscow Science Publ. 1967) p. 181

A. N. Zaikin, A. M. Zhabotinsky: Nature **225**, 535 (1970)
A. M. Zhabotinsky, A. N. Zaikin: J. Theor. Biol. **40**, 45 (1973)

A theoretical model accounting for the occurrence of spatial structures was first given by
A. M. Turing: Phil. Trans. Roy. Soc. **B 237**, 37 (1952)

Models of chemical reactions showing spatial and temporal structures were treated in numerous publications by Prigogine and his coworkers. See P. Glansdorff and I. Prigogine l.c. on page 307 with many references, and
G. Nicolis, I. Prigogine: *Self-organization in Non-equilibrium Systems* (Wiley, New York 1977)

A review of the statistical aspects of chemical reactions can be found in
D. Mc Quarry: *Supplementary Review Series in Appl. Probability* (Methuen, London 1967)

A detailed review over the whole field gives the
Faraday Symposium 9: Phys. Chemistry of Oscillatory Phenomena, London (1974)

For chemical oscillations see especially
G. Nicolis, J. Portnow: Chem. Rev. **73**, 365 (1973)

9.2 *Deterministic Processes, Without Diffusion, One Variable*

9.3 *Reaction and Diffusion Equations*

We essentially follow
F. Schlögl: Z. Phys. **253**, 147 (1972),
who gave the steady state solution. The transient solution was determined by
H. Ohno: Stuttgart (unpublished)

9.4 *Reaction-Diffusion Model with Two or Three Variables: the Brusselator and the Oregonator*

We give here our own nonlinear treatment (A. Wunderlin, H. Haken, unpublished) of the reaction-diffusion equations of the "Brusselator" reaction, originally introduced by Prigogine and coworkers, l.c. For related treatments see
J. F. G. Auchmuchty, G. Nicolis: Bull. Math. Biol. **37**, 1 (1974)
Y. Kuramoto, T. Tsusuki: Progr. Theor. Phys. **52**, 1399 (1974)
M. Herschkowitz-Kaufmann: Bull. Math. Biol. **37**, 589 (1975)
The Belousov-Zhabotinsky reaction is described in the already cited articles by Belousov and Zhabotinsky.

The "Oregonator" model reaction was formulated and treated by
R. J. Field, E. Korös, R. M. Noyes: J. Am. Chem. Soc. **49**, 8649 (1972)
R. J. Field, R. M. Noyes: Nature **237**, 390 (1972)
R. J. Field, R. M. Noyes: J. Chem Phys. **60**, 1877 (1974)
R. J. Field, R. M. Noyes: J. Am. Chem. Soc. **96**, 2001 (1974)

9.5 *Stochastic Model for a Chemical Reaction Without Diffusion. Birth and Death Processes. One Variable*

A first treatment of this model is due to
V. J. McNeil, D. F. Walls: J. Stat. Phys. **10**, 439 (1974)

9.6 *Stochastic Model for a Chemical Reaction with Diffusion. One Variable*

The master equation with diffusion is derived by
H. Haken: Z. Phys. **B 20**, 413 (1975)

We essentially follow
C. H. Gardiner, K. J. McNeil, D. F. Walls, I. S. Matheson: J. Stat. Phys. **14**, 4, 307 (1976)

Related to this chapter are the papers by
G. Nicolis, P. Aden, A. van Nypelseer: Progr. Theor. Phys. **52**, 1481 (1974)
M. Malek-Mansour, G. Nicolis: preprint Febr. 1975

9.7 *Stochastic Treatment of the Brusselator Close to Its Soft Mode Instability*

We essentially follow
H. Haken: Z. Phys. **B 20**, 413 (1975)

9.8 *Chemical Networks*

Related to this chapter are

G. F. Oster, A. S. Perelson: Chem. Reaction Dynamics. Arch. Rat. Mech. Anal. **55**, 230 (1974)

A. S. Perelson, G. F. Oster: Chem. Reaction Dynamics, Part II; Reaction Networks. Arch. Rat. Mech. Anal. **57**, 31 (1974/75)

with further references.

G. F. Oster, A. S. Perelson, A. Katchalsky: Quart. Rev. Biophys. **6**, 1 (1973)

O. E. Rössler: In *Lecture Notes in Biomathematics*, Vol. 4 (Springer, Berlin-Heidelberg-New York 1974) p. 419

O. E. Rössler: Z. Naturforsch. **31a**, 255 (1976)

10. Applications to Biology

10.1 *Ecology, Population Dynamics*

10.2 *Stochastic Models for a Predator-Prey System*

For general treatments see

N. S. Goel, N. Richter-Dyn: *Stochastic Models in Biology* (Academic Press, New York 1974)

D. Ludwig: In *Lecture Notes in Biomathematics*, Vol. 3: Stochastic Population Theories, ed. by S. Levin (Springer, Berlin-Heidelberg-New York 1974)

For a different treatment of the problem of this section see

V. T. N. Reddy: J. Statist. Phys. **13**, 1 (1975)

10.3 *A Simple Mathematical Model for Evolutionary Processes*

The equations discussed here seem to have first occurred in the realm of laser physics, where they explained mode selection in lasers (H. Haken, H. Sauermann: Z. Phys. **173**, 261 (1963)). The application of laser-type equations to biological processes was suggested by

H. Haken: Talk at the Internat. Conference *From Theoretical Physics to Biology*, ed. by M. Marois, Versailles 1969

see also

H. Haken: In *From Theoretical Physics to Biology*, ed. by M. Marois (Karger, Basel 1973)

A comprehensive and detailed theory of evolutionary processes has been developed by M. Eigen: Die Naturwissenschaften **58**, 465 (1971). With respect to the analogies emphasized in our book it is interesting to note that Eigen's "Bewertungsfunktion" is identical with the saturated gain function (8.35) of multimode lasers.

An approach to interpret evolutionary and other processes as games is outlined by

M. Eigen, R. Winkler-Oswatitsch: *Das Spiel* (Piper, München 1975)

An important new concept is that of hypercycles and, connected with it, of "quasi-species"

M. Eigen, P. Schuster: Naturwissensch. **64**, 541 (1977)

10.4 *A Model for Morphogenesis*

We present here a model due to Gierer and Meinhardt cf.

A. Gierer, M. Meinhardt: Biological pattern formation involving lateral inhibition. Lectures on Mathematics in the Life Sciences **7**, 163 (1974)

H. Meinhardt: The Formation of Morphogenetic Gradients and Fields. Ber. Deutsch. Bot. Ges. **87**, 101 (1974)

H. Meinhardt, A. Gierer: Applications of a theory of biological pattern formation based on lateral inhibition. J. Cell. Sci. **15**, 321 (1974)

H. Meinhardt: preprint 1976

10.5 *Order Parameters and Morphogenesis*

We present here unpublished results by H. Haken and H. Olbrich.

11. Sociology

We present here Weidlich's model.

W. Weidlich: Collective Phenomena **1**, 51 (1972)
W. Weidlich: Brit. J. math. stat. Psychol. **24**, 251 (1971)
W. Weidlich: In *Synergetics*, ed. by H. Haken (Teubner, Stuttgart 1973)

The following monographs deal with a mathematization of sociology:

J. S. Coleman: *Introduction to Mathematical Sociology* (The Free Press, New York 1964)
D. J. Bartholomew: *Stochastic Models for Social Processes* (Wiley, London 1967)

12. Chaos

12.1 *What Is Chaos?*

For mathematically rigorous treatments of examples of chaos by means of mappings and other topological methods see

S. Smale: Bull. A.M.S. **73**, 747 (1967)
T. Y. Li, J. A. Yorke: Am. Math. Monthly **82**, 985 (1975)
D. Ruelle, F. Takens: Commun. math. Phys. **20**, 167 (1971)

12.2 *The Lorenz Model. Motivation and Realization*

E. N. Lorenz: J. Atmospheric Sci. **20**, 130 (1963)
E. N. Lorenz: J. Atmospheric Sci. **20**, 448 (1963)

Historically, the first papers showing a "strange attractor". For further treatments of this model see

J. B. McLaughlin, P. C. Martin: Phys. Rev. **A12**, 186 (1975)
M. Lücke: J. Stat. Phys. **15**, 455 (1976)

For the laser fluid analogy presented in this chapter see

H. Haken: Phys. Lett. **53A**, 77 (1975)

12.3 *How Chaos Occurs*

H. Haken, A. Wunderlin: Phys. Lett. **62A**, 133 (1977)

12.4 *Chaos and the Failure of the Slaving Principle*

H. Haken, J. Zorell: Unpublished

12.5 *Correlation Function and Frequency Distribution*

M. Lücke: J. Stat. Phys. **15**, 455 (1976)
Y. Aizawa, I. Shimada: Preprint 1977

12.6 *Further Examples of Chaotic Motion*

Three-Body Problem:

H. Poincaré: *Les méthodes nouvelles de la méchanique céleste.* Gauthier-Villars, Paris (1892/99), Reprint (Dover Publ., New York 1960)

For electronic devices especially Khaikin's "universal circuit" see

A. A. Andronov, A. A. Vitt, S. E. Khaikin: *Theory of Oscillators* (Pergamon Press, Oxford-London-Edinburgh-New York-Toronto-Paris-Frankfurt 1966)

Gunn Oscillator:

K. Nakamura: Progr. Theoret. Phys. **57**, 1874 (1977)

Numerous chemical reaction models (without diffusion) have been treated by O. E. Roessler. For a summary and list of references consult

O. E. Roessler: In *Synergetics, A Workshop*, ed. by H. Haken (Springer, Berlin-Heidelberg-New York, 1977)

For chemical reaction models including diffusion see

Y. Kuramoto, T. Yamada: Progr. Theoret. Phys. **56**, 679 (1976)
T. Yamada, Y. Kuramoto: Progr. Theoret. Phys. **56**, 681 (1976)

Modulated chemical reactions have been treated by
K. Tomita, T. Kai, F. Hikami: Progr. Theoret. Phys. **57**, 1159 (1977)

For experimental evidence of chaos in chemical reactions see
R. A. Schmitz, K. R. Graziani, J. L. Hudson: J. Chem. Phys. **67**, 3040 (1977);
O. E. Roessler, to be published

Earth magnetic field:
J. A. Jacobs: Phys. Reports **26**, 183 (1976) with further references

Population dynamics:
R. M. May: Nature **261**, 459 (1976)

13. Some Historical Remarks and Outlook

J. F. G. Auchmuchty, G. Nicolis: Bull. Math. Biol. **37**, 323 (1975)
L. von Bertalanffi: Blätter für Deutsche Philosophie **18**, Nr. 3 and 4 (1945); Science **111**, 23 (1950); Brit. J. Phil. Sci. **1**, 134 (1950); *Biophysik des Fliessgleichgewichts* (Vieweg, Braunschweig 1953)
G. Czajkowski: Z. Phys. **270**, 25 (1974)
V. DeGiorgio, M. O. Scully: Phys. Rev. **A2**, 117a (1970)
P. Glansdorff, I. Prigogine: *Thermodynamic Theory of Structure, Stability and Fluctuations* (Wiley, New York 1971)
R. Graham, H. Haken: Z. Phys. **213**, 420 (1968); **237**, 31 (1970)
H. Haken: Z. Phys. **181**, 96 (1964)
M. Herschkowitz-Kaufman: Bull. Math. Biol. **37**, 589 (1975)
K. H. Janssen: Z. Phys. **270**, 67 (1974)
G. J. Klir: *The Approach to General Systems Theory* (Van Nostrand Reinhold Comp., New York 1969)
G. J. Klir, ed.: *Trends in General Systems Theory* (Wiley, New York 1972)
R. Landauer: IBM J. Res. Dev. **5**, 3 (1961); J. Appl. Phys. **33**, 2209 (1962); Ferroelectrics **2**, 47 (1971)
E. Laszlo (ed.): *The Relevance of General Systems Theory* (George Braziller, New York 1972)
I. Matheson, D. F. Walls, C. W. Gardiner: J. Stat. Phys. **12**, 21 (1975)
A. Nitzan, P. Ortoleva, J. Deutch, J. Ross: J. Chem. Phys. **61**, 1056 (1974)
I. Prigogine, G. Nicolis: J. Chem. Phys. **46**, 3542 (1967)
I. Prigogine, R. Lefever: J. Chem. Phys. **48**, 1695 (1968)
A. M. Turing: Phil. Trans. Roy. Soc. **B 234**, 37 (1952)

Subject Index